PHOTOPHYSIOLOGY

Current Topics in Photobiology and Photochemistry

Volume V

PHOTOPHYSIOLOGY

CURRENT TOPICS IN PHOTOBIOLOGY AND PHOTOCHEMISTRY

Edited by

Arthur C. Giese
Department of Biological Sciences
Stanford University, California

Volume V

1970

ACADEMIC PRESS · NEW YORK and LONDON

ACADEMIC PRESS, INC.
111 Fifth Avenue, New York, New York 10003

United Kingdom Edition published by
ACADEMIC PRESS, INC. (LONDON) LTD.
Berkeley Square House, London W1X 6BA

LIBRARY OF CONGRESS CATALOG CARD NUMBER: 63-16961

PRINTED IN THE UNITED STATES OF AMERICA

LIST OF CONTRIBUTORS

Numbers in parentheses indicate the pages on which the authors' contributions begin.

JAN AMESZ, *Biophysical Laboratory, State University, Leiden, The Netherlands* (97)

JOHN S. COOK, *University of Tennessee-Oak Ridge Graduate School of Biomedical Sciences and Biology Division, Oak Ridge, Tennessee* (191)

JOHN H. EPSTEIN, M.D., *Department of Dermatology, University of California School of Medicine, San Francisco, California* (235)

DAVID C. FORK, *Department of Plant Biology, Carnegie Institution of Washington, Stanford, California* (97)

L. I. GROSSWEINER, *Physics Department, Institute of Technology, Chicago, Illinois* (1)

J. WOODLAND HASTINGS, *Biological Laboratories, Harvard University, Cambridge, Massachusetts* (49)

E. A. OGRYZLO, *Department of Chemistry, University of British Columbia, Vancouver, Canada* (35)

ROBERT B. PAINTER, *Laboratory of Radiobiology, University of California Medical Center, San Francisco, California* (169)

H. RIPPS, *Department of Ophthalmology, New York University School of Medicine, New York, New York* (127)

R. A. WEALE, *Department of Physiological Optics, Institute of Ophthalmology, University of London, London, England* (127)

THÉRÈSE WILSON, *Department of Chemistry, Harvard University, Cambridge, Massachusetts* (49)

PREFACE

Active research in photochemistry and photobiology continues at an ever accelerated pace. Readers of previous volumes of *Photophysiology* have urged that the series be continued to enable them to keep abreast of the many fronts on which advances are being made. While *Photophysiology* is intended primarily for advanced students and researchers interested in the effects of light, it has reached a wider audience of those interested to know what is occurring in this borderline science.

To gain more insight into the needs and the ways to implement them, as well as to get advice on prospective authors, an advisory board has been formed, the members of which represent widely differing disciplines: Winslow Briggs (Harvard University), Roderick Clayton (Cornell University), William Hagins (National Institutes of Health), Philip Hanawalt (Stanford University), and Kendric Smith (Stanford Medical School). The board, meeting at the Photobiology Congress at Dartmouth University, decided, among other things, that the series should continue to cover a wide spectrum of fields from photochemistry and biophysics to photobiology of man. Treatises on single fields already occur in abundance, resulting often from conferences in limited areas of photobiology. The schedule of publication and order of chapters in each volume is to remain flexible and adaptable to needs. Each volume is planned to include some photochemistry and photobiology of plants, microorganisms, and animals, including man.

To emphasize the broad coverage, which to some is not implicit in the title *Photophysiology,* a subtitle—"Current Topics in Photobiology and Photochemistry"—was adopted with this volume.

We are indebted to those who have made suggestions and to the contributors to the volume, without whose industry and patience the series could not have developed. Continuing suggestions from the readers will help keep the series in accord with needs.

ARTHUR C. GIESE
March, 1970

CONTENTS

CHAPTER 4. SPECTROPHOTOMETRIC STUDIES OF THE MECHANISM OF PHOTOSYNTHESIS

DAVID C. FORK AND JAN AMESZ

CHAPTER 5. THE PHOTOPHYSIOLOGY OF VERTEBRATE COLOR VISION

H. RIPPS AND R. A. WEALE

CHAPTER 6. THE ACTION OF ULTRAVIOLET LIGHT ON MAMMALIAN CELLS

ROBERT B. PAINTER

CONTENTS OF OTHER VOLUMES

VOLUME I

VOLUME II

VOLUME III

VOLUME IV

PHOTOPHYSIOLOGY

Current Topics in Photobiology and Photochemistry

Volume V

Chapter 1

FLASH PHOTOLYSIS RESEARCH IN PHOTOBIOLOGY

L. I. Grossweiner

Physics Department, Illinois Institute of Technology, Chicago, Illinois

1. Introduction

The flash photolysis technique was invented two decades ago by G. Porter and R. G. W. Norrish (Porter, 1950), who conceived the idea of initiating a photochemical reaction with an intense light flash and identifying the initial photolysis products by taking optical absorption spectra of the irradiated system using a delayed light flash as the spectroscopic source. The early work led to the positive identification of numerous unstable intermediates, whose existence had been postulated only on the basis of indirect evidence, e.g., vibrationally-excited gas phase free radicals, small aromatic free radicals and triplet states, and inorganic radical-ions in solution. The potential applications of flash photolysis to photobiology were recognized almost immediately and early workers reported transient spectra from chlorophyll derivatives, retinal pigments, and sensitizing dyes which were identified with triplet states and free radicals (see the reviews of Livingston, 1966, and Grossweiner, 1966). However, the identification of intermediates *per se* is only one aspect of mechanism determination and must be augmented with information about the faster processes by which the transients are formed, the photochemical and thermal reactions by which they disappear, reaction kinetics, quantum yields, and the thermochemical parameters. Detailed investigations based on flash photolysis measurements plus other pertinent information (e.g., fluorescence and phosphorescence efficiencies and lifetimes, ESR, permanent product characterization) have led to reasonably complete descriptions of "simple" reactions such as the photolysis of aqueous phenol, although each advance typically opens as many questions as it solves.

Several fundamental types of chemical processes have been discovered or confirmed with flash photolysis, notably intermolecular triplet state energy transfer in solution, electron transfer and quenching reactions of triplet states, and the photochemical generation of solvated electrons. The possible role of such processes in photobiology is highly speculative at present. The principal objective of this chapter is to review the current status of two types of photochemical reactions in which flash photolysis research has made major contributions. The first category comprises dye-sensitized photooxidations, which are directly pertinent to the molecular stages in photodynamic action and may parallel the primary processes in natural phenomena promoted by visible light. An attempt has been made to correlate the available information on sensitization by a single dye, eosin Y, in free solution and when in the complexed or aggregated state. In addition, the recent research on sensitization by methylene blue and flavin pigments is reviewed. The other topic considered in detail is the photoionization of aromatic molecules in condensed systems, which has been the subject of recent research in connection with the possible biological implications. The first section of this chapter summarizes the recent "state-of-the-art" advances in flash photolysis techniques, which suggest the directions that future research is likely to follow. In drawing upon the published literature an attempt has been made to emphasize those aspects which can be synthesized into areas of general agreement. Therefore, the citations are illustrative rather than inclusive and undoubtedly reflect the particular interests of the author.

2. New Experimental Techniques in Flash Photolysis

The experimental aspects of flash photolysis have been reviewed by a number of authors including Porter (1959), Grossweiner (1966),

Livingston (1966), and Boag (1968). It is a remarkable tribute to the original design of Porter and Norrish (Porter, 1950) that it remains practical for many applications, particularly when the intermediates persist for microseconds or longer. However, this order of time response is inadequate for the investigation of shorter-lived transients, such as excited singlet states and some metastable states (e.g., the benzene triplet), and considerable effort is being expended towards the development of submicrosecond flash photolysis systems. The computer handling of flash photolysis data is another important direction of recent work, with the objectives of reducing the laborious manual effort required for the scanning of spectroscopic plates and oscilloscope traces, and improving the analysis of kinetics data in terms of the associated reaction mechanisms. The advantages and difficulties encountered in the applications of pulsed lasers to flash photolysis deserve special note, in view of the unusual characteristics of this new type of light source. The basic flash photolysis method in which light energy is injected as a high intensity pulse followed by fast optical detection of intermediates has been extended to various types of primary excitation (e.g., ionizing particles, heat, electrical impulses) and detection methods (e.g., EPR, electrical conductivity, polarography). Although this review emphasizes "optical" flash photolysis, the pertinent improvements in technique developed in connection with the related methods will be considered briefly.

2.1 High Speed Flash Photolysis

The time resolution of a flash photolysis system depends upon the duration of the photolysis flash and the time interval required before detection system recovers from electrical and optical interference. A correlation of the published characteristics of gas-filled lamps made by Porter (1960) shows that the energy dissipated varies approximately with the inverse-second power of the $(1/e)$ flash duration. A typical source can discharge 100 joules per centimeter of arc length in 10 μsec, but only 1 joule per centimeter in 1 μsec. For intermediates of long duration compared to the flash, the photometric sensitivity ΔOD is limited only by the total absorbed light energy according to:

$$(Wn/A)(f\bar{\lambda}\bar{\phi})\epsilon = 1.19 \times 10^5 \, \Delta\text{OD} \tag{1}$$

Where W (joules/cm) is the flash lamp input energy, n is the number of optical passes, A (cm^2) is the cross section of the reaction cuvette, f is the fraction of electrical energy converted to absorbed light, $\bar{\lambda}$ (nm) is the average absorbed wavelength, $\bar{\phi}$ is the average quantum yield, and $\epsilon(M^{-1}\text{cm}^{-1})$ is the molar absorbance of the transient at the measuring wavelength. For example, with $W = 100$, $n = 1$, $\bar{\lambda} = 400$, $\bar{\phi} = 0.1$,

$A = 1$, and $f = 0.01$, a usable detectivity of $\Delta OD = 0.01$ can resolve weakly absorbing species corresponding to $\epsilon = 30$. Thus, a strong light flash can generate measurable concentrations of almost any long-lived transient with modest requirements on the geometry of light collection, quantum yield, and detection sensitivity. However, when the transient lifetime is less than the flash duration, then the maximum yield is limited also by the rise time of the light flash. In the simple case where the transient decays with an exponential halftime τ(sec), W in Eq. (1) must be replaced by $E\tau$, where E(watts/cm) is the power input. In the previous example if the flash pulse builds up at 10 megawatts/cm, a species of 1 μsec lifetime will attain only 10% the yield of the long-lived intermediate. Furthermore, the absorption measurement must be made within several microseconds of the flash peak. The operating characteristics of typical fast arc, spark, and laser sources reported in the literature are summarized in Table I.

For species persisting for $>10^{-6}$ sec the selection of the detection system is largely a matter of convenience. Typically, spectroflash lamp sources with photographic plate detection are used for surveying the spectral region of interest while kinetics measurements are made with a constant monitoring lamp and photoelectric detection. However, broad wavelength spectra can be obtained with a photoelectric rapid scanning spectrophotometer (Pimentel, 1968) and a streak camera can be used to measure spectral changes in time at better than microsecond resolution. In general, photoelectric methods are preferable for flash photolysis because of higher photometric sensitivity by a factor of $\sim 10^2$, and because there is no necessity to correct for spectral sensitivity as with photographic emulsions. The major limitation in the submicrosecond range is the electrical noise in the detector-amplifier system, particularly the "shot" noise at the photodetector cathode. The well-known formula for r.m.s. shot noise current: $\bar{\imath}_n = \sqrt{2eiB}$ (i is the cathode current, e is the electron charge, and B is the bandwidth) given by Schottky (1918) leads to the following approximate relationship:

$$(\delta i/\bar{\imath}_n) = (\delta i/i_0)\sqrt{(i_0 T_r/e)} \tag{2}$$

where δi is the change in cathode current caused by the optical signal, i_0 is steady cathode current excited by the monitoring lamp, and T_r is the rise time of the detection system. The quantity $\delta i/\bar{\imath}_n$ is the "signal-to-noise" factor (S/N) which limits the accuracy possible in the measurement of optical density charges, while $\delta i/i_0$ determines the photometric sensitivity (see Fig. 1).

Assuming $S/N \sim 20$ is required for accurate flash photolysis work,

TABLE I
SHORT DURATION LIGHT SOURCES FOR FLASH PHOTOLYSIS

Charging voltage (kv)	Input energy (joules)	Time constant (μsec)	Input power (megawatts)	Effective length (cm)	General type	Reference
10	7500	21	360	10	annular lamp	Claesson and Lindqvist (1957)
3.2	0.5	0.2	2.5	0.1	open air gap	Fischer (1957)
10	100	2	50	4	4 argon lamps	Linschitz and Sarkanen (1958)
11	109	5	22	10	annular lamp	Charlson *et al.* (1960)
33	4850	0.2	24,000	20	z-pinch discharge	Niemann and Klenert (1968a,b)
		0.02	20 (output)	~1	Q-switched ruby laser	Wolff *et al.* (1967)
	6600	0.03	10 (output)	~1	Q-switched ruby laser	Danziger *et al.* (1967)
		0.02	4 (output)	1.25	Q-switched ruby laser with frequency doubler	Porter and Topp (1968)
100	1500	0.05	30,000	~1	spark generator	D'Alessio *et al.* (1968)
		0.03	200 (output)	~1	Q-switched ruby laser with frequency doubler	Novak and Windsor (1968)

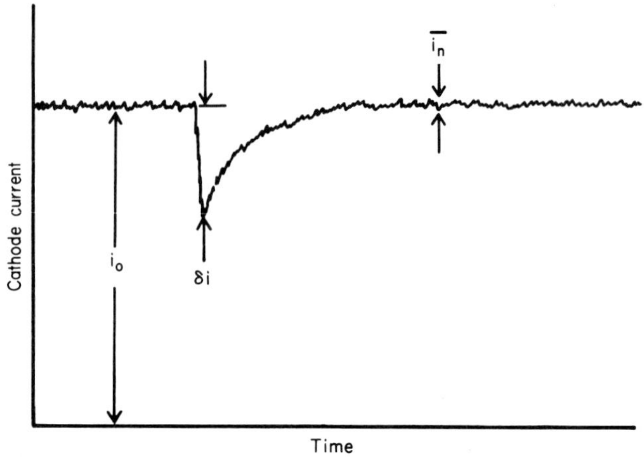

FIG. 1. Photodetector cathode current in flash photolysis: i_0 is the steady level established by the monitoring lamp, δi is the change in cathode current caused by the formation of the transient product, and $\bar{\imath}_n$ is the r.m.s. noise current. The photometric sensitivity is determined by $\delta i/i_0$ while the signal-to-noise parameter is $\delta i/\bar{\imath}_n$.

the photometric sensitivity is limited by the strength of the steady signal i_0. Taking a typical value of maximum photocurrent in a vacuum photomultiplier of 0.1 μamp, Eq. (2) leads to the following results:

T_r (sec)	10^{-4}	10^{-6}	10^{-8}
$\delta i/i_0$	0.003	0.03	0.3

Thus, conventional single flash methods with photoelectric detection are totally useless for flash photolysis in the nanosecond region. Solid state photodetectors can handle higher cathode currents without saturation, but would require very intense monitoring lamps which may be unsuitable for flash photolysis applications.

One possible solution of the noise problem in nanosecond flash photolysis is the use of the photographic method, which necessitates short duration photolysis flash and spectroflash lamps with a means of synchronization. Porter and Topp (1968) devised an ingenious procedure in which a 20 nanosec photolysis flash is obtained from a Q-switched ruby laser and converted to 3470 Å at 10% efficiency with a crystal frequency doubler. Part of the laser pulse is focused on a high-voltage air spark charged just below the breakdown point. The light pulse induces breakdown at a time controlled by the path distance from the laser to the spark gap, which provides a 13 nanosec spectroflash at a minimum 1 nanosec delay. A similar technique was employed by Novak and Windsor (1968) with the additional feature that kinetic measurements are obtained by replacing the spectrograph with a low-dispersion prism focused on the photocathode of an image-converter tube. A time resolved spectrum is produced on the phosphor screen by sweeping the electron beam from 20 nanosec to 100 μsec full scale. Porter and Topp (1969) improved the technique by utilizing the fluorescence of a dye excited by the laser flash as the spectroflash source. The same workers used the laser-pumping flash as the photoelectric monitoring source, which provided sufficient intensity for single-shot photoelectric detection at a time response of several nanoseconds. An alternative procedure is the use of periodic photolysis flashes with signal averaging instead of a single strong flash. Suppose that the photoelectric signal in a short time "sample" during the transient lifetime produces an electric charge δq which is stored in a capacitor and the process is repeated M times. The final signal voltage across the capacitor will be proportional to $M\delta q$ while the shot noise will be proproportional to $\sqrt{M}$, with a net gain in S/N by $\sqrt{M}$. This sequence must be repeated for successive time samples in order to reconstruct the actual signal. The application of these methods in flash photolysis has been pioneered by Witt and his co-workers (Witt, 1967a). In a recent arrangement (Wolff *et al.*, 1967) a giant pulse laser was operated repetitively over a period of minutes and the photoelectric signal measured with a sampling oscilloscope followed by an integrating circuit. The output of the integrator was displayed on a chart recorder which made it possible to measure 0.1% transmission charge in 10^{-8} sec. Pulsed lasers are also convenient photolysis lamps in the microsecond lifetime range where noise problems are not severe. For example, Danziger *et al.* (1967) studied the photochemistry of methylene blue using a 30 nanosec, 0.5 joule pulse from a Q-switched ruby laser and were able to convert the ground state dye almost completely to intermediates, which greatly simplifies the decay analysis. Furthermore, continuous lasers can act as efficient monitoring lamps because of the high spectral intensity and parallel light emission. A helium gas laser (6328 Å) was used as the monitoring lamp for the hydrated electron by Anbar and Hart (1967) in pulse radiolysis and would serve as well in flash photolysis. In fact, a useful beam can be confined to 1 mm diameter so that the nonuniformity of energy deposition across a reaction cell can be determined with a laser probe.

2.2 Use of Electronic Computers

Electronic computers can serve the two general functions in flash photolysis of automated data handling and the solving of differential rate equations. The first is not trivial by any means because point-by-point analysis of spectroscopic plates and oscilloscope photographs is time consuming and subject to error. The digitalization of a microdensitometer output is feasible with an analog-to-digital converter, so that the wavelength calibrations and emulsion sensitivity calculations can be performed by a digital computer. Overlapping absorption spectra can be resolved automatically by extrapolating the resolved portions of the bands so as to fit the data according to appropriate statistical criteria. An automatic scanning device which digitalizes Polaroid photographs of oscilloscope traces has been constructed at the Argonne National Laboratory (Sauer, 1966).

The interpretation of flash photolysis data in terms of reaction mechanism requires the identification of the transient spectra and the quantitative fitting of the measured concentration changes to appropriate rate equations. The analysis of kinetics is relatively direct when the buildup and decay functions follow the simple forms obtained by integration of the differential rate equations such as pure first-order and second-order reactions. However, many cases of practical interest cannot be solved exactly and one may resort to either integration of approximate rate equations or approximate analytical solutions of the correct rate equations. An alternative procedure which deserves much greater application than it has received is the computer solution of the rate equations, because the accuracy possible with modern machines is quite adequate for most experimental work. For example, a digital computer program prepared by Schmidt (1966) for pulse radiolysis or flash photolysis calculates the time dependence of the concentrations of 21 species, which engage in a maximum of 30 first-order and second-order reactions in terms of the rate constants. An alternative procedure which has been applied in the author's laboratory is to solve the differential rate equations with an electronic analog computer. Essentially, an analog computer consists of a set of operational amplifiers plus auxiliary components which can be wired so as to perform the operations of addition, subtraction, multiplication, division, and integration on time-varying voltages. By appropriate scaling, the dependence of different voltages on machine time simulates the changes of the chemical concentrations in real time, so that the rate constants can be determined by fitting the computer solutions to the experimental data. In a recent investigation of the eosin-sensitized oxidation of several aromatics, Chrysochoos and Grossweiner (1968) solved a set of seven first-order and second-order reactions with a small analog computer in order to determine the rate constants for triplet quenching and electron-transfer reactions. In another application. Youtsey and Grossweiner (1969a) matched the intensity profiles of the photolysis flash and spectroflash lamps with the appropriate time delay, so that the entire flash photolysis experiment involving photooxidation of aqueous *p*-cresol could be simulated on the analog computer. This procedure made it possible to determine rate constants and quantum yields for the photochemical hydrated electron under conditions where its lifetime was comparable to the duration of the light flashes.

2.3 Other Techniques

In addition to the advances in fast light pulses and computerized data analysis, a number of interesting experimental methods have been reported which are pertinent to the biological applications of flash photolysis. Porter and Strauss (1966) constructed a microbeam apparatus for flash photolysis on a microscope stage with samples 50 to 250 μm square and from 5 to several hundred microns thick. Holmström and Stening (1964) devised a novel irradiation technique in which fluorescent dye filters are used to increase the efficiency of optical pumping in the visible region. The sample is enclosed in a double wall, cylindrical cell with the active solution filter in the coaxial outer chamber. The dye is selected so that the sample receives direct illumination from the flash lamp in the long wavelength region plus the fluorescence emission from the filter excited by shorter wavelength light not absorbed by the sample. In a typical run using aqueous rhodamine B as the sample ($\lambda_{max} = 550$ nm) and alkaline fluorescein as the active filter [$\lambda_{max} = 494$ nm(abs); $\lambda_{max} = 518$ nm(fl)] the efficiency of rhodamine B bleaching by the flash lamp was improved by a factor of $\sim$5. More recent work has shown that organic dye solutions are capable of laser action when pumped with sufficiently intense light pulses, in which the broad fluorescence-band narrows remarkably and becomes coherent (Schafer *et al.*, 1966). Although the initial work was done using giant-pulse laser excitation, Sorokin *et al.* (1968) induced laser action in a number of organic dyes with a coaxial flashlamp of the Claesson and Lindqvist (1957) type. An interesting property of dye lasers of potential value in flash photolysis work is the shift of the emission wavelength with dye concentration. Extension of the Holström and Stening (1964) active filter technique to the laser geometry might lead to a highly efficient system for flash photolysis investigations of dyes and other colored organic molecules.

Recent improvements in technique have led to more reliable values of initial quantum yields. Bowers and Porter (1966, 1967) used a highly reproducible arrangement with potassium ferrioxalate actinometry to determine the absolute triplet quantum yields for chlorophyll solutions and several dyes and aromatics. Similar information was obtained by

Horrocks and Wilkinson (1968) for aromatic hydrocarbons using a novel method in which *relative* measurements of the triplet yields and fluorescence efficiencies in the presence of a fluorescence quencher, such as xenon gas, are combined to give absolute triplet yields. In related work, Kikuchi *et al.* (1968a,b) investigated E-type and P-type delayed fluorescence with a flash technique, in which the fluorescence intensity and triplet state concentrations were determined in a single apparatus.

The parallel aspects of flash photolysis and pulse radiolysis can be exploited in both types of measurements. In a photochemical process the free radicals and triplet states are generated directly by light absorption in the solute and also by reactions of the initial solutes species with the solvent or added substrates. On the other hand, in pulse radiolysis almost all energy is absorbed by the bulk solvent except in highly concentrated solutions. However, the resulting radicals and electronic excitation can generate free radicals and triplet states from added solutes, which may be the same as with photochemical excitation. For example, the same fluorescein semioxidized and semireduced radicals are formed by flash photolysis (Lindqvist, 1960) and pulse radiolysis (Cordier and Grossweiner, 1968) although by quite different mechanisms. Since the subsequent reactions do not depend on how the intermediates are formed, a choice can be made as to which technique is most convenient or reliable in a given case. This situation is illustrated by the recent determination of the extinction coefficients of the triplet-triplet transitions in several aromatic compounds made by Land (1968), in which the triplet states were formed indirectly by pulse radiolysis in cyclohexane solutions containing benzophenone. The aromatic triplet energy was transferred to benzophenone, followed by hydrogen abstraction from the solvent to give the ketyl radical of known extinction coefficient, which made possible an accurate determination of the aromatic triplet extinction coefficients. In another example (Grossweiner and Rodde, 1968) triplet eosin was formed with a light flash while the hydrated electron was generated with a delayed pulse of 33 mev. The fast reaction between these species was found to be chemiluminescent, and could be followed either by the decrease of the triplet dye optical absorption or light emission from the electronically excited dye.

3. Dye Sensitization

The natural phenomena in photobiology induced by visible light are sensitized photochemical processes at the molecular level; e.g., photosynthesis, phototropism, photoperiodism, phototaxis, and photoreception. The harmful effects of photodynamic action and light-induced carcinogenesis also fall into this broad class of reactions, in which the initial light absorption takes place in a colored sensitizer while the transformation of interest occurs in the "colorless" substrate. The flash photolysis method is well suited for the study of sensitized reactions, because the system can be irradiated with a visible light flash absorbed only by the sensitizer and the resulting transient spectra help identify the initial products of both the excited sensitizer and the substrate. Early flash photolysis results showed that the irradiation of chlorophyll derivatives and retinal pigments in solution generates short-lived triplet states and radical intermediates (e.g., the review of Grossweiner, 1966). Unfortunately, intermediates that can be attributed to the primary act are not obtained *in vivo* so that the biological significance of such species remains an open question. One possible approach involves altering the biological substrate in such a way that primary transients are formed; e.g., Porter and Strauss (1966) observed chlorophyll triplet spectra in plant leaves by treating with a cationic detergent and then encapsulating in glucose. This general technique has been applied to Chlorella and spinach chloroplasts by Witt and co-workers (Witt, 1967b) in an extensive series of investigations which have led to ten different types of absorption changes identified with triplet states, electron-transfer intermediates, and ion-transfer intermediates. Although studies of this type cannot be expected to "explain" photosynthesis or retinal vision, they provide useful information about the possible primary steps which assist in the evaluation of postulated physiological mechanisms. From this viewpoint, flash photolysis research on nonnatural photosensitizers such as dyes can be equally valuable, particularly in that the photochemistry may be less complex.

3.1 Sensitization by Eosin Y

It is interesting to select a particular organic dye and summarize the information that has been obtained about its sensitizing properties from physical and chemical measurements including flash photolysis. The dye selected for this survey is eosin Y (the di-sodium salt of 2′,4′,5′,7′-tetrabromofluorescein) which is a well-known photochemical and photodynamic sensitizer. Four aspects of the sensitization mechanism will be considered: primary excitation and disposition of absorbed light energy; fast interactions in dilute solutions; initial photochemical reactions with low molecular weight substrates; the effect of aggregation and complexing.

The optical spectrum of eosin shows strong absorption bands in the green and ultraviolet regions (518 mm and 255 nm in water) with overlapping, weaker bands between them. Beer's law is not obeyed except in very dilute solutions (Soederberg, 1913). This is indicative of aggregation. It is generally assumed that a shoulder near 490 nm(D band) which grows relative to the principal maximum(M band) is due to the dimer but this has not been proven. Illumination in any of the absorption bands leads to a strong green fluorescence (541 nm in water) attributed to emission from the thermally-equilibrated, first excited-singlet state (S_1) of the conjugated system. (The notation follows the energy diagram of Seliger and Morton [1968] in Vol. IV of this series.) The lifetime of state S_1 is approximately 5 nanoseconds (Maercks, 1968) which rules out its participation in diffusional chemical reactions with other than high concentration substrates. In viscous solvents such as glycerol, the strong fluorescence is accompanied by a weak, long-lived luminescence (Boudin, 1930). Parker and Hatchard (1961) demonstrated that it results from "delayed flourescence" when thermal excitation of the metastable triplet state T_1 induces population of S_1 followed by the allowed fluorescence transition to S_0. They also observed a long-lived weak emission at longer wavelengths (714 nm in ethanol) attributed to phosphorescence when T_1 decays to the ground state S_0. The absorption spectrum of the triplet state was obtained by Zwicker and Grossweiner (1963) with the flash photolysis method, corresponding to the allowed transitions $T_2 \leftarrow T_1$, $T_3 \leftarrow T_1$, etc. More recent flash photolysis work of Bowers and Porter (1967) has led to 0.71 ± 0.10 for the triplet quantum yield of aqueous eosin and 0.19 for the corresponding fluorescence quantum yield. The approach of the sum to unity shows that the radiationless transition $S_1 \rightarrow S_0$ is inefficient compared to fluorescence and the intersystem crossing $S_1 \rightarrow T_1$.

3.1.1 Anaerobic Photosensitization

A considerable body of work indicates that the lowest triplet state T_1 is responsible for the sensitized photochemical and photobiological reactions of eosin. This state has a long natural lifetime ($\sim$1 msec), because the inactivating $T_1 \rightarrow S_0$ transition is forbidden, which favors participation in diffusional and energy transfer processes. Kasche and Lindqvist (1965) demonstrated that triplet eosin is quenched by reacting with itself and ground state dye in the bimolecular processes: $T_1 + T_1 \rightarrow T_1 + S_0$ $T_1 + S_0 \rightarrow 2S_0$. The competing transfer of an electron between interacting molecules leads to the formation of the semi-reduced radical (R) and the semioxidized radical (X): $T_1 + T_1 \rightarrow \mathrm{R} + \mathrm{X}$; $T_1 + S_0 \rightarrow \mathrm{R} + \mathrm{X}$. The EPR measurements of Bubnov *et al.* (1959) indicate that R has the semiquinone structure with the unpaired electron on the central carbon atom, while a phenoxyl radical structure has been postulated for X by Kasche and Lindqvist (1965). The same dye intermediates may be formed when triplet eosin reacts with added oxidants and reductants.

For example, triplet eosin is reduced to the semiquinone by oxidizing phenol (Grossweiner and Zwicker, 1959, 1961), *p*-cresol, tyrosine and tryptophan (Grossweiner and Zwicker, 1963), *p*-phenylenediamine (Kasche and Lindqvist, 1965), allythiourea (Ohno *et al.*, 1966), aniline, resorcinol, *p*-sulfanilic acid, β-naphthol, and *p*-bromophenol (Chrysochoos and Grossweiner, 1968) and hydrogen peroxide (Youtsey and Grossweiner, 1969b). On the other hand, triplet eosin is oxidized to X by reducing ferricyanide (Kasche and Lindqvist, 1965) and *p*-benzoquinone (Ohno *et al.*, 1966). In these sensitized redox reactions the triplet energy of the substrate is higher than that of eosin and energy transfer is unfavorable. However, eosin and anthracene have approximately the same triplet energies and spin transfer can take place: $E^T + A \rightleftharpoons A^T + E$, where E and A refer to eosin and anthracene respectively. This reversible energy transfer was demonstrated by Kikuchi *et al.* (1968b) who flashed the system (in ethanol) with green light to excite E^T and then measured the yields of E^T and A^T by following the delayed fluorescence of each species. This case is particularly interesting because triplet eosin decays by the "E type" delayed fluorescence: $E^T \rightarrow E^*$ (E^* is the excited singlet state) while triplet anthracene decays via the biphotonic "P type" delayed fluorescence: $A^T + A^T \rightarrow A^* + A$.

The dye intermediates can be formed also by exposing aqueous solutions to ionizing radiation. The pulse radiolysis of anaerobic eosin solutions leads to the formation of R by the reaction with the hydrated electrons: $S_0 + e^-_{aq} + H^+ \rightarrow R$, and the formation of X by the attack of hydroxyl radicals: $S_0 + OH\cdot \rightarrow X + OH^-$ (Chrysochoos *et al.*, 1967). The reduction reaction was investigated for the triplet state by first exciting the dye with a light flash and then exposing the solution to a pulse of 33 mev electrons, in the presence of formate to scavenge hydroxyl radicals (Grossweiner and Rodde, 1968). The process is sufficiently energetic to leave the product in an excited state (possibly the dimer triplet) and the reaction can be followed by means of the chemiluminescent emission of eosin. The effect of complexing on the decay of the semiquinone was first observed in pulse radiolysis (of fluorescein) where the apparent bimolecular rate constant of the disproportionation reaction: $R + R \rightarrow S_0 + L$ diminishes with increasing dye concentration (Cordier and Grossweiner, 1968) (L is the leuco base of the dye). It was proposed that R complexes with the dye: $R + S_0 \rightleftharpoons (RS)$ so that only the equilibrium fraction of free semiquinone is available for disproportionation. A similar effect was found for eosin by pulse radiolysis and flash photolysis (Grossweiner, 1968a).

The overall scheme of an eosin-sensitized photochemical reaction in dilute anaerobic solution is summarized in Fig. 2, where QH represents

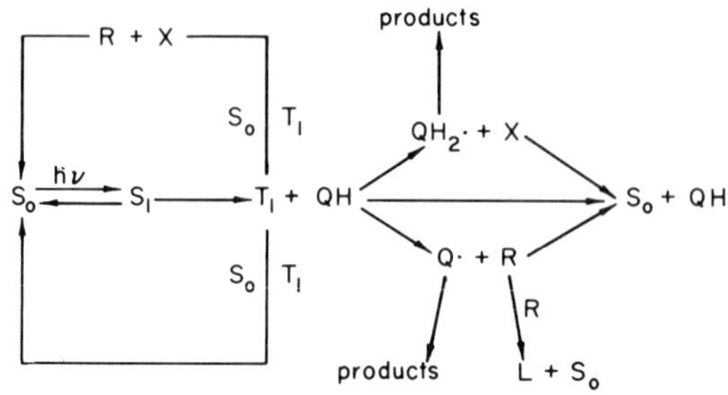

FIG. 2. Primary sensitization mechanism for aqueous eosin under anaerobic conditions. The dye species indicated are the ground state S_0, the excited singlet state S_1, the lowest triplet state T_1, the semireduced radical (semiquinone) R, the semioxidized radical X, and the leuco base L. The species QH represents a reactive substrate which can be half-oxidized to give Q· or half-reduced to give QH_2. The dismutation of R in the absence of a substrate is not shown.

an available substrate. Except at very high light intensities and low dye concentrations, the lifetime of T_1 is controlled by the $T_1 + S_0$ "self-quenching" reaction so that the quantum efficiency for the photolysis of QH will vary with the dye concentration. In addition, nonlinear intensity effects may be involved under flash irradiation conditions or with very intense steady lamps because of $T_1 + T_1$ quenching. It should be noted that the reaction of T_1 with QH and the back reaction of the resulting radicals may be fast processes, so that the overall effect is equivalent to "physical" quenching of T_1 by QH. However, the distinction between quenching and a fast, reversible electron-transfer reaction is important for the corresponding photodynamic effects because only the second case can lead to a primary process in which the initial reaction of the excited dye is with the substrate.

3.1.2 Aerobic Photosensitization

It is known that oxygen can react directly with triplet eosin via several mechanisms. Kasche (1967) has shown that flash photolysis of aqueous eosin in the presence of oxygen leads to triplet quenching and the formation of semioxidized dye: $T_1 + O_2 \rightarrow X + O_2^-$. The subsequent oxidation of the substrate by X will regenerate the dye: $QH + X \rightarrow Q\cdot + S_0$, so that the dye acts to transfer an electron from QH to oxygen. It might be assumed that the competing quenching reaction leaves a free oxygen molecule in the unexcited triplet state: $T_1 + O_2 \rightarrow S_0 + O_2$ but this is probably not the case. Beginning with the early work of Kautsky (1939) a number of experiments have shown that photosensitized oxidations can take place even when the dye and substrate are physically separate (e.g., see Bourbon and Schnuriger, 1967). The labile species proposed to explain that these effects include electronically-excited oxygen in the singlet state, ground-state oxygen with high vibrational excitation, and gaseous O_2^-. Recent work strongly supports the formation of excited singlet oxygen in dry systems (Schnuiger and Bourbon, 1968) and in solutions (Foote and Wexler, 1964) but does not necessarily rule out other possibilities.

The two oxygen states of interest are the $^1\Delta_g$ state at 0.98 ev above the ground state ($^3\Sigma_g^+$) and the $^1\Sigma_g^+$ state at 1.63 ev. The possibility of populating either state depends on the triplet energy of the sensitizer and the reaction conditions. Kearnes *et al.* (1967a) measured the distribution of products in the dye-sensitized photooxygenation of cholest-4-en-3β-ol and related the enone:epoxy ketone product ratio to the triplet energy of the sensitizer. It was found that sensitizers with triplet energy below approximately 1.7 ev give a ratio of 1:4 (e.g., methylene blue, hematoporphyrin, chlorin e_6) with a gradual increase to a ratio of 3:1 for high triplet energy sensitizers (e.g., acridine orange, triphenylene). Eosin is located at the start of the saturation high ratio region which indicates that it can sensitize the formation of both excited oxygen states. It was pointed out by Kearns *et al.* (1967b) that the much shorter lifetime of the higher energy $^1\Sigma_g^+$ state compared to the $^1\Delta_g$ state will lead to a strong dependence of the product ratio on the concentration of the substrate, which was observed with eosin where the ratio varied from 1:5 at 82 μM to 3:1 for 13 mM cholest-4-en-3β-ol. Another possibility which must be considered is enone production via direct hydrogen abstraction by the triplet state, in which case the product ratio would depend on the oxygen concentration. The product distribution for eosin was unchanged from 0.2 to 10 atm oxygen which eliminates the competing excited dye-substrate reaction in this particular system. However, this does not rule out the possibility of direct hydrogen abstraction (or electron transfer) with more readily oxidized substrates such as phenols and aromatic amines, even in aerobic solutions. Furthermore, the formation of the dye-oxygen complex (moloxide of the type proposed by Schenck (1948) must be considered. In a recent investigation of the dye-sensitized photooxidation of leucofluorescein (FH_2) (Koizumi and Usui, 1968; Usui *et al.*, 1969) with methylene blue, thionine, eosin, and acridine as photosensitizers, it was found that the results support the excited oxygen reaction: $FH_2 + {}^1O_2 \rightarrow FH\cdot + HO_2\cdot$. However, bleaching of the dye takes place also in aerobic solutions with low quantum yield and the authors favor the formation of a charge-transfer complex as the intermediate: $T_1 + O_2 \rightarrow S^+ \ldots O_2^- \rightarrow$ products.

3.1.3 Effect of Interactions on Photosensitization

The reactions of eosin discussed thus far involve dilute solutions and relatively low molecular weight substrates. In order to extend the results to biological phenomena it is necessary to consider the influence of dye to dye and dye to substrate interactions. It is known that binding and aggregation of dyes may lead to extensive changes in the optical properties and sensitizing ability (e.g., the review of Bellin, 1965) which might be highly significant in the photobiological system. The aspects which appear to be most basic are the influence of the interactions on the formation of the sensitizer triplet state and its subsequent reactions with oxygen and the biological substrate. The theory of McRae and Kasha (1964) suggests that the triplet quantum yield can be enhanced by dye to dye interactions such as occur in dimers or oriented aggregates, because the allowed inactivating transitions from S_1 to S_0 are suppressed in certain geometrical arrangements. The prediction is supported by photochemical experiments showing that the quantum yield increases when the sensitizer is bound compared to the free state. Bellin and Oster (1957) reported that the eosin-sensitized photoreduction of ascorbic acid is increased fivefold when the dye is bound to the inert, water soluble polymer polyvinylpyrrolidone. It was found also that the self-quenching of eosin fluorescence takes place at lower dye concentrations in the presence of the polymer, which is attributed to the reaction: $S_1 + S_0 \rightarrow T_1 + S_0$. This process differs from the molecular exciton model of McRae and Kasha in that the excitation is associated with a particular dye molecule and not a large group of molecules in the aggregate. However, the end result of higher triplet yield because of excited-singlet state interactions is common to both theories. Youtsey and Grossweiner (1967) found that the fluorescence efficiency of eosin is lower when the dye is multiply-bound to human serum albumin, compared to free and singly-bound eosin (i.e., less than 1 eosin molecule per protein molecule) which is explained by dye to dye interactions of the same type.

Eosin is also an efficient solid state sensitizer. For example, it sensitizes the surface photoconductivity of zinc oxide to visible wavelengths when the dye is on the form of adsorbed molecules, monolayers, or aggregated layers (e.g., the review of Grossweiner, 1968b). According to the mechanism of Weigl (1963), surface oxygen molecules trap zinc oxide conduction electrons in the dark and lower the vacuum electrical conductivity. Under the action of light the excited dye molecule transfers an electron to the conduction band, while the trapped electron as O_2^- occupies the hole and releases the oxygen molecule. This is an example of the generalized electron acceptor molecule (EAM) model proposed by Grossweiner (1968b) in which optical excitation of the sensitizer reverses a dark electron transfer from the substrate to the EAM, which may be the dye itself or an auxiliary surface agent. Many photochemical and photobiological sensitizers are also solid-state sensitizers. For example, the photoconductive zinc oxide-resin binder systems used in the Electrofax process are sensitized by xanthene, triphenylmethane, thiazole, thiazine, azine, and acridine dyes (Greig, 1962), which are the classes of photodynamic dyes active in the degradation of DNA (Bellin and Grossman, 1965). On the other hand, the azo dyes appear to be photodynamically inactive and have never been reported as sensitizers in electrophotography.

3.2 Sensitization by Other Dyes

Bourdon and Schnuriger (1967) proposed a classification of sensitizing dyes based on the photochemical reactivity of the excited dye. According to their scheme, *Class I* dyes have a higher reactivity for oxygen than for good reducing agents and are almost always photodynamic. They include the dyes of quinonoid structure such as xanthenes, thiazines and acridines. In general, *Class I* dyes are fluorescent in solution, phosphorescent in rigid media, and may be photo-reduced. The *Class II* dyes react readily with hydrogen donors by abstraction in the excited state in preference to reaction with oxygen. They include the anthraquinone vat dyes and derivatives of indanthrone, flavanthrone, and pyranthrone. *Class III* dyes may be photodynamic and also are active phototenderers; i.e., the degradation of dyed textile fibers caused by exposure to visible light. The same authors suggest that the essential difference between the two classes is that the π–π^* triplet state is the lowest metastable state of the *Class I* dyes while the n–π^* triplet state is lower than the π–π^* triplet state in the *Class II* dyes (Bourdon and Schnuriger, 1967). The dyes of anthraquinone structure which behave more like *Class I* in their relative reactivity with oxygen (*Class III*) may be those in which substitution lowers the π–π^* triplet level below the n–π^* level. It is apparent from the brief summary of eosin photosensitization that the mechanism can be quite complex for relatively simple chemical systems and strongly dependent on the specific conditions. Two other classes of dyes which have been studied in detail comparable to the fluoresceins are the thiazines, particularly thionine and methylene blue, and the lumiflavin derivatives, particularly riboflavin and riboflavin-5′-phosphate (FMN). These dyes are effective photochemical and photobiological sensitizers and have been investigated with a wide variety of techniques.

3.2.1 Methylene Blue

The earlier flash photolysis results obtained with methylene blue suggest that the initial sensitization processes closely parallel those of eosin, i.e., light absorption in the visible band leads to population of the triplet state via the excited singlet state; the triplet state reacts with the various reductants to form the semiquinone, which in turn undergoes dismutation leading to the leuco base. The experimental support for this mechanism was reviewed by Grossweiner (1966). One of the interesting complexities in the study of aqueous methylene blue involves the spectra, dissociation constants, and chemical reactivities of the different aggregation and protonation states. The dye itself exists in three states which can be denoted as: MBH_2^{3+} ($pK_a = -5.1$), MBH^{2+} ($pK_a = 0.0$), and MB^+ so that a multiplicity of states might be expected in the intermediates. Matsumoto (1964) studied the flash photolysis of methylene blue at different pH in the presence of ferrous ion, *p*-phenylenediamine, and hydroquinone as reductants and identified the spectra of four semiquinone states: $M\dot{B}H_3^{3+}$ (in conc. H_2SO_4), $M\dot{B}H_2^{2+}$ (in 0.1 *N* H_2SO_4), $M\dot{B}H^+$ (pH 3–8), and $M\dot{B}$ (pH >9). These assignments were confirmed in a pulse radiolysis investigation made by Keene *et al.* (1965; 1968) in which methylene blue was irradiated with 2 μsec pulses of 4 mev in the presence of formate to convert OH· and H· from water decomposition to CO_2^- or COOH depending on the acidity. The reduction of the dye by e_{aq}^- and oxidized formate led to spectra of the semiquinone protonation states and also accurate measurements of their dismutation rate constants. The pulse radiolysis of aqueous methylene blue in the absence of formate gives the semiquinone absorption at 420 nm (due to e_{aq}^- attack) and a band near 520 nm which was attributed to the semioxidized radical $M\dot{B}^{2+}$ from the reaction with OH·. The assignments have been expanded by recent laser flash photolysis experiments of Danziger *et al.* (1967) in which it was possible to convert 5 μM methylene blue entirely into transients by means of a 0.5 joule, 30 nanosec giant pulse ruby flash. The results confirm the earlier identification of the triplet state with a 30 μsec lifetime band at 415–420 nm in neutral or alkaline solution (Kato *et al.*, 1964; Matsumoto, 1964). However, a new 2 μsec transient was observed near 550 nm which was identified with the triplet state of the dimer from its dependence on initial dye concentration. The 520 nm absorption of 140 μsec lifetime was found to result from the decay of the dimer triplet state and cannot be the monomeric form of the semioxidized dye. In fact, the description given by Kato *et al.* (1964) of a "loosely-combined pair of semiquinone and a half-oxidized state" appears to be correct and leads to a structure of the type: MB^{2+} - - -$M\dot{B}$, where the 420 nm band is due to the semiquinone moiety and the 520 nm band is due to the semioxidized moiety. An investigation of methylene blue dimerization in mixed solvents (Blandamer *et al.*, 1967) shows that complexing is strongly favored in pure water at low temperatures. It is suggested that the water structure plays a special role in stabilizing aromatic anion and cation dimers, a factor which may be of considerable significance in the interpretation of biological photosensitization.

3.2.2 Flavin Pigments

The flavin pigments are of considerable interest because of their enzymatic function as hydrogen carriers in flavoproteins. They are also effective photochemical sensitizers, which have motivated flash photolysis research directed towards clarifying the mechanism of the oxidation-reduction reactions and the structure of the intermediates. The early work of Holmström (1962, 1964) on FMN showed that irradiation in deaerated aqueous solutions leads to semiquinone formation and that the yield is enhanced in the presence of ethylenediamine tetraacetic acid (EDTA) as a reductant. A triplet state precursor to the semiquinone was postulated on the basis of quenching by iodide ion. Similar results were obtained by Terenin *et al.* (1965) with riboflavin and luminflavin in pyridine or glycerol solutions using diphenylamine as the reductant. The transient spectra showed both the semiquinone absorption and that of the diphenylamine radical cation. Thus, the earliest flash photolysis work suggests that flavin reduction in these systems takes place in a two-step, one-electron process, although the triplet spectra were not detected. An investigation of aqueous luminflavin (Tegner and Holmström, 1966) led to the identification of a short-lived band at 650 nm with the triplet state. In a more recent study of luminflavin flash photolysis, Knowles and Roe (1968) resolved the triplet bands at 375 nm and 640 nm and the semiquinone bands at 347 nm and 490 nm (pH 6.9). It was established that luminflavin has a reactive triplet state by demonstrating triplet to triplet energy transfer from lumiflavin to acridine. However, no direct evidence was found to support the assumption that the triplet state is the precursor of the semiquinone in the flavin-sensitized oxidation of guanylic acid in aqueous solution. Green and Tollin (1968a) made an extensive investigation of flavin photoreduction in nonaqueous, hydroxylic solvents including primary alcohols, ethylene glycol, propane and butane diols, and glycerole, and concluded that the lowest triplet state of lumiflavin is the reactive intermediate in hydrogen abstraction either from the solvent or from added reductants such as EDTA. However, in the case of riboflavin or FMN it is proposed that an *intra*molecular hydrogen atom transfer takes place in the triplet state (from the 2′ hydroxyl group of the ribityl side chain) to form a biradical which is the effective oxidizing species. The sequence was summarized by: $FMN + h\nu \rightarrow FMN_S^* \rightarrow FMN_T^* \rightarrow \cdot FMN\cdot$; $\cdot FMN\cdot + AH_2 \rightarrow (FMNH\cdot + AH\cdot) \rightarrow FMNH_2 + A$, where $\cdot FMN\cdot$ is the biradical and AH_2 is the reductant. The sensitization act in this case is essentially a two-electron transfer, in contrast to the two-stage, one-electron transfer sequence proposed for aqueous systems. An extension of this work to the glycerol/lumiflavin/*p*-benzoquinone system (Green and Tollin, 1968b) displayed complex behavior involving the semiquinones of the flavin and benzoquinone and the oxidized solvent radical. Similar results were obtained with FMN except that the fully reduced flavin is reoxidized by benzoquinone: $FMNH_2 + BQ \rightarrow FMNH\cdot + BQH\cdot \rightarrow FMN + BQH_2$, where BQ, BQH·, and BQH_2 denote benzoquinone, the semiquinone, and hydroquinone, respectively. This sequential electron transfer from glycerol through flavin to quinone is consistent with the postulated role of flavin in mitochondrial electron transport. However, the specific role of the flavin triplet state remains an open question which undoubtedly will be the subject of additional work.

4. Photoionization of Aromatics in Solution

In most work on organic photochemistry it has been assumed that the primary photochemical act involves bond splitting, photoisomerization, or the formation of reactive molecular excited states. However, recent flash photolysis research has demonstrated that photoionization of aromatics by ejection of a solvated electron may take place in polar fluids, which is of direct interest in photobiology because many processes involve oxidation-reduction steps in which the electron itself might be an intermediary.

4.1 Rigid Media

It has been known since the early work of G. N. Lewis and his associates (Lewis and Lipkin, 1942; Lewis and Bigeleisen, 1943) that the near-ultraviolet irradiation of oxidizable molecules in a rigid hydrocarbon medium leads to the free radical products of one-electron oxidation. The mechanism has been the subject of much speculation because the photon energy of the absorbed light is typically 3 to 4 ev smaller than the gas phase photoionization potential (PIP) of the molecule. For example, in the oxidation of N,N,N′, N′-tetramethyl-*p*-phenylenediamine (TMPD) to Wurster's Blue, the conversion at liquid air temperature can be effected with 3.5 ev photons, (Lewis and Bigeleisen, 1943) although the gas phase PIP is 6.6–6.7 ev. This reaction in 3-methylpentane has been studied extensively by Albrecht and co-workers, based on measurements of phosphorescence (Cadogan and Albrecht, 1965), photoconductivity (Johnson and Albrecht, 1966a,b), polarized photoselection (McClain and Albrecht, 1966), flash photoconductivity (Pilloff and Albrecht, 1968), and quantitative spectrophotometry (Cadogan and Albrecht, 1968). The accumulated evidence indicates that photoionization involves a two-photon process, in which the lowest triplet state of TMPD absorbs the second photon. This mechanism is supported by the emission measurements of Gibbons *et al.* (1965), who reported that the thermoluminescence yield from toluene, phenol, aniline, naphthalene, phenanthrene, fluorene, diphenylmethane, perylene, and biphenyl after 2537 Å irradiation in 3-methylpentane at 77°K is proportional to the square of the photolyzing light intensity. The mechanism proposed by the latter workers can be summarized as: $S_0 + h\nu_1 \rightarrow S_1 \rightarrow T_1$; $T_1 + h\nu_2 \rightarrow S^{**} \rightarrow S^+ + e^-$ where S^{**} is a highly excited state resulting from light absorption by the triplet state.

The optical studies have been accomplished by ESR investigations which lead to the same triplet-triplet absorption mechanism for the production of free radicals; e.g., for aromatic amines and naphthalene in frozen ethanol (Siegel and Eisenthal, 1965); α-naphthol in frozen methanol (Shimokoshi *et al.*, 1967); purine, adenosine, and guanosine in ice at 77°K (Hélène *et al.*, 1966), and tyrosine, polytyrosine, tryptophan, and their peptides in ice at 77°K (Santus *et al.*, 1968). The ESR technique can distinguish unambiguously between triplet states and free radicals, and in conjunction with phosphorescence measurements shows the conversion kinetics between these intermediates. The aromatic triplet lifetimes are typically the order of seconds at 77°K so that the biphotonic process is quite feasible at ordinary light intensities. However, the situation is more complex at higher temperatures where the triplet lifetimes are shorter and diffusional processes can take place.

4.2 Fluid Meida

The first indication that ultraviolet irradiation of an aromatic in water at room temperature leads to photoionization follows from early flash photolysis experiments showing that both neutral phenol and the phenolate anion give the same intermediate (Grossweiner and Mulac, 1959), which was identified as the neutral phenoxyl radical (Land *et al.*, 1961). Jortner *et al.* (1963) proposed that the electron is ejected from phenolate as a hydrated electron (e^-_{aq}), based on scavenging experiments in which irradiation of aqueous phenolate sensitizes the reduction of nitrous oxide to molecular nitrogen: $N_2O + e^-_{aq} + H^+ \rightarrow N_2 + OH\cdot$. The photochemical generation of the hydrated electron was confirmed in a series of investigations in the author's laboratory showing that flash photolysis of oxidizable aromatics in aqueous solution leads to the broad, red absorption band of e^-_{aq} plus longer-lived spectra, identified with the aromatic radical products (Grossweiner *et al.*, 1963; Swenson *et al.*, 1963;

Dobson and Grossweiner, 1965; Grossweiner and Joschek, 1965; Joschek and Grossweiner, 1966; Chrysochoos and Grossweiner, 1968). Photoionization was found to be a significant initial process for benzene derivatives with electron-donating substituents (e.g., phenol, anisole, aniline, thiphenol, *p*-xylene), aryl mono-carboxylic acids, and 5-member ring heterocyclics (e.g., benzofuran, thiophene, pyrrole, and indole derivatives). Many of the same molecules can be photoionized in rigid hydrocarbon media at low temperatures, which suggests that the biphotonic mechanism with triplet intermediary may apply. However, there are important differences between the two cases because the triplet lifetimes of single aromatic compounds in fluid water are very short, the strong coulombic attraction between the ejected electron and the radical ion is shielded by the high dielectric constant of water, and direct electron transfer from the excited state to the medium may be facilitated by hydrogen bonding in the unexcited molecule. An early attempt to detect a nonlinear dependence of the initial e^-_{aq} yield from *p*-cresol on the photolysis flash intensity led to negative results (Dobson and Grossweiner, 1965). More recently, Youtsey and Grossweiner (1969a) repeated this experiment using analog computed simulation of the flash lamp intensity profile to correct for e^-_{aq} growth and decay during the light flash and again a linear intensity dependence gave the best fit to the data. However, this does not necessarily prove that electron ejection requires only the absorption of one photon, because a short-lived precurser to the hydrated electron can lead to overall linear intensity dependence at high light intensities. For example, consider the initial step: $A + h\nu_1 \rightarrow A^*$ followed by the two competing processes: $A^* \rightarrow A$; $A^* + h\nu_2 \rightarrow A^+ + e^-_{aq}$. A kinetics analysis (Siegel and Eisenthal, 1965) shows that the rate of e^-_{aq} formation is a square function of light intensity at low intensities (i.e., when the dark decay of A^* is much higher than the rate of second photon absorption) and a linear function of light intensity at limiting high light intensities. Thus, the linear flash photolysis result does not prove that the biphotonic process is inoperative at ordinary light intensities.

Efforts have been made to investigate the TMPD case in fluid organic solvents in order to clarify the photoionization mechanism. Yamamoto *et al.* (1966) observed a transient of 82 μsec lifetime after flash photolysis of TMPD in liquid paraffin at 20°C, different from the very long-lived Wurster's Blue ($TMPD^+$) absorption obtained in polar solvents which they identified as the triplet state. Potashnik *et al.* (1969) extended the flash irradiation measurements to fluid solution of isopentane-isopropanol at —145°C and identified a long-lived emission as the delayed fluorescence resulting from triplet-triplet annihilation (P-type). The addition of oxygen led to triplet quenching as expected, but the yield of $TMPD^+$ was not completely suppressed even in high oxygen concentrations. Nevertheless, the light intensity measurements under oxygenated conditions gave the square dependence characteristic of a biphotonic process. The efficiency of the nontriplet biphotonic oxidation process is considerably higher in more polar solvents such as methanol and increases with increasing temperature. These authors suggest that the second photon is absorbed by a charge transfer state formed between the solute and the solvent, extending a mechanism proposed by Grossweiner and Joschek (1965) for aqueous solutions. On the other hand, in deaerated isopentane-isopropanol solutions the yield of $TMPD^+$ was almost linear in light intensity but very susceptible to purification of the solvent. The proposed mechanism in this case involves single-photon oxidation of the TMPD triplet state by very smal amounts of impurity, a process investigated in detail by Linschitz *et al.* (1967). Thus, there appear to be at least three photooxidation mechanisms which can be operative under appropriate conditions: (a) the biphotonic process with triplet state intermediary in low-temperature rigid media; (b) the monophotonic process with triplet intermediary in the presence of trace impurities in fluid, nonpolar solvents; (c) the biphotonic process with charge-transfer intermediary in polar solvents, particularly at higher temperatures.

4.3 The Primary Act

The scavenging experiments of Jortner *et al.* (1963) on phenolate were analyzed in terms of the Noyes cage scavenging theory (Noyes, 1955) in which the π–π^* excited state "dissociates" to give the phenoxyl radical and hydrated electron in the solvent cage. The theory postulates that the diffusive separation of the "geminate" co-products competes with secondary cage recombination, so that the maximum quantum yield for scavenger conversion is taken to be the primary dissociation yield in the excited state. The analysis predicts that the quantum yield of reacted scavenger should depend on the square root of the scavenger concentration, which was observed for the scavenging of e^-_{aq} from phenolate by N_2O. The extension of this work to β-naphthol by Ottolenghi (1963) showed that photolysis of the neutral molecule leads to both hydrogen atoms and hydrated electrons. A comparison of the scavenger quantum yields with fluorescence efficiency data (Weller, 1952) indicates that the hydrated electron is ejected from the thermally equilibrated fluorescent state. A similar mechanism was proposed by Youtsey (1968) for the flash photolysis of aqueous *p*-cresol, in which electron ejection and splitting of the O—H bond are competitive processes, both of which lead to the same *p*-methylphenoxyl radical. In a recent paper Czapski and Ottolenghi (1968) criticize the solvent cage model as applied to photochemical scavenging because it leads to inconsistent rate constants for the reactions between the scavenger and e_{aq}^-. They propose that the observed dependence of quantum yield on scavenger concentration can be explained in terms of homogeneous back reactions in the bulk solution and they demonstrate that this is the case in the photolysis of aqueous iodide ion. The flash photolysis method was applied to determine the effect of N_2O concentration on the initial radical yields. In the case of phenolate ion, the phenoxyl radical yield does not change when the N_2O partial pressure is increased from 0.015 atm to 1 atm. The lower concentration leads to a maximum e_{aq}^- lifetime in the bulk $\sim$1 μsec so that the homogeneous reaction can account for the observed scavenging of e_{aq}^- by N_2O under flash photolysis conditions. The maximum e_{aq}^- lifetime at 1 atm N_2O is $\sim$10 nanosec, which shows that detection of a solvent cage effect would require much higher scavenger concentrations than have been employed.

The preliminary results of Czapski and Ottolenghi suggest that the intermediate state from which the electron is ejected is not available for reaction with added scavengers under ordinary experimental conditions. However, recent work of Steiner and Kirby (1969) on the fluorescence quantum yields and lifetimes of indole derivatives in the presence of quenchers shows a remarkable correlation between quenching efficiency and the reactivity of the quencher with hyrated electrons. For example, a correlation coefficient of 0.82 ± 0.05 was found for aqueous acetyltryptophan (pH 5) fluorescence quenching by 23 different compounds of widely different chemical types, including inorganic anions and cations, sulfur containing compounds, amino acids, and uncharged aliphatics. The parallel decrease of the fluorescence quantum yield and lifetime indicate that radiationless deactivation of the excited state is the dominant mechanism (excluding several cases in which complexing was found) and it is proposed that quenching results from the transfer of an electron from the excited state to the acceptor molecule. A similar competition between fluorescence and e_{aq}^- formation was observed by Ottolenghi (1963) for β-naphthol and Youtsey (1968) for *p*-cresol. The process parallels case (b) above except that the "impurity" is the quencher, present in much higher concentrations to permit reaction with the excited singlet state. The common factor between the different photooxidation mechanisms appear to be that a monophotonic process is possible when the excited electron is transferred from either the triplet state or the fluorescent state to an electron acceptor molecule. However, in the absence of impurities or quenchers a second photon excitation is required to transfer the electron from the excited molecule or its subsequent intermediate state to the medium as a hydrated electron.

4.4 Biological Implications

The role of the hydrated electron in photobiology remains a speculative subject. A major problem is the high reactivity of hydrated electrons with proteins and nucleic acids. For example, the bimolecular rate constant for the reaction of e_{aq}^- with DNA is $\sim 10^{13}$ liters/mole-sec (Scholes *et al.*, 1965) which makes it difficult experimentally to distinguish the rapid formation and reaction of e_{aq}^- from other primary processes such as bond splitting and triplet energy transfer. The ultraviolet inactivation of aqueous enzymes is a phenomenon in which the hydrated electron may be involved, because the aromatic amino acids are readily photoionized in free solution. This would not be a major factor in enzymes of high cystine content according to the theory of McLaren and co-workers (McLaren and Luse, 1961; Luse and McLaren, 1963; McLaren and Hidalgo-Salvatierra, 1964), in which the quantum yield for enzyme inactivation is the sum of the quantum efficiencies for destruction of the light-absorbing amino acid residues. However, if cooperative interactions take place between the aromatic amino acids and cystine as proposed by Augenstein and co-workers (Augenstein, 1953, 1956; Augenstein and Ghiron, 1961; Grist *et al.*, 1965; Augenstein and Riley, 1964; Risi *et al.*, 1967) then the hydrated electron may act as a sensitizing intermediary. Evidence for such interactions derives from the photochemical studies of Dose (1968a,b) who reported that the addition of tryptophan and tyrosine to a lesser extent enhances the reduction of the disulfide bond of free cystine. In a recent flash photolysis investigation (Grossweiner and Usui, 1969) it was found that the cystine electron adduct is formed either by irradiating indole in the presence of cystine or by irradiating indole in the presence of cysteine and nitrous oxide. The proposed mechanism parallels the pulse radiolysis study of cystamine by Adams *et al.* (1967) which led to the following reactions for the two cases: (a) *reductive path* $RSSR + e_{aq}^- \rightarrow RSSR^-$; (b) *oxidative path* $e_{aq}^- + N_2O + H^+ \rightarrow OH\cdot + N_2$; $OH\cdot + RSH \rightarrow RS\cdot + H_2O$; $RS\cdot + RSH \rightleftharpoons RSSR^- + H^+$ $(R = NH_2CH_2CH_2)$ except that e_{aq}^- is formed by photoionization of indole. Although this finding appears to explain the results of Dose with free cystine, it does not imply that hydrated electrons must be involved in ultraviolet enzyme inactivation because other competing processes may be more efficient.

The first flash photolysis studies of nucleic acid derivativatives have been reported recently. Herbert *et al.* (1968) observed a transient band at 320 nm by flash photolysis of orotic acid at pH 5. The kinetics indicate that the intermediate reacts with orotic acid and is quenched by oxygen, which led to the identification with the triplet state. It is proposed that the bimolecular reaction of the triplet state with the unexcited molecule leads to dimerization, in agreement with the mechanism of Greenstock *et al.* (1967). A combined flash photolysis and pulse radiolysis investigation of 5-bromouracil by Danziger *et al.* (1968) indicates the occurrence of at least two primary photochemical reactions, which cannot lead to dimerization or hydration products because they require irradiation below 240 nm. Flash photolysis of BrU in acidic solution in the presence of Br^- led to the Br_2^- absorption at 360 nm (Grossweiner and Matheson, 1957), which is attributed to the homolytic cleavage of the C-Br bond. A band at 330 nm formed only in alkaline solutions was identified with the semioxidized radical of BrU anion resulting from ejection of a hydrated electron. The assignment is based on the doubling of the band height by the presence of N_2O and the formation of the same band by pulse radiolysis in the presence of N_2O, which are explained by the reaction of OH· with the BrU anion in both cases. These preliminary experiments suggest that the photolysis mechanism of the pyrimidine bases in aqueous solution is quite complex, involving triplet formation, bond cleavage, and electron ejection to a relative extent strongly dependent on the irradiation conditions. However, the system appears to be amenable to the combined attack of flash photolysis, pulse radiolysis, and product analysis and one would expect considerable progress during the next few years.

5. Discussion and Summary

The most important parameters in flash photolysis are the quantum yield for the production of the intermediates, their lifetime under the experimental conditions, and the detection sensitivity in an accessible spectral region. At the present state of development it is possible to study virtually any encounter-limited free radical reaction in solution with a submicrosecond light flash and to detect many aromatic π–π^* and n–π^* triplet states. The Q-switched, giant pulse laser source of nanosecond duration should resolve all metastable states and the fluorescent states of many molecules, while the picosecond mode locked laser (Rentzepis, 1968) makes it possible to study vibrational relaxation of excited states and other processes with time constants $\sim 10^{-12}$ sec. In cases where the detection of the intermediates from optical absorption spectra is not feasible, it may be possible to substitute another physical property such as electrical conductivity (Witt, 1967a) or EPR (Smaller *et al.*, 1968). However, the fastest measurements of this type require 10^{-5}–10^{-6} sec which limits the applicability. On the other hand, the initiation of reactions by the rapid injection of ionizing radiation from a microwave linear accelerator or pulsed Van de Graaf generator can be effected in several nanoseconds with present equipment, and will attain the picosecond region if the electron bunching substructure of the linear accelerator output can be utilized. Since many triplet state and free radical photochemical intermediates can be formed also with pulse radiolysis, the two techniques are mutually complimentary and can be correlated to advantage.

The detailed studies of dye sensitization in fluid solution show that the triplet state of the sensitizer invariably is the reactive intermediate. The same conclusion follows for photodynamic sensitization, although it is based on less direct evidence; e.g., the reviews of Spikes (1968) and Grossweiner (1969). However, in solid state sensitization it is generally accepted that the excited singlet state of the sensitizer transfers either charge or energy to the substrate (West and Carroll, 1966). The primary sensitization act is not an equilibrium process so that the conventional thermodynamic parameters cannot be applied to evaluate the energetics. A comparison of different dyes in solution can be made, following the approach of Chrysochoos and Grossweiner (1968), in which the ability of the triplet state to accept an electron is proportional to (E_T–F$E_S°$), where $E_S°$ is the reduction potential of the dye, E_T is the (positive) phosphorescence energy and F is the Faraday constant. Roughly, the dyes with the least negative catholic half-wave potentials should be the best electron acceptors, leading to oxidizing ability in the order: methylene blue, riboflavin, eosin. In the case of aqueous eosin this energy parameter equals 35 kcal/mole, so that triplet eosin should be able to oxidize molecules whose first oxidation potentials are less negative than —1.5 volts on the U.S. scale. The "apparent potentials" of Fieser (1930) indicate that $E°$ for phenol, aniline, β-naphthol, and resorcinol ranges from —1.2 to —1.3 volts, which is consistent with the flash photolysis results of Chrysochoos and Grossweiner (1968). A similar estimation shows that triplet eosin can effect the one-electron oxidation of hydrogen peroxide as observed by Youtsey and Grossweiner (1969b).

From the molecular orbital viewpoint, the reduction of the triplet dye involves the addition of an electron to the lowest empty molecular orbital of the *pi* electron system, accompanied by the transition of an electron from the triplet state to the vacancy in the highest filled molecular orbital. Conversely, oxidation of the excited dye requires the removal of the most energetic electron, so that corresponding energy parameter is ($E^* + \mathrm{F}E^{\circ}_{S+}$) where E°_{S+} is the reduction potential of the semioxidized form of the dye and E^* is the energy of the excited state. Unfortunately, anodic potentials of sensitizing dyes are not available although, in general, the best sensitizers of photoreduction reactions should be the dyes with the lowest PIP and highest excitation energy. The available data indicates that the PIP of dyes is typically 7.2–7.4 ev (Terenin and Vilessov, 1964), so that E^* probably is the more important factor. Excluding the triplet-triplet and triplet-dye electron-transfer reactions, photosensitized reductions are less frequent in solution. However, the significantly higher energy of the excited singlet state compared to the triplet state may explain why the former is the intermediary in solid state sensitizations, in which the excited dye donates an electron to the conduction band of the semiconductor substrate.

An important consideration in both dye sensitization and photoionization of aromatics in solution involves the utilization of the absorbed light quanta. Bowers and Porter (1967) have shown that the total of the triplet state quantum yield plus the fluorescence efficiency is close to unity for the fluorescein dyes in aqueous solution and also several polycyclic hydrocarbons in nonpolar organic solvents, which indicates that radiationless decay modes of the excited singlet state (including fast reversible photolysis steps) are minor in these cases. However, the situation is quite different with the ultraviolet photolysis of aromatics in aqueous solution where hydrated electron formation yields are significant, e.g., 0.23 for phenolate at 2537 Å (Jortner *et al.*, 1963) and 0.092 for β-naphtholate at 214–360 nm (Ottolenghi, 1963). Furthermore, in the case of the neutral phenolic molecules, splitting of the O—H bond can take place as well, probably from the vibrationally excited or "hot" excited singlet state. A reasonable sequence of events subsequent to light absorption by an aromatic molecule is bond splitting before 10^{-12} to 10^{-13} sec, thermal stabilization of the excited singlet state after 10^{-11} sec, and partition of the thermally-equilibrated excited singlet state into fluorescence, radiationless inactivation, intersystem crossing to the triplet state, and the sequence which leads to the photochemical solvated electron from 10^{-7} to 10^{-9} sec. As discussed in Section 4, electron release is possible also via absorption of a second photon in the triplet state or a fast reaction of the triplet state with an oxidizing impurity, but the principal mechanism in polar media at room temperature appears to involve the nontriplet pathway. This already complex situation may be further complicated by proton transfer between the excited singlet state and the medium, to an extent dependent on the acidity and lifetime of the excited state (Weller, 1952), or the formation of an excimer, i.e., a stoichiometric complex of the excited molecule with the solvent (Walker *et al.*, 1966; 1967). Correlated investigations of the fluorescence properties with flash photolysis can help sort out the relative importance of the competing primary processes, because each technique provides direct information which can only be inferred from the other.

There is an element of truth in the generalization that the simplest molecules have the most complicated photochemistry. For example, sensitization by the fluorescein dyes has been clarified to a considerable extent, while important questions remain unanswered in the mechanism of halide ion photolysis. Apart from the obvious factor that a deeper level of understanding is sought for in the smaller molecules, there are physical reasons why intramolecular processes in large molecules may be more amenable to experiment. In the case of polycyclic aromatic hydrocarbons and dyes, the electronic properties in the zeroeth order can be correlated with the length of the conjugated chain by comparisons between homologous series of molecules. The concept of the chromophore and the auxochrome is useful in such molecules, to identify spectral and intensity changes with particular functional groups. The shift of the electronic transitions to lower energies facilitates the measurement of absolute flourescence and phosphorescence efficiences, while the longer-lived triplet states can be investigated with flash photolysis. In addition to such experimental advantages, there is an important photochemical implication in the "color" of highly conjugated systems, because the photon energy in the longest wavelength absorptions may be too low to break the weakest bonds in the molecule. Thus anaerobic aqueous eosin solutions are completely stable to irradiation in the 518 nm band, although photolysis can be effected with ultraviolet light. The spectral selectivity has an important consequence in the photochemistry of pyrimidine bases, because the radical products observed by flash irradiation below the 260 nm absorption can be isolated from the precursors of the dimer and hydrate products, which are believed to be excited states.

Although these factors tend to offset the chemical complexity of larger molecules, the fundamental problem remains that *in vitro* photochemical reactions may have little relationship to the true function of the molecule in nature. For example, the primary photochemical act in the biological system may be relatively insignificant in the free molecule or buried in the "noise." It is absolutely essential that quantum yields in the biological process be defined (a nontrivial problem) and measured to ensure that results obtained with model systems are relevant to the basic biological problem. The alteration of the primary mechanisms in the free molecule caused by intermolecular interactions can be explored by introducing molecules which can accept excitation energy or act as electron donors and acceptors. At a higher level of organization, flash photolysis experiments can be carried out with molecular biological systems including aqueous enzymes and nucleic acids, particularly to determine whether the initial transient spectra parallel the results obtained with the chromophores in free solution. The extension of flash photolysis to unmodified cellular systems has not been successful thus far. However, recent results obtained with chlorophyll in detergent treated plant leaves (Porter and Strauss, 1966), chlorophyll in altered chloroplasts at low temperature (summarized by Witt, 1967b) and cytochrome *c* in bacterial cell suspensions at low temperature (summarized by Chance *et al.*, 1967) indicate that this technique will be of increasing importance.

Acknowledgment

The author is pleased to acknowledge the support of the National Institutes of Health on Grant No. EY-516 during the preparation of this article.

References

Adams, G. E., McNaughton, G. S., and Michael, B. D. (1967). *In* "The Chemistry of Ionization and Excitation" (G. R. A. Johnson and G. Scholes, eds.), pp. 281–293. Taylor and Francis, London.

Anbar, M., and Hart, E. J. (1967). *J. Am. Chem. Soc.* **71,** 4163.

Augenstein, L. (1953). *In* "Symposium on Information Theory in Biology" (H. Quastler, ed.), pp. 119–122. Univ. of Illinois Press, Urbana.

Augenstein, L. (1956). *In* "Symposium on Information Theory in Biology" (H. Yockey, ed.), pp. 287–291. Pergamon Press, New York.

Augenstein, L., and Ghiron, C. A. (1961). *Proc. Natl. Acad. Sci. U. S.* **47,** 1733.

Augenstein, L., and Riley, P. (1964). *Photochem. Photobiol.* **3,** 353.

Bellin, J. S. (1965). *Photochem. Photobiol.* **4,** 33.

Bellin, J. S., and Grossman, L. I. (1965). *Photochem. Photobiol.* **4,** 45.

Bellin, J. S. and Oster, G. (1957). *J. Am. Chem. Soc.* **79,** 2461.

Blandamer, M. J., Brivati, J. A., Fox, M. F., Symons, M. C. R., and Verma, G. S. P. (1967). *Trans. Faraday Soc.* **63,** 1850.

Boag, J. W. (1968). *Photochem. Photobiol.* **8,** 565.

Boudin, S. (1930). *J. Chim. Phys.* **27,** 285.

Bourdon, J., and Schnuriger, B. (1967). *In* "Physics and Chemistry of the Organic Solid State" (D. Fox, M. M. Labes, and A. Weissberger, eds.), Vol. III, pp. 59–131. Wiley (Interscience) New York.

Bowers, P. G., and Porter, G. (1966). *Proc. Roy. Soc. (London)* **A296,** 435.

Bowers, P. G., and Porter, G. (1967). *Proc. Roy. Soc. (London)* **A297,** 348.

Bubnov, N. N., Kibalko, L. A., Tsepalov, V. F., and Shliapintokh, V. Ya. (1959). *Opt. i Spektroskopiya* **7,** 117.

Cadogan, K. D., and Albrecht, A. C. (1965). *J. Chem. Phys.* **43,** 2550.

Cadogan, K. D., and Albrecht, A. C. (1968). *J. Phys. Chem.* **72,** 929.

Chance, B., DeVault, D., Legallais, V., Mela, L., and Yonetani, T. (1967). *In* "Fast Reactions and Primary Processes in Chemical Kinetics" (S. Claesson, ed.), pp. 437–464. Wiley (Interscience), New York.

Charlson, R. J., Harrison, H., and Hardwick, R. (1960). *Rev. Sci. Instr.* **31,** 46.

Chrysochoos, J., and Grossweiner, L. I. (1968). *Photochem. Photobiol.* **8,** 193.

Chrysochoos, J., Ovadia, J., and Grossweiner, L. I. (1967). *J. Phys. Chem.* **71,** 1629.

Claesson, S., and Lindqvist, L. (1957). *Arkiv Kemi* **12,** 1.
Cordier, P., and Grossweiner, L. I. (1968). *J. Phys. Chem.* **72,** 2018.
Czapski, G., and Ottolenghi, M. (1968). *Israel J. Chem.* **6,** 75.
D'Alessio, J. T., Pantaleone, A., and Zampach, J. (1968). *Photochem. Photobiol.* **7,** 583.
Danziger, R. M., Bar-Eli, K. H., and Weiss, K. (1967). *J. Phys. Chem.* **71,** 2633.
Danziger, R. M., Hayon, E., and Langmuir, M. E. (1968). *J. Phys. Chem.* **72,** 3842.
Dobson, G., and Grossweiner, L. I. (1965). *Trans. Faraday Soc.* **61,** 708.
Dose, K. (1968a). *Photochem. Photobiol.* **7,** 671.
Dose, K. (1968b). *Photochem. Photobiol.* **8,** 331.
Fieser, L. F. (1930). *J. Am. Chem. Soc.* **52,** 5204.
Fischer, H. (1957). *J. Opt. Soc. Am.* **47,** 981.
Foote, C. S., and Wexler, S. (1964). *J. Am. Chem. Soc.* **86,** 3880.
Gibbons, W. A., Porter, G., and Savadatti, M. I. (1965). *Nature* **206,** 1355.
Green, M., and Tollin, G. (1968a). *Photochem. Photobiol.* **7,** 129.
Green, M., and Tollin, G. (1968b). *Photochem. Photobiol.* **7,** 155.
Greenstock, C. L., Brown, I. H., Hunt, J. W., and Johns, H. E. (1967). *Biochem. Biophys. Res. Comm.* **27,** 431.
Greig, H. G. (1962). U. S. Patent No. 3,052,540.
Grist, K., Taylor, T., and Augenstein, L. (1965). *Radiation Res.* **26,** 198.
Grossweiner, L. I. (1966). *In* "Advances in Radiation Biology" (L. G. Augenstein, R. Mason, M. R. Zelle, eds.), Vol. 2, pp. 83–133. Academic Press, New York.
Grossweiner, L. I. (1968a). *In* "Advances in Chemistry Series" No. 81, Vol. 1, pp. 309–320. American Chemical Society, Washington, D. C.
Grossweiner, L. I. (1968b). *Photochem. Photobiol.* **8,** 411.
Grossweiner, L. I. (1969). *Photochem. Photobiol.* **10,** 183.
Grossweiner, L. I., and Joschek, H. I. (1965). *In* "Advances in Chemistry Series, No. 50, Solvated Electron" (R. F. Gould, ed.), pp. 279–288. American Chemical Society, Washington, D. C.
Grossweiner, L. I., and Matheson, M. S. (1957). *J. Phys. Chem.* **61,** 1089.
Grossweiner, L. I., and Mulac, W. A. (1959). *Radiation Res.* **10,** 515.
Grossweiner, L. I., and Rodde, A. F., Jr. (1968). *J. Phys. Chem.* **72,** 756.
Grossweiner, L. I., and Usui, Y. (1969). *Photochem. Photobiol.* **11,** 53.
Grossweiner, L. I., and Zwicker, E. F. (1959). *J. Chem. Phys.* **31,** 1141.
Grossweiner, L. I., and Zwicker, E. F. (1961). *J. Chem. Phys.* **34,** 1411.
Grossweiner, L. I., and Zwicker, E. F. (1963). *J. Chem. Phys.* **39,** 2774.
Grossweiner, L. I., Swenson, G. W., and Zwicker, E. F. (1963). *Science* **141,** 805.
Hélène, C., Santus, R., and Douzou, P. (1966). *Photochem. Photobiol.* **5,** 127.
Herbert, M. A., Hunt, J. W., and Johns, H. E. (1968). *Biochem. Biophys. Res. Comm.* **33,** 643.
Holmström, B. (1962). *Bull. Soc. Chim. Belg.* **71,** 869.
Holmström, B. (1964). *Photochem. Photobiol.* **3,** 97.
Holmström, B., and Stening, B. (1964). *Photochem. Photobiol.* **3,** 55.
Horrocks, A. R., and Wilkinson, F. (1968). *Proc. Roy. Soc.* (*London*) **A306,** 275.
Johnson, G. E., and Albrecht, A. C. (1966a). *J. Chem. Phys.* **44,** 3162.
Johnson, G. E., and Albrecht, A. C. (1966b). *J. Chem. Phys.* **44,** 3179.
Jortner, J., Ottolenghi, M., and Stein, G. (1963). *J. Am. Chem. Soc.* **85,** 2712.
Joschek, H. I., and Grossweiner, L. I. (1966). *J. Am. Chem. Soc.* **88,** 3261.
Kasche, V. (1967). *Photochem. Photobiol.* **6,** 643.
Kasche, V., and Lindqvist, L. (1965). *Photochem. Photobiol.* **4,** 923.

Kato, S., Morita, M., and Koizumi, M. (1964). *Bull. Chem. Soc. Japan* **37,** 117.
Kautsky, H. (1939). *Trans. Faraday Soc.* **35,** 216.
Kearns, D. R., Hollins, R. A., Khan, A. U., Chambers, R. W., and Radlick, P. (1967a). *J. Am. Chem. Soc.* **89,** 5455.
Kearns, D. R., Hollins, R. A., Khan, A. U., Chambers, R. W., and Radlick, P. (1967b). *J. Am. Chem. Soc.* **89,** 5456.
Keene, J. P., Land, E. J., and Swallow, A. J. (1965). *In* "Pulse Radiolysis" (M. Ebert, J. P. Keene, A. J. Swallow, and J. H. Baxendale, eds.), pp. 227–245. Academic Press, London.
Keene, J. P., Land, E. J., and Swallow, A. J. (1968). *J. Chim. Phys.* **65,** 371.
Kikuchi, K., Kokubun, H., and Koizumi, M. (1968a). *Bull. Chem. Soc. Japan* **41,** 1545.
Kikuchi, K., Kokubun, H., and Koizumi, M. (1968b). *Z. Phys. Chem. N. F.* **62,** 75.
Knowles, A., and Roe, E. M. F. (1968). *Photochem. Photobiol.* **7,** 421.
Koizumi, M., and Usui, Y. (1968). *Tetrahedron Letters* No. 57, 6011.
Land, E. J. (1968). *Proc. Roy. Soc.* (*London*) **A305,** 457.
Land, E. J., Porter, G., and Strachan, E. (1961). *Trans. Faraday Soc.* **57,** 1885.
Lewis, G. N., and Bigeleisen, J. (1943). *J. Am. Chem. Soc.* **65,** 520; 2419, and 2424.
Lewis, G. N., and Lipkin, D. (1942). *J. Am. Chem. Soc.* **64,** 2801.
Lindqvist, L. (1960). *Arkiv Kemi* **16,** 79.
Linschitz, H., and Sarkanen, K. (1958). *J. Am. Chem. Soc.* **80,** 4826.
Linschitz, H., Ottolenghi, M., and Bensasson, R. (1967). *J. Am. Chem. Soc.* **89,** 4592.
Livingston, R. (1966). *In* "Actions Chimiques et Biologiques des Radiations" (M. Haissinsky, ed.), Vol. 9, pp. 1–35. Masson, Paris.
Luse, A. D., and McLaren, A. D. (1963). *Photochem. Photobiol.* **2,** 343.
McClain, W. M., and Albrecht, A. C. (1966). *J. Chem. Phys.* **44,** 1594.
McLaren, A. D., and Hidalgo-Salvatierra, O. (1964). *Photochem. Photobiol.* **3,** 349.
McLaren, A. D., and Luse, R. A. (1961). *Science* **134,** 836.
McRae, E. G., and Kasha, M. (1964). *In* "Physical Processes in Radiation Biology" (L. Augenstein, R. Mason, and B. Rosenberg, eds.), pp. 17–42. Academic Press, New York.
Maercks, O. (1968). *Z. Physik.* **109,** 685.
Matsumoto, S. (1964). *Bull. Chem. Soc. Japan* **37,** 491.
Niemann, E. G., and Klenert, M. (1968a). *Appl. Opt.* **7,** 295–299.
Niemann, E. G., and Klenert, M. (1968b). *J. Phys. Chem.* **72,** 3766.
Novak, J. R., and Windsor, M. W. (1968). *Proc. Roy. Soc.* (*London*) **A308,** 95.
Noyes, R. M. (1955). *J. Am. Chem. Soc.* **77,** 2042.
Ohno, T., Kato, S., and Koizumi, M. (1966). *Bull. Chem. Soc. Japan* **39,** 232.
Ottolenghi, M. (1963). *J. Am. Chem. Soc.* **85,** 3557.
Parker, C. A., and Hatchard, C. G. (1961). *Trans. Faraday Soc.* **57,** 1894.
Pilloff, H. S., and Albrecht, A. C. (1968). *J. Chem. Phys.* **49,** 4891.
Pimentel, G. C. (1968). *Appl. Phys.* **7,** 2155.
Porter, G. (1950). *Proc. Roy. Soc.* (*London*) **A200,** 284.
Porter, G. (1959). *Radiation Res. Supp.* **1,** 479.
Porter, G. (1960). *Z. Elektrochem.* **64,** 59.
Porter, G., and Strauss, G. (1966). *Proc. Roy. Soc.* (*London*) **A295,** 1.
Porter, G., and Topp, M. R. (1968). *Native* **220,** 1228.
Porter, G., and Topp, M. R. (1969). *Proc. Roy. Soc.* (*London*) (in press).
Potashnick, R., Ottolenghi, M., and Benasson, R. (1969). *J. Phys. Chem.* **73,** 1912.
Rentzepis, P. M. (1968). *Photochem. Photobiol.* **8,** 579.

Risi, S., Dose, K., Rathinasamy, T. K., and Augenstein, L. (1967). *Photochem. Photobiol.* **6,** 423.

Santus, R., Hélène, C., and Ptak, M. (1968). *Photochem. Photobiol.* **7,** 341.

Sauer, M. C., Jr. (1966). Argonne National Laboratory Report 7113.

Schafer, F. P., Schmidt, W., and Volze, J. (1966). *Appl. Phys. Letters,* **9,** 306.

Schenck, G. O. (1949). *Naturwissenchaften* **35,** 28.

Schmidt, K. H. (1966). Argonne National Laboratory Report 7199.

Schnuriger, B., and Bourbon, J. (1968). *Photochem. Photobiol.* **8,** 361.

Scholes, G., Shaw, P., Wilson, R. L., and Ebert, M. (1965). *In* "Pulse Radiolysis" (M. Ebert, J. P. Keene, A. J. Swallow, and J. H. Baxendale, eds.), pp. 151–164. Academic Press, London.

Schottky, W. (1918). *Ann. Physik.* **57,** 541.

Seliger, H. H., and Morton, R. A. (1968). *In* "Photophysiology" (A. C. Giese, ed.), Vol. 4, pp. 253–314. Academic Press, New York.

Shimokoshi, K., Mori, Y., and Tanaka, I. (1967). *Bull. Chem. Soc. Japan* **40,** 302.

Siegel, S., and Eisenthal, K. (1965). *J. Chem. Phys.* **42,** 2494.

Smaller, B., Remko, J. R., and Avery, E. C. (1968). *J. Chem. Phys.* **48,** 5174.

Soederberg, B. (1913). *Ann. d. Phys.* **41,** 381.

Sorokin, P. O., Lankard, J. R., Moruzzi, V. L., and Hammond, E. C. (1968). *J. Chem. Phys.* **48,** 4726.

Spikes, J. D. (1968). *In* "Photophysiology" (A. C. Giese, ed.), Vol. 3, pp. 33–64. Academic Press, New York.

Steiner, R. F., and Kirby, E. P. (1969). *J. Phys. Chem.* **73,** 4130.

Swenson, G. W., Zwicker, E. F., and Grossweiner, L. I. (1963). *Science* **141,** 1042.

Tegner, L., and Holmström, B. (1966). *Photochem. Photobiol.* **5,** 223.

Terenin, A., and Vilessov, F. (1964). *In* "Advances in Photochemistry" (W. A. Noyes, Jr., G. S. Hammond, and J. N. Pitts, Jr., eds.), Vol. 2, pp. 385–421. Wiley (Interscience), New York.

Terenin, A., Tachin, V., and Shakhverdov, P. (1965). *Photochem. Photobiol.* **4,** 505.

Usui, Y., Iwanaga, C., and Koizumi, M. (1969). *Bull. Chem. Soc. Japan* **42,** 1231.

Walker, M. S., Bednar, T. W., and Lumry, R. (1966). *J. Chem. Phys.* **45,** 2355.

Walker, M. S., Bednar, T. W., and Lumry, R. (1967). *J. Chem. Phys.* **42,** 1020.

Weigl, J. W. (1963). *In* "Photographic Science" (W. F. Berg, ed.), pp. 345–355. Focal Press, London.

Weller, A. (1952). *Z. Electrochem.* **56,** 662.

West, W., and Carroll, B. H. (1966). *In* "The Theory of the Photographic Process" (C. E. K. Mees and T. H. James, eds.), pp. 233–277. Macmillan, New York.

Witt, H. T. (1967a). *In* "Fast Reactions and Primary Processes in Chemical Kinetics" (S. Claesson, ed.), pp. 81–97. Wiley (Interscience), New York.

Witt, H. T. (1967b). Ibid. 261–310.

Wolff, Ch., Buchwald, H., Ruppel, H., and Witt, H. T. (1967). *Naturwissenschaften* **54,** 489.

Yamamoto, N., Nakato, Y., and Tsubomura, H. (1966). *Bull. Chem. Soc. Japan* **39,** 2603.

Youtsey, K. J. (1968). Thesis, Illinois Institute of Technology (unpublished).

Youtsey, K. J., and Grossweiner, L. I. (1967). *Photochem. Photobiol.* **6,** 721.

Youtsey, K. J., and Grossweiner, L. I. (1969a). Unreported result.

Youtsey, K. J., and Grossweiner, L. I. (1969b). *J. Phys. Chem.* **73,** 447.

Zwicker, E. F., and Grossweiner, L. I. (1963). *J. Phys. Chem.* **67,** 549.

Chapter 2

PHYSICAL PROPERTIES OF SINGLET OXYGEN

E. A. Ogryzlo

Department of Chemistry, University of British Columbia, Vancouver, Canada

1. Introduction

The importance of molecular oxygen in countless processes in the earth's "biosphere" has long been recognized. However, it is only during the last five years that it has become increasingly apparent that one of its electronically excited states is involved in a number of these processes. The most significant observations are the following. Visible and ultraviolet light absorbed by objects usually results in electronic excitation of the constituent molecules. Because molecular oxygen permeates most materials it is a common quencher of such electronic excitation. Especially in the case of "triplet states," such quenching often leads quite efficiently to the formation of electronically excited oxygen molecules. These molecules, which are referred to as singlet oxygen because of their spin multiplicity, not only have an extraordinarily long radiative lifetime and can suffer many collisions without losing their energy, but can form complexes with numerous molecules giving them an indefinite lifetime since the complex can now "regenerate" the singlet oxygen for reaction when it is required. The chemical reactivity of this singlet state of oxygen appears to be highly specific, and it would not be surprising to find that life processes have made considerable use of its unique characteristics.

In this article we will review the information which has been obtained by physical methods regarding the two electronically excited states of oxygen which may be important in the above processes.

2. Theoretical Description

The three relevant electronic levels of O_2 are shown in Fig. 1. The singlet delta state ($O_2(^1\Delta_g)$) lies 22.5 kcal above the triplet sigma ground state ($O_2(^3\Sigma_g^-)$), while the singlet sigma state ($O_2(^1\Sigma_g^+)$) lies 37.5 kcal above the ground state. (Other states lie more than 98 kcal above ground.) Part of the uniqueness of molecular oxygen undoubtedly stems from the fact that the three lowest electronic levels all arise from the same configuration, i.e., having two electrons in the pair of degenerate antibonding orbitals (Mulliken, 1932). As in most many-electron problems there is unfortunately not one unique set of acceptable orbitals which can be used to describe the states. For example, the orbitals that are useful for establishing the term symbols for these states are the Π_{+1}^* and Π_{-1}^* orbitals shown in Fig. 2(a). They have sharp values ($\pm$h) of orbital angular momentum along the internuclear axis and are there-

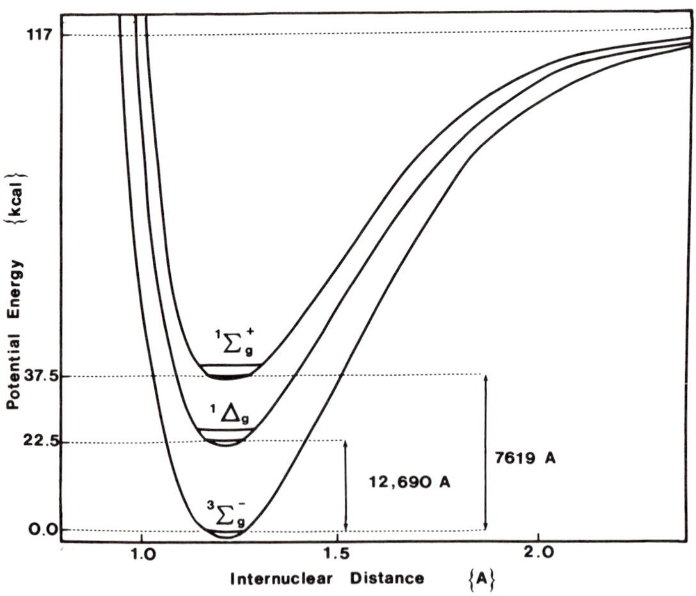

FIG. 1. Potential energy curves for the three lowest electronic states of O_2.

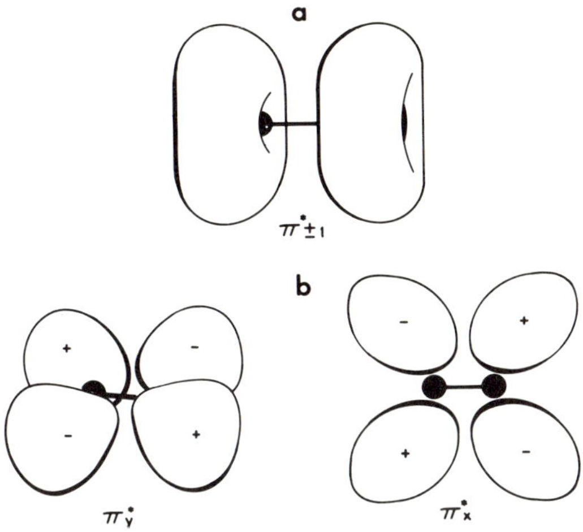

FIG. 2. (a) Contour surfaces of the probability distribution function ($\Psi^*\Psi$) for the Π^*_{+1} (and Π^*_{-1}) orbitals of O_2. (b) Contour surfaces of Ψ for Π^*_x and Π^*_y orbitals of O_2.

fore useful in describing the interaction of the molecule with a magnetic field. The following "orbital wave functions" for the three relevant levels of O_2 (which satisfy the Pauli principle) are then obtained by assigning electrons (1) and (2) to these two orbitals. Ignoring the spin functions and all inner electrons:

$$\begin{array}{lll} & \Pi^*_{+1} & \Pi^*_{-1} \\ {}^1\Sigma_g{}^+ = \dfrac{1}{\sqrt{2}}\,(\Pi^*_{+1}(1)\Pi^*_{-1}(2) + \Pi^*_{+1}(2)\Pi^*_{-1}(1)) & \uparrow & \downarrow \\ {}^1\Delta_g = \begin{cases}\Pi^*_{+1}(1)\Pi^*_{+1}(2) \\ \Pi^*_{-1}(1)\Pi^*_{-1}(2)\end{cases} & \begin{matrix}\uparrow\downarrow \\ \ \end{matrix} & \begin{matrix}\ \\ \uparrow\downarrow\end{matrix} \\ {}^3\Sigma_g{}^+ = \dfrac{1}{\sqrt{2}}\,(\Pi^*_{+1}(1)\Pi^*_{-1}(2) - \Pi^*_{+1}(2)\Pi^*_{-1}(1) & \uparrow & \uparrow \end{array}$$

This assignment (which is repeated somewhat pictorially with circles and arrows on the right hand side) has led some workers to suggest that since the ${}^1\Sigma_g{}^+$ state appears to have two electrons in separate orbitals it should show reactions similar to the ground state, i.e., those characteristics of a "diradical." On the other hand, both components of the ${}^1\Delta_g$ state appear to have the electrons paired in a single orbital, suggesting it should undergo reactions like ethylene—with which it is isoelectronic. Such an analysis is, however, definitely erroneous.

A somewhat more satisfactory analysis can be obtained by using the alternative set of "real orbitals Π^*_x and Π^*_y shown in Fig. 2(b). These are related to the complex functions used previously by the relations:

$$\Pi^*_{+1} = \Pi^*_x + i\,\Pi^*_y$$
$$\Pi^*_{-1} = \Pi^*_x - i\,\Pi^*_y$$

With these new basis functions the orbital part of the wave functions for the relevant states of O_2 become:

$$\begin{array}{lll} & \Pi^*_x & \Pi^*_y \\ {}^1\Sigma_g^+ = \frac{1}{\sqrt{2}}(\Pi^*_x(1)\Pi^*_x(2) + \Pi^*_y(1)\Pi^*_y(2)) & & \uparrow\downarrow \\ {}^1\Delta_g = \begin{cases} \frac{1}{\sqrt{2}}(\Pi^*_x(1)\Pi^*_x(2) - \Pi^*_y(1)\Pi^*_y(2)) \\ \frac{1}{\sqrt{2}}(\Pi^*_x(1)\Pi^*_y(2) + \Pi^*_x(2)\Pi^*_y(1)) \end{cases} & \begin{array}{c}\uparrow\downarrow \\ \uparrow\end{array} & \begin{array}{c} \\ \downarrow\end{array} \\ {}^3\Sigma_g^- = \frac{1}{\sqrt{2}}(\Pi^*_x(1)\Pi^*_y(2) - \Pi^*_x(2)\Pi^*_y(1)) & \uparrow & \uparrow \end{array}$$

These functions not only constitute an equally valid description of the free molecule but, more significantly, are the starting point for a description of the effect of any perturbation which destroys the cylindrical symmetry of the molecule—such as the approach of a reactive molecule. The effect of such a perturbation is to mix the ${}^1\Sigma_g^+$ state with the first component of ${}^1\Delta_g$. The effect of such a perturbation is always to stabilize the ${}^1\Delta_g$ component at the expense of the ${}^1\Sigma_g^+$ state, (Griffiths, 1964). Though such an effect has never been shown to be large for any non-reactive interaction, this fact together with the required correlation with the ground state of the product which will be discussed later always makes this (${}^1\Delta_g$) component the most favourable state for concerted addition reactions when these are symmetry allowed. Since the other ${}^1\Delta_g$ component appears to have its electrons in separate orbitals, this state should display a "diradical" character similar to the triplet ground state. On the other hand, the ${}^1\Sigma_g^+$ state does not display this "diradical" configuration and, as we have seen above, may display a repulsive interaction with most species. Before it is concluded from this analysis that $O_2({}^1\Sigma_g^+)$ is a singularly unreactive species, it should be noted that the mixing of this state with a ${}^1\Delta_g$ component on collision would also be expected to bring about the electronic relaxation of $O_2({}^1\Sigma_g^+)$ to $O_2({}^1\Delta_g)$ (initially with vibrational excitation). We shall see later that this is indeed the conclusion reached from experimental results. It may, therefore, prove difficult to verify this predicted unreactivity of $O_2(^1\Sigma_g^+)$.

3. Formation of Singlet Oxygen by Energy Transfer

The efficient quenching of triplet (phosphorescent) states, and in some cases, excited singlet (fluorescent) states of organic molecules by molecular oxygen has been known for many years. It is rather remarkable that it was in 1931 that Kautsky and de Bruijn (1931) proposed that the efficient quenching of phosphorescent (triplet) states by oxygen results in the formation of excited (singlet) oxygen molecules and that these are the reactive species in a certain class of oxidation reactions. Though his proposal was not embraced by the scientific community and other mechanisms of oxygen quenching and photooxidation were in vogue for varying periods, there is now strong theoretical (Kawaoka *et al.*, 1967) and experimental (Foote, 1968) support for Kautsky's mechanism. In our current language its essential features are the following. The absorption of light by molecules normally raises them to excited singlet states which can be different at different wavelengths. In almost all cases, electronic and vibrational relaxation rapidly brings the molecules to the lowest vibrational levels of the lowest excited singlet state from which fluorescence is often observed. Quenching of this state by oxygen is not very effective because of the short radiative lifetime (10^{-6}–10^{-8} secs) of most such states. However, radiationless conversion (intersystem crossing) to the triplet states (manifold) may occur within that lifetime. Relaxation to the lowest vibrational levels of the lowest triplet state then also occurs extremely rapidly. The long radiative lifetimes of triplet states (usually 10^{-3} sec) makes them much more susceptible to quenching by other molecules and for most such states a very small concentration of oxygen can efficiently eliminate the phosphorescence. Theoretical arguments (Kawaoka *et al.*, 1967) suggest that such oxygen induced triplet–singlet relaxation is facilitated by the fact that the triplet character of oxygen makes the overall process spin allowed. Furthermore, the process becomes most efficient when least electronic energy needs to be degraded into vibrational-rotational excitation in the interacting pair. Consequently, the most favorable quenching process is one in which the electronic excitation of the organic molecule is transferred to oxygen, producing $O_2(^1\Sigma_g^+)$ or $O_2(^1\Delta_g)$. If more than 37.5 kcal is available then the former would be expected to be the dominant product. However, since $O_2(^1\Sigma_g^+)$ is easily relaxed to $O_2(^1\Delta_g)$ by collisions with most molecules (Arnold *et al.*, 1968) the only experimentally observable consequence has been the formation of $O_2(^1\Delta_g)$.

Direct spectroscopic observation of the formation of $O_2(^1\Delta_g)$ by energy transfer has been reported for benzene (Snelling, 1968), naphthalene (Kearns *et al.*, 1969), and several naphthalene derivative (Wasserman *et al.*, 1969). The only rate constant available for the formation of $O_2(^1\Delta_g)$ in such a quenching reaction is 3×10^{10} 1 moles^{-1} sec^{-1} for benzene which is within a factor of 10 of collision frequency.

4. Radiative Transition Probabilities

Molecular oxygen possesses several extremely weak absorption bands in the near infrared which can be used to calculate the radiative lifetimes of the two singlet states of oxygen. From the intensity of the absorption band at 12,690 Å a radiative lifetime of 1.07 hours (half-life = 45 minutes) has been calculated (Badger *et al.*, 1965) for the $^1\Delta_g$ state. Similarly, from its absorption band at 7,619 Å the radiative lifetime of the $^1\Sigma_g^+$ state has been calculated to be 7 seconds (Childs and Mecke, 1931) (half-life = 5 seconds). These transitions have been identified as "magnetic dipole" transitions which have transition probabilities many orders of magnitude smaller ($\approx 10^{-8}$) than those for allowed electric dipole transitions. In a collision with another molecule, electric dipole transitions can be induced in these molecules. However, because the perturbation due to the second molecule is normally relatively small, the radiative lifetime of such a collision complex is of the order of 1 second. Because the $^1\Sigma_g^+$ half-life is already of this order of magnitude its radiative lifetime is not much affected by other gases. However, in the presence of O_2, N_2 and CO_2, the radiative lifetime of $O_2(^1\Delta_g)$ can be shortened appreciably (Badger *et al.*, 1965). The following equation summarizes the quantitative relationship between the *radiative* lifetime (τ_r) and the pressure of these gases (in atmospheres).

$$\tau_r = \frac{1.07}{1 + 3.8P(O_2) + 3.0P(CO_2) + 0.7P(N_2)} \text{ hours}$$

τ_r should not be confused with the actual lifetime of the species which is discussed in the next section and is normally much shorter.

There is an additional radiative process which has been observed more recently. This is the cooperative process in which two excited molecules pool their energies to produce a photon with the combined energy of both molecules. The process gives rise to bands at 6340 Å and 7030 Å when both molecules are $O_2(^1\Delta_g)$, (Bader and Ogryzlo, 1964) and a band at 4800 Å when the pair is $O_2(^1\Delta_g) \cdot O_2(^1\Sigma_g^+)$ (Khan and Kasha, 1966).

The rate constants for these processes are comparable with those for the pressure induced radiative processes discussed above (see Section 6). However, since the excited molecule concentrations are normally very much smaller than the concentrations of other molecules, these processes do not make a significant contribution to the radiative lifetimes of $O_2(^1\Delta_g)$ and $O_2(^1\Sigma_g^+)$. Nevertheless, the radiation from the first of the above two processes provides a convenient (because it occurs in the visible) and sensitive (because it is proportional to the square of the concentration) method of monitoring the singlet oxygen concentration in the flow system. The details of these processes will be discussed in Section 6.

5. Collisional Quenching and Lifetimes

Under laboratory conditions the actual lifetimes (τ) of $O_2(^1\Sigma_g^+)$ and $O_2(^1\Delta_g)$ are found to be much shorter than the radiative lifetimes (τ_r) discussed in the last section.

5.1— $O_2(\Sigma^1{}_g^+)$

The rate of deactivation of $^1\Sigma_g^+$ with a number of different molecules has been determined with "discharge-generated" molecules (Arnold *et al.* 1968). There is good theoretical (see Section 1) and experimental (Arnold *et al.*, 1968) evidence for believing that the process can be written:

$$O_2(^1\Sigma_g^+) + \text{Quencher} \rightarrow O_2(^1\Delta_g) + \text{Quencher}$$

i.e., the relaxation is to $^1\Delta_g$. The experimental results are summarized in Fig. 3 where the rate constants and quenching efficiencies are plotted against the boiling points of the quencher. It can be seen that the correlation is only approximate. A better correlation is not found with any other single molecular property. This suggests that a number of factors may be important in the process. However, with some reservations, an order of magnitude estimate of the quenching efficiency can be made for an unstudied molecule by noting its boiling point. For example, the solvents used in most photooxidation studies would most probably quench $^1\Sigma_g^+$ in about 10^2 collisions. This would shorten its lifetime under these conditions to about 10^{-10} sec.

5.2— $O_2(^1\Delta_g)$

Compared with the $^1\Sigma_g^+$ state, the $^1\Delta_g$ shows remarkable resistance to deactivation. Because of this stability, direct measurement of quenching rates in discharge-flow systems has been found difficult. Recently, how-

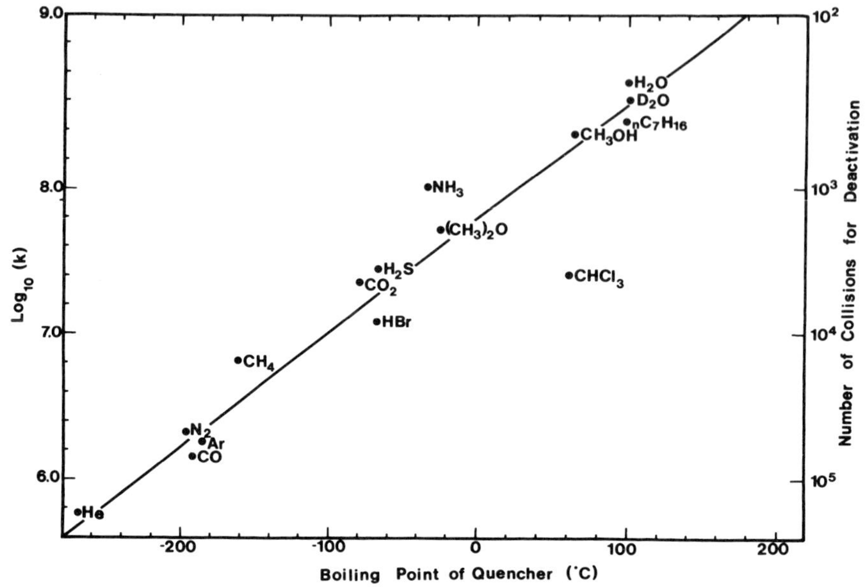

FIG. 3. Quenching constants for $O_2(^1\Sigma^+{}_g)$ in units of l moles^{-1} sec^{-1} as a function of the boiling point of the quencher. The approximate number of gas kinetic collisions necessary to cause the electronic relaxations are given on the right hand size of the figure.

ever, the following rate constants have been reported (Findlay *et al.*, 1969).

$$O_2(^1\Delta_g) + \text{Quencher} \rightarrow O_2(^3\Sigma_g^-) + \text{Quencher}$$

$$k(C_6H_6) = 2.6 \times 10^3 \text{ l moles}^{-1} \text{ sec}^{-1} (\approx 5 \times 10^7 \text{ collisions})$$

$$k(N_2) < 2 \text{ l moles}^{-1} \text{ sec}^{-1} (>6 \times 10^{10} \text{ collisions})$$

$$k(O_2) = 1.3 \times 10^3 \text{ l moles}^{-1} \text{ sec}^{-1} (\approx 10^8 \text{ collisions})$$

The fact that O_2 is much more efficient than N_2 at quenching $^1\Delta_g$ (whereas it isn't for $^1\Sigma_g^+$) is consistent with the fact that the $^1\Delta_g$ relaxation involves a change in multiplicity. The process can be written:

$$^3\{O_2(^1\Delta_g) \cdot O_2(^3\Sigma_g)\} \rightarrow {}^3\{O_2(^3\Sigma_g^-) \cdot O_2(^3\Sigma_g^-)\}$$

which is now "spin-allowed" since the initial and final complex can both have the same multiplicity, (3).

The difference between the quenching efficiency of N_2 and C_6H_6 suggests that there may exist a correlation with boiling point similar to that found with $^1\Sigma_g^+$. In that case the same line (Fig. 3) can be used to estimate quenching rates for $^1\Delta_g$ simply by multiplying the values of k by 10^{-5}. It should not be concluded from the above observations that "boiling point" is a reliable indication of quenching efficiency when the nature of the species differs. For example, it has been reported by Ouannès and Wilson, (1968) that tertiary amines are very efficient quenchers of $^1\Delta_g$. The reason for this strong "interaction" is not clear, although it has been suggested by these workers that "charge-transfer" interactions may be responsible. There also is evidence that carotene is an effective quencher of singlet oxygen (Foot and Denny, 1968) although a direct test of this effect with discharge-generated $^1\Delta_g$ has not been reported.

6. Energy Pooling

Three types of energy pooling processes have been either observed or proposed. Type I is called "cooperative emission" or "radiative pooling." This process gives rise to the visible emission bands of oxygen shown in Fig. 4 (Gray and Ogryzlo, 1969). The bands at 6340 Å and

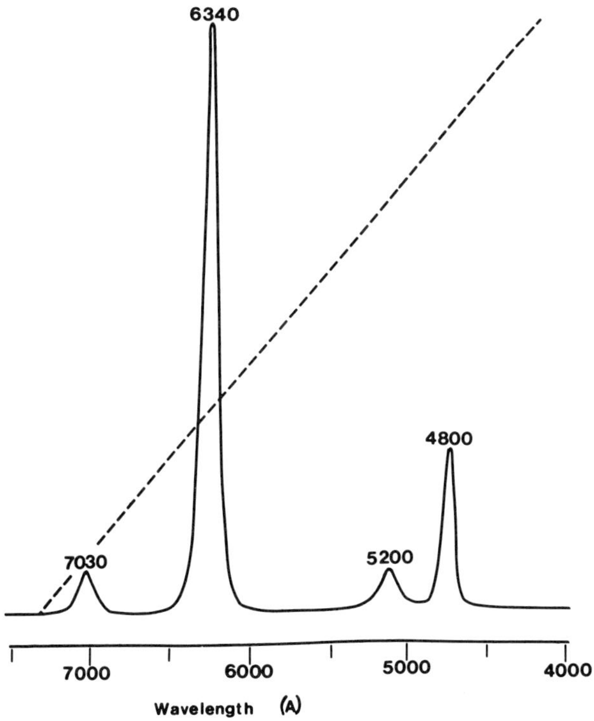

FIG. 4. Emission from a stream of singlet oxygen molecules at about 100°K. The spectrum was obtained with the IP21 photomultiplier and a Bausch and Lomb grating monochrometer with 1 mm slits. Total pressure ≈10 torr. The photomultiplier response is shown with a dotted line.

7030 Å which are of comparable intensity under all conditions have been assigned to the process:

$$O_2(^1\Delta_g) + O_2(^1\Delta_g) \rightarrow O_2(^3\Sigma_g^-) + O_2(^3\Sigma_g^-) + h\nu$$

i.e., to a simultaneous electronic transition in two molecules with the excitation of both molecules combined in a single photon (Arnold *et al.*, 1964). The 6340 Å peak represents the 0-0 vibrational band and the 7030 Å band the 0–1 band. The relative magnitudes of these two bands and the absence of other members of this progression is consistent with the nature of the transition (Dianov-Klokov, 1959). It has been demonstrated that the two $^1\Delta_g$ molecules are not bound to one another but are probably simply in a state of collision (Arnold *et al.*, 1965). The rate constant for the process is 0.03 l moles^{-1} sec^{-1} (Falick and Mahan, 1967) which means that the process occurs on about 1 collision in 10^{12} between $^1\Delta_g$ molecules. The chemical generation of $O_2(^1\Delta_g)$ has been detected by this red emission (Browne and Ogryzlo, 1964).

The emission at 4800 Å and 5200 Å only appears with a significant intensity at low temperatures (Khan and Kasha, 1966; Gray and Ogryzlo, 1969). It has been assigned to the process:

$$O_2(^1\Sigma_g^+) + O_2(^1\Delta_g) \rightarrow O_2(^3\Sigma_g^-) + O_2(^3\Sigma_g^-) + h\nu$$

The luminescence at 4800 Å which accompanies the low temperature decomposition of some photoperoxides can probably be attributed to the species (Canva *et al.*, 1969).

Type II energy pooling has been called "energy disproportionation." The most important example for singlet oxygen is the reaction (Young and Black, 1965)

$$O_2(^1\Delta_g) + O_2(^1\Delta_g) \rightarrow O_2(^1\Sigma_g^+) + O_2(^3\Sigma_g^-)$$

In this process the two $^1\Delta_g$ molecules pool their energy raising one molecule to the $^1\Sigma_g^+$ state while the other relaxes to the ground state. The rate constant for this process is 1.5×10^3 moles^{-1} sec^{-1} (Arnold and Ogryzlo, 1967). The occurrence of this reaction means that $^1\Sigma_g^+$ will be produced in any system containing $^1\Delta_g$. However, the quenching constants for $^1\Delta_g$ with other molecules are of the same order of magnitude (see Section 5.2.). Consequently $^1\Sigma_g^+$ will be produced efficiently only when the concentration of $^1\Delta_g$ is comparable with other molecules in the system.

Nevertheless, such processes may be important in some chemiluminescent reactions. It has been shown (Ogryzlo and Pearson, 1968) that the chemiluminescence of a solution of H_2O_2–Cl_2 and violanthrone probably involves the reaction

$$O_2(^1\Delta_g) + {}^3(\text{violanthrone}) \rightarrow O_2(^3\Sigma_g^-) + {}^1(\text{violanthrone})^*$$

i.e., in an energy disproportionation process the violanthrone triplet is raised to an excited singlet state. Since other aromatic molecules (anthracene and rubrene, for example) also show emission from their excited singlet state when exposed to singlet oxygen (Gray and Ogryzlo, 1969), such processes may account for the often observed weak luminescence from oxygen producing systems (Stauff and Lohmann, 1964).

Type III pooling which is called "cooperative transfer" or "termolecular pooling" (Ogryzlo and Pearson, 1968) is exemplified by the process

$$O_2(^1\Delta_g) + O_2(^1\Delta_g) + A \rightarrow O_2(^3\Sigma_g^-) + O_2(^3\Sigma_g^-) + \overset{*}{A}$$

in which there is a simultaneous transfer of the excitation energy of two oxygen molecules to a third molecule. It has been proposed that such processes (involving O_2 $(^1\Sigma_g^+)$ as well) can account for numerous chemiluminescent reactions (Khan and Kasha, 1966). Though there is, as yet, no concrete evidence that such processes play a significant part in any known system, some recent observations by Canva *et al.* (1969) on the luminescence of systems containing singlet oxygen and aromatic molecules suggest the need for retaining this possibility.

7. Stabilization of Singlet Oxygen

It can be concluded from the data presented in Section 5.2 that even under the most favourable conditions the lifetime of "free $O_2(^1\Delta_g)$" does not exceed 10^{-3} sec in a condensed phase. If some other species were capable of "trapping" singlet oxygen in an "unrelaxable" state, from which it could be regenerated, its potential as an oxidizing agent would be greatly enhanced. This possibility was first demonstrated with diphenylanthracene (Wasserman and Scheffer, 1967). In this case a stable transannular peroxide can be formed by exposing diphenylanthracene to singlet oxygen. Then, by heating the peroxide to temperatures above about 80°C typical singlet oxygen reactions can be observed suggesting that the species is released under these conditions. Unfortunately, it cannot be stated with complete confidence that the reaction is not a direct reaction with the transannular peroxide rather than with free $O_2(^1\Delta_g)$. A similar result was obtained with rubrene. Since the formation of such transannular peroxides is best described in more chemical terms, the details are given in the next chapter. A discussion of why singlet oxygen is regenerated by such peroxides is given by Khan and Kearns (1968).

More recently Canva *et al.* (1969) obtained evidence that singlet oxygen can be trapped by the following molecules: rhodamine B, eosin,

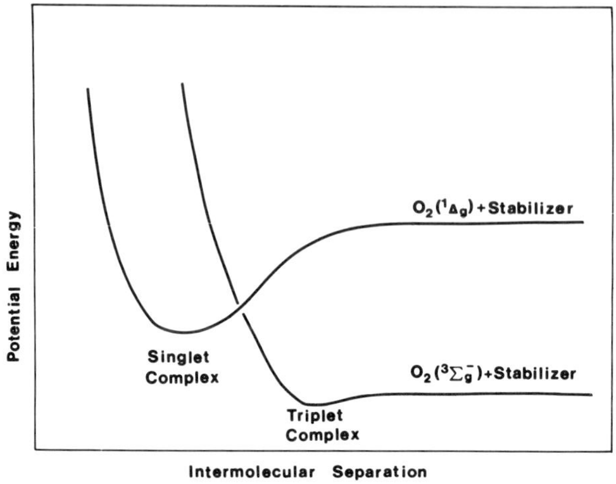

FIG. 5. Schematic representation of potential energy curves which could account for stabilization of singlet oxygen in complexes.

erythrosin, rose bengal, acridine orange, methylene blue, pyrene, 1-2 benzopyrene, anthracene, and perylene. Exposure of these compounds to photochemically and chemically generated singlet oxygen at temperatures of about –60°C produced an intermediate which regenerated singlet oxygen on warming. These observations have been confirmed with discharge-generated $O_2(^1\Delta_g)$ (Ogryzlo and Pattenden, 1970). Since the intermediate is stable for periods greater than the unperturbed radiative half-life, the singlet complex must be the ground electronic state, and the potential energy curves probably have a form such as that shown in Fig. 5. This requires a sufficiently large repulsive interaction between $O_2(^3\Sigma_g^-)$ and the "stabilizing" molecule to place it energetically above the singlet complex at the equilibrium intermolecular separation of the singlet complex. This is not inconsistent with the theoretical analysis of Section 2. In view of the effect such complexes can have on the lifetime and mobility of singlet oxygen, they may be important intermediates in oxidation processes.

ACKNOWLEDGMENT

The research for this paper was supported by the National Research Council of Canada, Grant No. A935 and by the Defense Research Board of Canada, Grant No. 9530–31

REFERENCES

Arnold, S. J., and Ogryzlo, E. A. (1967). *Can. J. Phys.* **45**, 2053.
Arnold, S. J., Ogryzlo, E. A., and Witzke, H. (1964). *J. Chem. Phys.* **40**, 1769.

Arnold, S. J., Browne, R. J., and Ogryzlo, E. A. (1965). *Photochem. and Photobiol.* **4,** 963.

Arnold, S. J., Kubo, M., and Ogryzlo, E. A. (1968). *In* "Advances in Chem., Ser. 77, Oxidation of Organic Compounds, #3" (R. F. Gould, ed.), pp. 133–142. American Chemical Society, Washington, D. C.

Bader, L. W., and Ogryzlo, E. A. (1964). *Discussions Faraday Soc.* **37,** 46.

Badger, R. M., Wright, A. C., and Whitlock, R. F. (1965). *J. Chem. Phys.* **43,** 4345.

Browne, R. J., and Ogryzlo, E. A. (1964). *Proc. Chem. Soc.* April, **117**.

Canva, J., Balny, C., Douzou, P., and Bourdon, J. (1969). *C. R. Acad. Science, Paris* **268,** 1027.

Childs, W. H. J., and Mecke, R. (1931). *Z. Physik* **68,** 344.

Dianov-Klokov, V. I. (1959). *Opt. Spectr.* **6,** 290.

Falick, A. M., and Mahan, B. H. (1967). *J. Chem. Phys.* **47,** 4778.

Findlay, F. D., Fortin, C. J., and Snelling, D. R. (1969). *Chem. Phys. Letters* **3,** 204.

Foote, C. S. (1968). *Accounts Chem. Res.* **1,** 104–110.

Foote, C. S., and Denny, R. W. (1968). *J. Am. Chem. Soc.* **90,** 6233.

Gray, E. W., and Ogryzlo, E. A. (1969). *Chem. Phys. Letters* **3,** 658.

Griffiths, J. S. (1964). *In* "Oxygen in Animal Organisms" (F. Dickens and E. Neil, eds.), pp. 141–148. MacMillan, New York.

Herzberg, G. (1950). "Spectra of Diatomic Molecules." Van Nostrand, Princeton, New Jersey.

Kautsky, H., and de Bruijn, H. (1931). *Naturwissenschaften* **19,** 1043.

Kawaoka, K., Khan, A. U., Kearns, D. R. (1967). *J. Chem. Phys.* **46,** 1842.

Kearns, D. R., Khan, A. U., Duncan, C. K., and Maki, A. H. (1969). *J. Am. Chem. Soc.* **91,** 1039.

Khan, A. U., and Kasha, M. (1966). *J. Am. Chem. Soc.* **88,** 1574.

Khan, A. U., and Kearns, D. R. (1968). *In* "Advances in Chem., Ser. 77, Oxidation of Organic Compounds, #3" (R. F. Gould, ed.), pp. 143–168. American Chemical Society, Washington, D. C.

Mulliken, R. S. (1932). *Rev. Mod. Phys.* **4,** 1–86.

Ogryzlo, E. A., and Pattenden, T. W. E. (1970). (To be published).

Ogryzlo, E. A., and Pearson, A. E. (1968). *J. Phys. Chem.* **72,** 2913.

Ouannès, C., and Wilson, C. (1968). *J. Am. Chem. Soc.* **90,** 6577.

Slater, J. C. (1963). "Quantum Theory of Molecules and Solids, Vol. I, Electronic Structure of Molecules." McGraw-Hill, New York.

Snelling, D. R. (1968). *Chem. Phys. Letters* **2,** 346.

Stauff, J., and Lohman, F. (1964). *Z. Physik. Chem.* **40,** 123.

Wasserman, H. W., and Scheffer, J. R. (1967). *J. Am. Chem. Soc.* **89,** 3073.

Wasserman, E., Kuck, V. J., Delavan, W. M., and Yager, W. A. (1969). *J. Am. Chem. Soc.* **91,** 1040.

Young, R. A., and Black, G. (1965). *J. Chem. Phys.* **42,** 3240.

Chapter 3

CHEMICAL AND BIOLOGICAL ASPECTS OF SINGLET EXCITED MOLECULAR OXYGEN

Thérèse Wilson and J. Woodland Hastings

Department of Chemistry and Biological Laboratories, Harvard University, Cambridge, Massachusetts

1. Introduction

1.1 Oxidation by Molecular Oxygen: Free Radical Reactions, Photooxygenation

It has long been appreciated that in the presence of light and oxygen, organic compounds can undergo two main types of reactions: radical mediated "autoxidations" and photooxygenations (see Calvert and Pitts, 1966).

In the first, light in the ultraviolet is generally required. Its action has a catalytic type of effect, such that a large number of organic molecules may be oxidized for each quantum of light absorbed (i.e., the quantum yield is high).

The vast majority of seemingly spontaneous degradative processes involving oxygen are reactions of this type, involving free-radicals (peroxy ROO· or alkoxy RO·) in a chain mechanism initiated by light. Examples are the formation of hydroperoxides and peroxides from alcohols, aldehydes, ketones, acids and hydrocarbons; the drying of oils and the degradation of rubber. When the concentration of these peroxides reaches a certain level, a relatively slow oxidation can accelerate and become a combustion resulting in very simple products: H_2O, CO, and CO_2. Because a detailed understanding of the mechanism of these processes has now been achieved, they can often be effectively controlled, initiated, or inhibited.

The second type of reaction is usually encountered in the presence of a dye or some natural pigment; often one which is highly fluorescent. These reactions utilize the lower energy photons of visible light absorbed by the dye, and the quantum yields are unity or less. Although the fluorescence can be partly quenched, the dye is not destroyed in the reaction where it serves as a "sensitizer." For example, in the presence of methylene blue and red light, anthracene will take up one molecule of oxygen to form an endoperoxide. This is a typical example of photooxygenation.

Until the last five years, the role of light and sensitizer in the mechanism of photooxygenation remained ambiguous. A large body of evidence now points to the intermediacy of molecular oxygen in an excited spectroscopic state (singlet state) of long lifetime, having a low excitation energy compared to the energies of most chemical bonds. The excitation energy of the dye is transferred to oxygen, forming an excited oxygen molecule which attacks the substrate.

This article will review the chemistry of processes in which excited molecular oxygen participates. After a brief survey of the nature of singlet oxygen and of the physical chemical evidence which led to its discovery, we shall outline some details of the methods of production and detection of singlet oxygen and discuss its typical reactions. The main emphasis will be on the basic processes involved, with some speculation concerning the possible role of singlet oxygen in biological and biochemical processes.

A vast number of papers have recently appeared on the subject of singlet oxygen, which seems relevant to the problems of air polution, atmospheric spectroscopy and chemiluminescence as well as organic and biological chemistry. Besides a highly selective reactivity and long lifetime, its ability to participate in unusual energy pooling processes confers a unique interest to singlet oxygen.

1.2 Nature of Singlet Oxygen

From a chemical standpoint, two significant features of the oxygen molecule are: (*a*) its triplet ground state (designated $^3\Sigma_g^-$) which is responsible for its paramagnetism and its diradical-like properties; (*b*) the existence of two low-lying electronically excited singlet states, designated $^1\Delta_g$ and $^1\Sigma_g^+$ with **22.5** and **37.5** kcal of excitation energy respectively (Fig. 1). Because transitions between the ground state and these two excited singlet states are spin-forbidden, the absorption

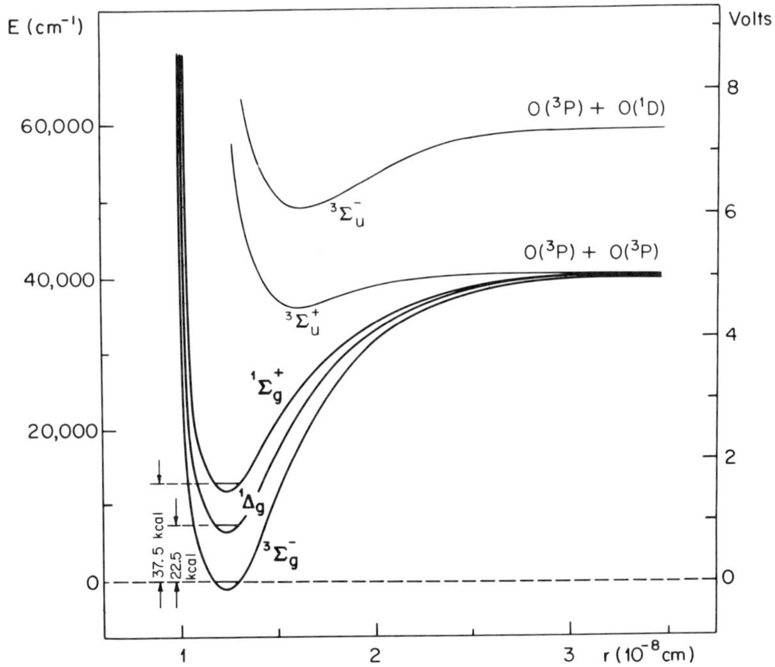

FIG. 1. Potential energy diagram for the lower states of the oxygen molecule.

coefficients are very small. However, absorption leading to these $^1\Delta_g$ and $^1\Sigma_g^+$ excited states can be measured in solution if the O_2 is dissolved under high pressure (Evans, 1969). Moreover the emission from molecules which are in these singlet excited states is unlikely, e.g., their radiative lifetimes are unusually long: 1 hour for the $^1\Delta_g$ and 7 seconds for the $^1\Sigma_g^+$.

All three states can be considered to be formed from two ground state ^3P atoms; they differ in the way two of their sixteen electrons occupy the two degenerate (e.g., of equal energy) antibonding π_x^* and π_y^* orbitals.

In the ground state, $^3\Sigma_g^-$, the two electrons have parallel spins and hence give a triplet level, but their orbital angular momenta are in opposite directions and therefore cancel, making the state a Σ state. The two upper states are both singlet, i.e., the electron spins are paired. In the lower of these two states, $^1\Delta_g$, the two electrons may be described as rotating about the interatomic axis in the *same* direction, yielding a resultant component of angular momentum of 2 units; i.e., a Δ state. Since this rotation can be clockwise or counterclockwise, this level is doubly degenerate. The upper level, $^1\Sigma_g^+$ has zero angular momentum because the electrons are rotating in *opposite* directions. The reader is referred to the article by E. A. Ogryzlo (Chapter 2) for a more rigorous and detailed discussion of the spectroscopic and physical properties of singlet oxygen.

1.3 Discovery and Rediscovery

The destructive effect of light in the presence of a dye and oxygen—the photodynamic action—was well-known in a variety of biological systems long before the chemistry of any simple photooxygenation reaction was established. Some of these reactions were themselves of interest from a biological standpoint. For example, certain endoperoxides were discovered to have the property of regenerating the oxygen molecule upon heating: Moureu *et al.* (1926) showed that the red hydrocarbon rubrene gives a colorless peroxide when exposed to light in air, but becomes red and highly fluorescent again upon heating. A formal analogy

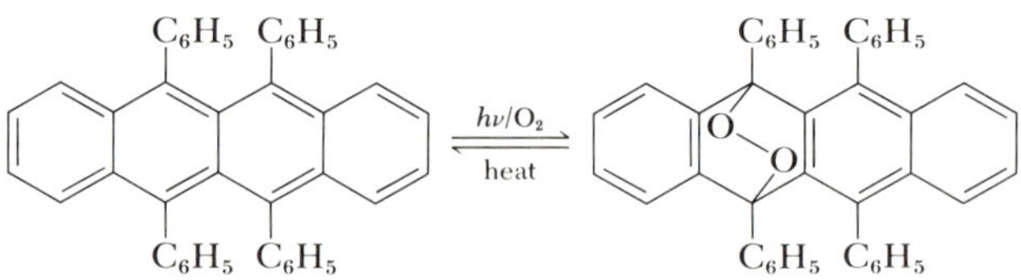

between this type of reaction and the reversible binding of oxygen with hemoglobin was obvious. Thus photooxygenation became the object of a great deal of exploratory work from a biological standpoint, while at the same time the relation between quenching of fluorescence by oxygen and photooxidation was investigated.

The most incisive studies were those of Kautsky (Kautsky and de Bruyn, 1931). From a long series of experiments he concluded that the light energy absorbed by a sensitizing dye was transferred to oxygen (thereby quenching the dye fluorescence) giving an "active" metastable excited oxygen O_2^* which attacked the substrate, e.g., the acceptor (A). Thus,

$$\text{Dye} + h\nu \rightarrow \text{Dye}^*$$
$$\text{Dye}^* \rightarrow \text{Dye} + h\nu \text{ (fluorescence)}$$
$$\text{Dye}^* + O_2 \rightarrow \text{Dye} + O_2^*$$
$$A + O_2^* \rightarrow AO_2$$

A mechanism of this type was indicated first because direct energy transfer from the sensitizing dye to the substrate (A) had to be ruled out, since the absorption bands of many substrates lie at shorter wavelengths than that of the dye used; i.e., the excited dye molecules do not have sufficient energy to populate the excited states of these acceptors. Such is the case in the methylene blue sensitized photooxidation of anthracene mentioned earlier. Obviously some intermediate has to participate in the reaction between the excited dye, oxygen and substrate. Even in the apparently simpler case where the substrate acts as its own sensitizer, it was shown that the quantum yield increases with the concentration of substrate (Bowen and Steadman, 1934). Thus a direct reaction between the primary excited state of the substrate (acceptor) and oxygen must be ruled out in this case too: a second molecule of substrate has to participate in the reaction.

Kautsky (1939) elegantly demonstrated the validity of this attractively simple hypothesis by showing that photooxidation could occur even in a system having dye and substrate bound to *physically separate* grains of silica gel; only gaseous oxygen could possibly diffuse and come in contact successively with both.

The fact that the Kautsky mechanism was disregarded is difficult to explain in view of his striking experiments. It might in part be attributed to the fact that there still existed at the time a great deal of confusion about the fate of light absorbed by polyatomic molecules—singlet and triplet states, fluorescence and phosphorescence, modes of radiationless conversion of electronic energy, etc. Likewise, the existence of the two singlet species of oxygen, the $^1\Delta_g$ and $^1\Sigma_g^+$ at energy levels of 22.5 and 37.5 kcal above ground, obscured the discussion. Somehow the possible participation of the lower energy state was discounted; thus the effectiveness of a sensitizer absorbing at 800 nm (which is energetically incapable of exciting the higher state $^1\Sigma_g^+$) was considered incompatible with the singlet oxygen hypothesis (Gaffron, 1936), even though there was ample energy in this case for excitation of the lower singlet state. Finally, a number of examples of apparent dye specificity were known, and this was taken as evidence that a reaction involving both dye and substrate had to occur, in spite of Kautsky's proof to the contrary.

In any event, Kautsky's hypothesis was arbitrarily discarded and ignored for the next 25 years, for the benefit of a more elusive mechanism advocated first by Schönberg (1935) then by Schenck (see Gollnick and Schenck, 1964), where an unstable sensitizer-oxygen complex is the active intermediate. Thus the subject of "active" oxygen was curiously dormant and almost ignored for many years. Its "rediscovery" followed Seliger's observation that a red luminescent flash occurs upon mixing H_2O_2 and HClO. Seliger (1960) was testing the ability of a number of different oxidizing agents to stimulate the chemiluminescence of luminol when he noted that the two oxidants above give a red glow upon mixing. He reported that the spectrum of this emission, which had been first observed 33 years earlier by Mallet (1927) but had been largely forgotten, consisted of a rather narrow band centered at 634 nm. (See Khan and Kasha, 1970, for an excellent historical account.)

Khan and Kasha (1963) must certainly be credited with recognizing the importance of the observation, with pursuing the study and identifying the emission with an electronically excited state of oxygen. They confirmed Seliger's data and were able to record an additional band at 703 nm. Because the separation between those two bands (1567 cm^{-1}) corresponds within experimental error to the ground state vibrational frequency of molecular oxygen (1580 cm^{-1}), and since the O–O bond in the peroxide had been shown (Cahill and Taube, 1952) to remain intact during the reaction,

$$H_2O_2 + HClO \rightarrow O_2 + HCl + H_2O$$

they attributed the emission to an excited state of molecular oxygen. They tentatively assigned the 634 and 703 nm peaks to the 0,0 and the 0,1 bands of the ${}^1\Sigma_g^+ \rightarrow {}^3\Sigma_g^-$ transitions of O_2. The fact that the two observed bands were blue shifted from their corresponding gas-phase positions (at 762 and 865 nm respectively) was attributed to solvation.

Although this specific assignment was shortly shown to be incorrect by Ogryzlo and co-workers, Khan and Kasha were right in attributing the luminescence to a metastable excited singlet oxygen molecule, and their paper had the very great merit of reviving Kautsky's hypothesis concerning the role of singlet oxygen and of predicting its importance.

Independently, in studying the emission from a stream of gaseous oxygen after being subjected to electric discharge, Ogryzlo and coworkers observed the same two bands as those from the H_2O_2–HClO reaction (Arnold *et al.*, 1964). But since the known sharper bands from the 0,0 and 0,1 transitions in the $^1\Sigma_g^+ \rightarrow {}^3\Sigma_g^-$ system (at 762 and 865 nm) were also present in the spectrum, it was clearly not possible to attribute the 634 and 703 nm peaks to that system.

These authors also reinvestigated the hydrogen peroxide-hypochlorite chemiluminescence and recorded not only the 634 and 703 nm bands, but in addition they were able to observe weakly the 762 nm band from $^1\Sigma_g^+$ as well as the intense transition at 1270 nm corresponding to $^1\Delta_g \rightarrow {}^3\Sigma_g^-$. Evidently the 634 and 703 nm bands had to have the same origin in the solution luminescences and in the gaseous discharge, and could be said not to derive from the $^1\Sigma_g^+ \rightarrow {}^3\Sigma_g^-$ system.

Arnold *et al.* (1969) made the important assumption that the 634 and 703 nm bands originate from a dimer of two excited ($^1\Delta_g$) oxygen molecules undergoing simultaneous deactivation with emission of a single quantum combining the energies of two excited $^1\Delta_g$ molecules

$$2O_2(^1\Delta_g) \rightarrow 2O_2(^3\Sigma_g^-) + h\nu$$

(at 634 nm, if both product molecules are left in their ground vibrational levels; at 703 nm, if one product molecule is left in its first vibrational level) (Fig. 2). Indeed, twice the electronic energy of one $O_2(^1\Delta_g)$ correspond to 15,765 cm^{-1} or 634 nm; moreover, the bands are diffuse and their intensity appears to parallel the concentration of $O_2(^1\Delta_g)$ independently measured with a calorimetric detector in the gas phase studies (Browne and Ogryzlo, 1964; Bader and Ogryzlo, 1964).

Though unconventional, the hypothesis of simultaneous electronic transitions was not unprecedented. The visible *absorption* bands at 634 and 580 nm in liquid oxygen, which are responsible for its blue color, had previously been attributed to such double transitions, in which one photon simultaneously elevates *two* electrons on two different molecules to excited states (Ellis and Kneser, 1933; Dianov-Klokov, 1959). Thus Ogryzlo's work in emission spectroscopy beautifully confirmed this assignment in absorption.

$$2O_2(^3\Sigma_g^-) + h\nu_{634\text{nm}} \rightleftarrows 2O_2(^1\Delta_g)$$

Although this interpretation of the 634–703 nm bands was in dispute for some time, the evidence is now compellingly in its favor. It is well established, for example, that their intensity depends on the square of the $O_2(^1\Delta_g)$ concentration (Bader and Ogryzlo, 1964; Whitlow and Finlay, 1967; Arnold and Ogryzlo, 1967). It seems that the species emitting the double transition should be viewed, not as a loosely bound dimer, but as a simple collisional complex of two $O_2(^1\Delta_g)$ undergoing simultaneous deactivation to the ground state (Arnold *et al.*, 1965; see also Hoytink, 1969).

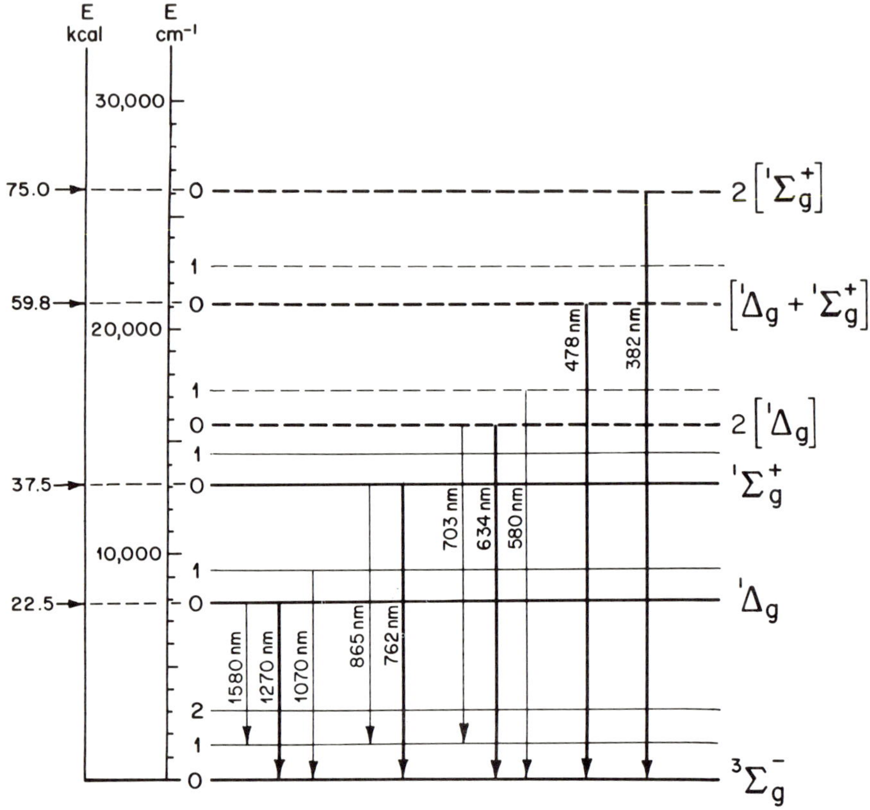

FIG. 2. Electronic and associated vibrational levels of the oxygen molecule in its ground and singlet excited states. The vertical arrows indicate some of the transitions observed in emission. The dotted horizontal lines correspond to the levels of the "molecular pairs" from which the observed simultaneous transitions originate.

There is also evidence that simultaneous electronic transitions may involve oxygen molecules in different excited singlet states. The absorption band in liquid oxygen at 478 nm corresponds to the simultaneous transition giving rise to one excited molecule of oxygen in the $^1\Sigma_g{}^+$ state and one in the $^1\Delta_g$ state.

$$2O_2(^3\Sigma_g{}^-) + h\nu_{478nm} \rightleftarrows O_2(^1\Sigma_g{}^+) + O_2(^1\Delta_g)$$

The same "mixed" double transition has been observed in emission also (Khan and Kasha, 1966).

The phenomenon of simultaneous transition may even occur between chemically different molecules. An absorption band at 350 nm in a chloroform solution of naphthalene and oxygen is attributed to a simultaneous transition populating the naphthalene triplet and exciting the oxygen molecule to the lowest singlet state (Dijkgraaf *et al.*, 1962).

$$^{1}\text{Naphth} + O_2(^{3}\Sigma_g^-) + h\nu_{345\text{nm}} \rightarrow {}^{3}\text{Naphth}^* + O_2(^{1}\Delta_g)$$

Thus the ground work had been laid, showing that excited oxygen in the metastable singlet state could be generated either by reaction of hydrogen peroxide with hypochlorite or by electrodeless discharge in a stream of oxygen. Foote was prompt to realize the implications of this result for dye-sensitized photooxygenation reactions and showed that substrates subjected to the H_2O_2–HClO system were converted to oxidation products identical with those obtained with light and an appropriate dye (Foote and Wexler, 1964).

Corey and Taylor (1964) approached the problem at precisely the same time but with singlet oxygen generated in a stream of gaseous oxygen by an electric discharge prior to bubbling it through the solution of substrate. Their findings were essentially identical, a striking result in view of the different methods of generating the excited oxygen. If Corey's and Foote's results differed somewhat in the type and reactivity of the acceptors successfully oxidized, it could be attributed to rather irrelevant differences in reaction conditions, such as solvents. The important fact demonstrated by these two investigators is that excited singlet oxygen can exist in solution and can carry out reactions that were previously thought to require light. They thus added a great weight of circumstantial evidence, if no direct proof as yet, to Kautsky's photooxygenation mechanism.

2. Laboratory Sources of Singlet Oxygen, Basic Reactions Involved

Several different methods of producing singlet oxygen have been mentioned, which we shall now discuss in greater detail.

2.1 Electrical Discharge

In principle, this method has the advantage of unambiguity; the singlet oxygen being generated externally, one should be able to study its properties like that of any other pure reagent. In practice, however, the conversion of ground state oxygen into excited oxygen is only partial. Thus, the singlet oxygen is always mixed with a large excess of ground state oxygen and often with traces of less desirable contaminants such as ozone and oxygen atoms.

An electrodeless microwave or radio frequency discharge is used (Bader and Ogryzlo, 1964; Arnold and Ogryzlo, 1967; Clyne *et al.*, 1965). It produces directly about 10% of oxygen atoms in their ground ^3P state and another 10% of excited O_2 molecules in the $^1\Delta_g$ state (Foner and Hudson, 1956; Bader and Ogryzlo, 1964). It seems that singlet oxygen molecules in the upper level (the $^1\Sigma_g{}^+$ state) result mostly from secondary processes involving O atoms immediately after the discharge, such as

$$O + O + M \rightarrow O_2(^1\Sigma_g{}^+) + M \qquad \text{(here the "third body" } M \text{ is } {}^3O_2)$$

although this point is in dispute (Young and Sharpless, 1963; Clyne *et al.*, 1965).

Ozone is formed in another three body reaction:

$$O + {}^3O_2 + M \rightarrow O_3 + M$$

and destroyed by the bimolecular process (Mathias and Schiff, 1964)

$$O_3 + O \rightarrow 2\,{}^3O_2.$$

By operating at low pressures of oxygen, of the order of a few millimeters or less and at room temperature, the steady concentration of ozone formed is apparently negligible.

Singlet oxygen is probably involved also in processes controlling the concentration of ozone in the stratosphere, such as

$$O_2(^1\Delta_g) + O_3 \rightarrow 2\,{}^3O_2 + O$$
$$O_2(^1\Sigma_g{}^+) + O_3 \rightarrow 2\,{}^3O_2 + O$$
$$O_3 + O \rightarrow O_2(^1\Sigma_g{}^+) + {}^3O_2$$

(see Schiff, 1969).

The easiest method for eliminating oxygen atoms from the discharged oxygen stream consists in distilling mercury into the discharge region and arranging to have a deposit of mercuric oxide immediately after the discharge[1] (Arnold *et al.*, 1965; Ogryzlo and Pearson, 1968).

The presence of residual oxygen atoms can be detected by introducing nitric oxide in the gas stream and looking for the appearance of the yellow air glow resulting from the reaction

$$NO + O \rightarrow NO_2 + h\nu$$

A sensitive method of oxygen atom titration based on this reaction has been described (Fontijn *et al.*, 1964).

The concentration of $O_2(^1\Delta_g)$ in the stream can be conveniently

[1] If the speed of pumping is not fast enough, this may require setting up an auxiliary temporary discharge downstream from that point. Oxygen atoms will tend to diffuse upstream from the discharge and react with mercury there. Once an HgO deposit is formed, the auxiliary discharge is turned off and the main discharge allowed to operate (Falick, 1967).

measured by the intensity of the 634 and 703 nm bands, corresponding to the simultaneous deactivation of two molecules according to the bimolecular process discussed earlier (Section 1.3 and Fig. 2).

$$O_2(^1\Delta_g) + O_2(^1\Delta_g) \rightarrow 2^3O_2 + h\nu_{634nm} \quad (1)$$

The intensity at 634 nm is proportional to the square of the $O_2(^1\Delta_g)$ concentration. The transition from a single $O_2(^1\Delta_g)$ occurs in the infrared (1270 nm), which is a far less convenient region of the spectrum to monitor.

Singlet oxygen in the $^1\Delta_g$ state is remarkably immune to collisional deactivation. In the usual conditions of discharge experiments (pressure of a few torr) its lifetime is determined mostly by wall deactivation (Arnold *et al.*, 1968). The rate constant of reaction (1) above is far too small to appreciably affect the concentration of $O_2(^1\Delta_g)$: only about one in 10^{11} collisions between two $O_2(^1\Delta_g)$ will lead to radiative deactivations at 634 nm.

Another rare but remarkable bimolecular reaction between two $O_2(^1\Delta_g)$ is a disproportionation yielding one oxygen molecule in the upper $^1\Sigma_g{}^+$ state, and one in the ground state.

$$O_2(^1\Delta_g) + O_2(^1\Delta_g) \rightarrow O_2(^1\Sigma_g{}^+) + {}^3O_2 \quad (2)$$
$$^1O_2(^1\Sigma_g{}^+) \rightarrow {}^3O_2 + h\nu_{762nm}$$

This process is responsible for the building up, in a gas stream free of oxygen atoms, of a steady concentration of $O_2(^1\Sigma_g{}^+)$ after the discharge, and thus the emission band at 762 nm (Arnold and Ogryzlo, 1967). In the gas phase at low pressure, the collision of two $O_2(^1\Delta_g)$ is actually more likely to result in emission at 762 nm via this mechanism than in emission at 634 nm via the double transition (Eq. 1). Most of the $O_2(^1\Sigma_g{}^+)$ will also be deactivated at the wall, but about 500 times faster than $O_2(^1\Delta_g)$ (Arnold and Ogryzlo, 1967).

$O_2(^1\Sigma_g{}^+)$ is considerably more easily quenched by foreign gases than $O_2(^1\Delta_g)$. No obvious relation is apparent between the efficiency of a gas as a quencher and its nature, other than the higher its boiling point, the better it quenches [the greater the perturbation it effects on $^1O_2{}^*$ (Arnold *et al.*, 1968)]. Water is very efficient: according to Ogryzlo, one in 10 collisions with water molecules will relax $O_2(^1\Sigma_g{}^+)$, most likely to the $^1\Delta_g$ state. Taking all of these considerations into account, the steady concentration of $^1\Sigma_g{}^+$ can be expected to be always very small, about 10^5 less than the $^1\Delta_g$ concentration.

The electrical discharge method has so far been used mostly for studying gas phase reactions. Adding an inert gas such as helium makes it possible to operate the discharge at higher total pressure without increasing the production of ozone or reducing the yield of singlet oxygen.[2]

The gaseous mixture rich in $^1O_2$ can then be bubbled through liquids. This method was first applied by Corey and Taylor (1964) using bromobenzene solutions of different typical oxidizable acceptors such as anthracene derivatives and conjugated dienes; it enabled them to demonstrate that the products of their reaction with singlet oxygen generated in this way were identical with the products obtained via dye-sensitized photooxygenation. This result immediately underscores the fact that the reaction of singlet oxygen with acceptor must be highly efficient since competing deactivating collisions reduce the effective lifetime of $O_2(^1\Delta_g)$ in solution to 10^{-3}–10^{-5} sec (see Ogryzlo, Chapter 2, this volume; Foote and Denny, 1968). If the acceptors are fairly unreactive and require long reaction times, it becomes crucial to choose a truly inert solvent. A low vapor pressure is also desirable in order to minimize the loss by evaporation.

2.2 Photochemical Oxygenation

The combined action of light and oxygen on suitable sensitizer-acceptor mixtures is of course the oldest technique for studying the reactions of singlet oxygen, and also in a general way the most convenient. Experimental methods and the vast literature have been the object of several recent review articles (Bowen, 1963; Gollnick and Schenck, 1967; Foote, 1968; Gollnick, 1968). Let us emphasize here the importance of using a monochromator or color filters to isolate a spectral region of the light source that matches the absorption of the sensitizer, in order to minimize side reactions and obtain unambiguous results.

The key step is the transfer of electronic energy from the excited "sensitizer" molecule S^* to a ground state oxygen molecule, thereby raising the latter to its excited singlet state (represented here by $^1O_2^*$ without further defining at this time its spectroscopic state).[3]

$$S + h\nu \rightarrow {}^1S^* \tag{3}$$

$$^1S^* \rightarrow S(+h\nu) \quad \text{internal conversion or fluorescence} \tag{4}$$

$$^1S^* \rightarrow {}^3S^* \quad \text{intersystem crossing} \tag{5}$$

$$^3S^* \rightarrow S(+h\nu) \quad \text{internal conversion or phosphorescence} \tag{6}$$

$$^3S^* + {}^3O_2 \rightarrow S + {}^1O_2^* \quad \text{energy transfer to form singlet oxygen} \tag{7}$$

[2] Probably because the ground state oxygen $^3O_2$ is a triplet, it is an especially efficient quencher of the $^1\Delta_g$ state.

[3] The acceptor can in fact often act as its own sensitizer. In that case, the need for some intermediate (singlet oxygen) is not immediately obvious. Kinetic studies of the photooxidation of rubrene showed that the peroxide is not formed in one step (see Section 1.3). Besides, rubrene can act as sensitizer of the photooxygenation of another acceptor (see below, p. 63).

In most dyes, the energy difference between the first excited singlet and triplet states (singlet–triplet splitting $\Delta E_{ST} \lesssim 10$ kcal) is much smaller than the excitation energy for $O_2(^1\Delta_g)$ and therefore there is no question that the excitation energy has to come from the triplet state of the sensitizer, as written above (7). In the readily photooxidizable polynuclear aromatics such as naphthalene, anthracene and naphthacene, ΔE_{ST} is about 30 kcal (see Turro, 1965, pp. 86 and 132), so that the energy transfer could conceivably occur from the singlet state $^1S^*$ via an equally allowed reaction:

$$^1S^* + {}^3O_2 \rightarrow {}^3S^* + {}^1O_2^*.$$

In any case, the much longer lifetime of the triplet state ($\sim 10^5$ longer than an excited singlet) makes it much more susceptible to collisional quenching and deactivation by energy transfer. Kinetic and spectroscopic evidence seem to confirm that (7) is the predominant mode of excitation of oxygen in all cases (Stevens and Algar, 1967; Snelling, 1968).

The relation between photoformation of singlet oxygen and phosphorescence quenching by oxygen is obvious. Oxygen is also known to enhance the singlet-triplet absorption (intersystem crossing) in organic molecules, which is the basis for a useful technique for locating triplet states (Evans, 1957, 1959). To explain this effect, Tsubomura and Mulliken (1960) have proposed a charge-transfer complex between the organic molecule and oxygen, which would relax the forbiddeness of the singlet-triplet transition. They also concluded that the paramagnetism of ground state oxygen was of no importance here (see Hoytink, 1969).

Thus quenching of phosphorescence by oxygen could occur through enhanced intersystem crossing, as well as through energy transfer yielding singlet oxygen. Kawaoka *et al.* (1967) have discussed in detail the relative importance of these two modes of quenching by oxygen. From quantum-mechanical calculations, they conclude that the latter is usually much faster than the first, mainly because less electronic energy needs to be degraded in vibrational energy (Frank-Condon factor). Kawaoka *et al.* (1967) predict that the quenching of phosphorescence by energy transfer to oxygen may well account for the experimentally observed rates, which can reach the diffusion controlled limit.

The question of which of the two singlet states of oxygen is generated by the energy transfer process has not yet received an unambiguous answer. Kearns *et al.* (1967a) suggested that this will depend on the triplet energy level of the sensitizer. A high energy sensitizer (with $E_T > 50$ kcal) would produce primarily $^1\Sigma_g^+$ oxygen, whereas evidently the quenching of a low energy sensitizer ($E_T < 38$ kcal) could only result in $O_2(^1\Delta_g)$; sensitizers with E_T between 38 and 50 kcal would produce increasing relative amounts of $^1\Sigma_g^+$ to $^1\Delta_g$ as their triplet energy increases. Different reactivities of the two singlet oxygens could lead to different products from the same reactive substrate. Theoretically correct and attractive as this may be, it must be kept in mind that $O_2(^1\Sigma_g^+)$ is so quickly relaxed by collisions to the $^1\Delta_g$ state that it is not very likely to have a chance of reacting. Ogryzlo also points out that $O_2(^1\Sigma_g^+)$ may actually be quite unreactive (Chapter 2).

Kearns *et al.* (1967a,b) attempted to explain on the basis of their "two singlet oxygen" hypothesis some unaccounted dye specificity effects in the sensitized photooxidation of cholest-4-en-3β-ol, where two products are obtained in a ratio which varies with the dye (Nickon and Mendelson, 1965). Primarily for the reason given above, this interpretation is questionable. The photochemistry of dyes is extremely complicated, which makes them less than desirable sensitizers from a mechanistic standpoint. Their chemical stability when exposed to strong light is far from perfect, as evidenced by their bleaching. Besides transferring their excitation energy to form singlet oxygen, they can, depending on their structure, participate with the acceptor in various less efficient competitive processes such as hydrogen or electron transfer, or react in these manners with oxygen itself, with possible formation of hydrogen peroxide or hydroperoxide and free-radical side reactions. The less reactive the acceptor is towards singlet oxygen, the more significant these side-processes will be. Using the photooxidation of the octalol I (below) as a model system for the cholestenol case, Foote and Wong (1969) have recently shown that one of the oxidation products (the enone III, attributed by Kearns *et al.* (1967a) to the reaction of the substrate with $O_2(^1\Sigma_g^+)$ was probably not a singlet oxygen reaction, but the result of a direct reaction between sensitizer and substrate.

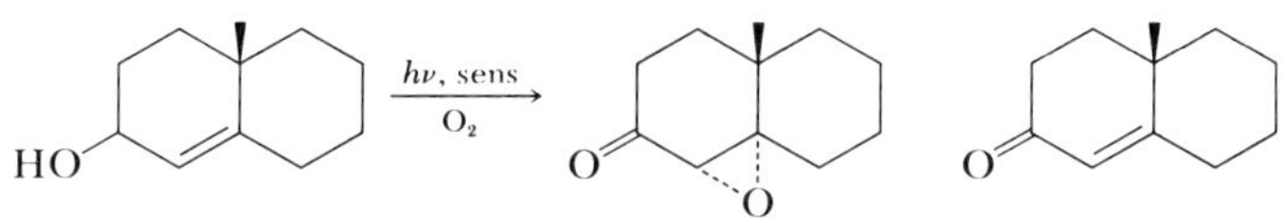

I II III

The possible direct reactions between excited dyes and substrates have recently been discussed by Foote (1969), in the context of photooxidation via singlet oxygen.

Until the work of Foote and of Corey, the prevalent interpretation of the photooxidation mechanism was that first suggested by Schönberg (1935), and later presented by Schenck with a detailed formalism. It assumed the intermediacy, not of singlet oxygen, but of a transient excited sensitizer-oxygen complex (a moloxide SOO), which later transfers oxygen to the acceptor (Gollnick and Schenck, 1964).

$$S^* + O_2 \rightarrow (SOO)$$

$$(SOO) + A \rightarrow AO_2 + S$$

This mechanism is evidently difficult to distinguish kinetically from Kautsky's mechanism, and it may well be that a weak complex between sensitizer and oxygen precedes the formation of free singlet oxygen. But whereas no moloxide has ever been isolated or physically characterized, there is now ample indirect and some direct evidence to implicate singlet oxygen. Besides, one would expect the relative rates of oxygenation of various acceptors to be different in direct or sensitized oxidation (the moloxide being either SOO or AOO), but this is not the case (Wilson, 1966). Relative rates of oxidation are independent of the sensitizer (Kopecky and Reich, 1965) as one can expect if the reactive species is, in all cases, singlet oxygen.

The most convincing argument against a moloxide intermediate is that photooxidation can occur even when acceptor and sensitizer are not in physical contact and only gaseous oxygen can diffuse from one to the other (Kautsky and de Bruyn, 1931; Kautsky, 1939). In a very elegant experiment, refining Kautsky's, Schnuringer and Bourdon (1968) showed that the photooxidation of rubrene or of 9,10-diphenylanthracene photosensitized by methylene blue or eosin took place when acceptor and sensitizer were separated by up to 500 Å of oxygen-permeable stearate salts films.

In direct photooxidation, where an aromatic hydrocarbon like 9,10-diphenylanthracene (DPA) is both the sensitizer and the acceptor, it may seem remarkable that here again energy transfer first takes place between the anthracene and oxygen

$$^3A^* + {}^3O_2 \rightarrow A + {}^1O_2^* \tag{7}$$

followed by

$$A + {}^1O_2^* \rightarrow AO_2 \tag{8}$$

instead of immediate reaction

$$^3A^* + {}^3O_2 \rightarrow AO_2 \tag{9}$$

It can indeed be shown that reaction (9) does not occur to any appreciable extent, by adding a reactive acceptor like tetramethylethylene (TME) (Wilson, 1966). The first singlet and triplet excited states of TME are too high to be excited at the wavelength used to excite DPA. Yet, if TME is present in high enough concentration it stops the oxidation of DPA by capturing all the ${}^1O_2^*$ formed by reaction (7):

$$\mathrm{TME} + {}^1\mathrm{O}_2{}^* \rightarrow \mathrm{TMEO}_2 \tag{10}$$

Consequently, the simple energy transfer (7) must occur much more rapidly, probably at greater distance, than reaction (9).

For its basic rather than practical interest, we must mention here that oxygenation can be caused by direct photochemical excitation of oxygen dissolved at high pressure in an inert organic solvent. Evans (1969) has recently reported the oxidation of 9,10-dimethylanthracene and 1,3-diphenylisobenzofuran dissolved in $CCl_2F\text{-}CClF_2$ at 2000 lb/sq. in., under intense irradiation at wavelengths corresponding to the transitions from ground state oxygen into ${}^1\Delta_g$, ${}^1\Sigma_g{}^+$ as well as to $({}^1\Delta_g)_2$ and $({}^1\Delta_g + {}^1\Sigma_g{}^+)$ states. Other wavelengths were ineffective; no sensitizers were present.

2.3 Chemical Oxygenation

2.3.1 Heterolytic Oxidation of Hydrogen Peroxide

This method has already been discussed at some length earlier, because of its importance in the singlet oxygen hypothesis for oxygenation (see Section 1.3). The decomposition of hydrogen peroxide in alkaline solutions by oxidants reacting by an electron pair mechanism yields molecular oxygen in its singlet excited state.

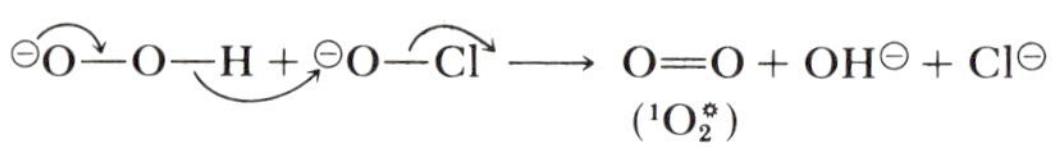

With sodium hypochlorite, Foote and Wexler (1964) were able to obtain preparative scale oxygenation products from a variety of acceptors such as olefins, dienes, dienones, and furans. The method is very simple: NaOCl is added dropwise to a chilled and stirred alkaline solution of the acceptor containing an excess of hydrogen peroxide, usually in methanol. Yields of up to 80% of oxidation products are reported when the acceptor is present in sufficiently high concentration (0.1 M) and when methanol, ethanol, or a mixture of methanol-t-butyl alcohol are used as solvents. For reasons not yet understood, much lower yields are obtained in solvents like acetonitrile, dioxane, or tetrahydrofuran (Foote *et al.*, 1968).

Obvious solubility problems arise with oxidizable substrates like polyacenes. Anthracene derivatives, for instance, give only extremely poor yields of products.

To obviate in part the solvent limitations, McKeown and Waters (1966), proposed an alternative method in which bromine replaces NaOCl, in a two-phase system:

$$^{\ominus}O{-}O{-}H + Br{-}Br \longrightarrow \underset{(^1O_2^*)}{O{=}O} + HBr + Br^{\ominus}$$

for example: an upper layer of DPA in chlorobenzene, a lower layer of stirred aqueous alkaline hydrogen peroxide to which bromine is slowly added. The singlet oxygen evolved rises through the upper layer where it reacts with the acceptor in 50% yields. The reaction of bromine and hydrogen peroxide emits a particularly intense red chemiluminescence (at 634 nm and 702 nm). However, side reactions due to bromine are likely to take place. Therefore at the present time, both Foote's and Waters' methods of chemical oxygenation appear to have only a limited practical value, but, as discussed earlier, their success had most important mechanistic implications.

2.3.2 Decomposition of Triphenylphosphite-ozone Adduct

Alkyl and arylphosphites react with ozone to give the corresponding phosphates and molecular oxygen. Thompson (1961) found that when the reaction is carried out at —70°, some triarylphosphites first form a metastable 1:1 adduct with ozone, which then decomposes upon heating into the phosphate and oxygen. Murray and Kaplan (1968) showed that in the case of the triphenylphosphite-ozone adduct, the oxygen evolved had a reactivity like that of singlet oxygen.

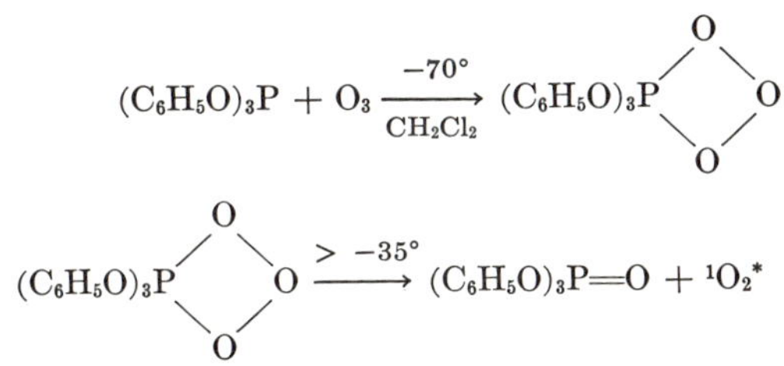

Thus, here is a convenient source of singlet oxygen susceptible to temperature control: drops of the ozonide are added to the acceptor solution at low temperatures; the reaction takes place on warming. There is direct spectroscopic (E. Wasserman *et al.*, 1968) and chemical (Murray and Kaplan, 1968) evidence that the oxygen evolving is in the singlet excited state, and the products obtained are consistent with this assumption. Nevertheless, Bartlett and Mendenhall (1968, 1970a) have shown that, at least in one case, a reactive olefin, tetramethylethylene, will react with the ozonide at temperatures well below that at which the ozonide starts decomposing. The oxidation product so obtained is identical with the photooxygenation product, as if a molecular displacement reaction occurred directly between acceptor and ozonide, mimicking free singlet oxygen.

The decomposition of ozonide is very susceptible to base catalysis; for example, moist pyridine will bring about the immediate release of singlet oxygen (Bartlett and Mendenhall, 1970b).

2.3.3 Thermal Decomposition of Transannular Peroxides of Polyacenes

Polynuclear aromatic hydrocarbons such as anthracene and naphthacene derivatives are among the best acceptors of singlet oxygen (see Sections 1.3 and 4.1). The product of their photooxidation is a transannular peroxide, which in some cases has the remarkable property of regenerating oxygen and the original polyacene when heated (see Section 1.3). In the case of 9,10-diphenylanthracene (DPA) Wasserman and Scheffer (1967) showed that the regenerated oxygen released

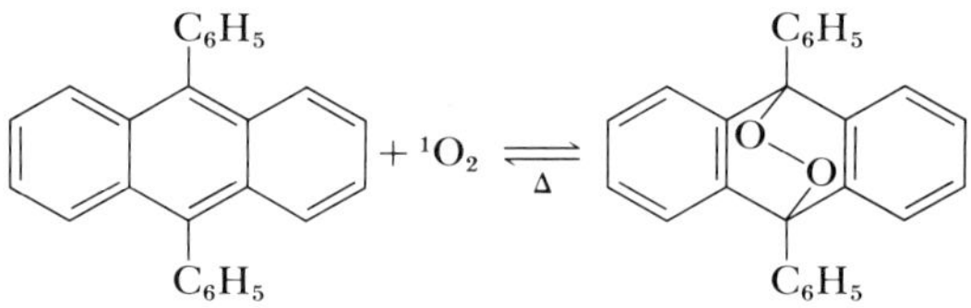

had singlet oxygen reactivity with different acceptors. Thus, here again is a temperature-controlled source of singlet oxygen. The peroxide of DPA decomposes only very slowly at temperatures below 100°. Wasserman treated mixed solutions of peroxide and acceptor (in a 2/1 ratio) in methylene chloride or benzene at refluxing temperatures for a day or so, and reports remarkably high yields of product.

The transannular peroxide of 1,4-dimethoxy-9,10-diphenylanthracene is already unstable at room temperature. In pyridine solution, this peroxide undergoes first order decomposition (activation energy $E_a = 19$ kcal) yielding near-quantitative amounts of oxygen and the anthracene. Such solutions can be used to peroxidize a good acceptor like 1,3-diphenylisobenzofuran (DPBF) or rubrene, but in preliminary experiments the yields of transfer of oxygen from the original anthracene peroxide to these acceptors is much lower than that obtained by Wasserman with the peroxide of 9,10-diphenylanthracene ($\sim$50% with DPBF, 15% with rubrene) (Wilson, 1969a).

3. Physical Evidence for Singlet Oxygen Involvement

When a chemical reaction is suspected of proceeding via singlet oxygen, the argument is most often based on the nature and relative yields of oxygenation products obtained from some typical ${}^1O_2^*$ acceptors (see Section 4), compared with the product distribution from the same acceptors subjected to singlet oxygen generated photochemically or by electron discharge. Such evidence can often be compelling but is only circumstantial. There do exist different direct physical methods of identifying singlet oxygen which we shall discuss briefly here in view of their theoretical significance.

3.1 Mass Spectrometric Evidence

Because $O_2(^1\Delta_g)$ carries 0.98 ev (22.5 kcal) of electronic excitation energy, correspondingly less energy is required for its ionization. Consequently, when a stream of oxygen enriched in $O_2(^1\Delta_g)$ is analyzed with a mass spectrometer, it gives a higher O_2^+ ion intensity at lower electron energy than a stream of normal oxygen. From the resulting shift in the appearance potential curve of O_2^+, one can calculate the concentration of $O_2(^1\Delta_g)$ in the gas stream (Foner and Hudson, 1956).

This technique requires a mass spectrometer with a sampling system especially designed to minimize the number of wall collisions on the way to the ionization chamber. The short lifetime of $^1\Sigma_g^+$ probably precludes its detection.

3.2 Calorimetric Evidence

Excited singlet oxygen molecules are efficiently deactivated on collision with a platinum wire coated with a catalyst (cobalt oxide) placed in the gas stream. An amount of heat corresponding to the $O_2(^1\Delta_g)$ excitation energy is thus liberated on the wire. This is the basis for an isothermal calorimetric method of measuring the concentration of $O_2(^1\Delta_g)$ (Elias *et al.*, 1959).

Both this calorimetric method and the mass spectrometric technique make use of the amount of excitation energy of the oxygen molecules. They do not provide unambiguous evidence of singlet oxygen, yet they have been instrumental in indicating its presence in discharged gaseous oxygen and in obtaining the first laboratory tests of its reactivity. For selective characterization of singlet oxygen, one has to turn to spectroscopy.

3.3 Electronic Emission Spectrum

The electronic emission spectrum of singlet oxygen has been discussed earlier (see Section 1.3) in some details. Bands at 478 nm ($^1\Delta_g + {}^1\Sigma_g^+$ double transition), 634 nm (2 $^1\Delta_g$ double transition), 762 nm ($^1\Sigma_g^+$) and 1.270 nm ($^1\Delta_g$) (and satellite bands corresponding to vibrational levels of these transitions) in a luminescence system constitute unambiguous evidence of singlet oxygen (Fig. 2).

3.4 Electron Paramagnetic Resonance Spectroscopy (EPR)

Because of the *orbital* angular momentum of the two π^* electrons (see Section 1.2), the excited oxygen molecule in the $^1\Delta_g$ state has a characteristic four-line EPR spectrum corresponding to about 10,000 gauss in the x-band (easy to distinguish from the EPR spectrum of ground state oxygen $^3\Sigma_g^-$ arising from the *spin* magnetic moment of the two π^* electrons) (Falick *et al.*, 1965; Wasserman *et al.*, 1968, 1969; Kearns *et al.*, 1969).

Electronic emission and EPR spectroscopy have recently proven the occurrence of singlet oxygen in different systems (see Table I). Of particular importance are the EPR studies of $O_2(^1\Delta_g)$ generated by photosensitization from naphthalene and naphthalene derivatives, because they demonstrate that the energy transfer from photoexcited sensitizer to oxygen is indeed a highly efficient process.

Unfortunately, all these physical methods seem restricted to the gas phase,[4] with the possible exception of emission spectroscpy: whether the red luminescence from the reaction of hypocholorite and hydrogen peroxide is emitted homogenously from the solution or from gas bubbles is apparently not known for sure (Khan and Kasha, 1964).

4. Chemical Reactivity

Photooxygenation and chemical reactions of singlet oxygen have been the subject of several recent review articles, which present a broad picture of the field from an organic chemist viewpoint. The two excellent articles by Foote (1968, 1969) emphasize the mechanistic aspects of the reactions whereas the earlier ones by Arbuzov (1965), Gollnick and Schenck (1967) and that by Gollnick (1968) are particularly valuable for their treatment of specific examples and thorough survey of the literature. We will therefore limit ourselves to a brief discussion of the three main types of reactions which occur between singlet oxygen and unsaturated systems.

4.1 1,4-Cycloaddition

Singlet oxygen adds directly to the 1,4-positions of conjugated dienes, heterocyclics such as furans and pyrroles, anthracenes and other polycyclic aromatic hydrocarbons to form endoperoxides. The reaction is similar

[4] The photoionization technique (McNeal and Cook, 1967) was overlooked in this discussion. Clark and Wayne (1969) used argon resonance radiation at 1049 Å and 1067 Å, which is able to ionize $O_2(^1\Delta_g)$ but not $^3O_2$, O or O_3. The ion current produced is measured. This technique is highly sensitive.

TABLE I
PHYSICAL EVIDENCE FOR SINGLET OXYGEN INVOLVEMENT

Source of $^1O_2$	Methods of $^1O_2$ detection			
	Mass spectrometry	Isothermal calorimetry	Emission spectroscopy	EPR
Electrical discharge in gaseous oxygen	Foner and Hudson (1956)	Elias *et al.* (1959); Bader and Ogryzlo (1964)	Orgyzlo and co-workers (see Sections 1.3 and 2.1)	Fallick *et al.* (1965)
Gas phase photosensitization			Emission of 1.270 nm, with benzene as sensitizer: Snelling (1968)	With naphthalene as sensitizer: Kearns *et al.* (1969); with naphthalene and derivatives as sensitizer: Wasserman *et al.* (1969)
Reaction of H_2O_2 with halogens or hypochlorite			Seliger (1960) Khan and Kasha (1963); Ogryzlo and co-workers; see Section 1.3	
Decomposition of triphenylphosphite ozonide				Wasserman *et al.* (1968)

to the Diels-Alder reaction, i.e., singlet oxygen acts as a dienophile. The natural product ascaridole is an endoperoxide formed in plants by chlorophyll-sensitized photooxidation of α-terpinene (Schenck and Ziegler, 1944; Schneck *et al.*, 1953; Corey and Taylor, 1964).

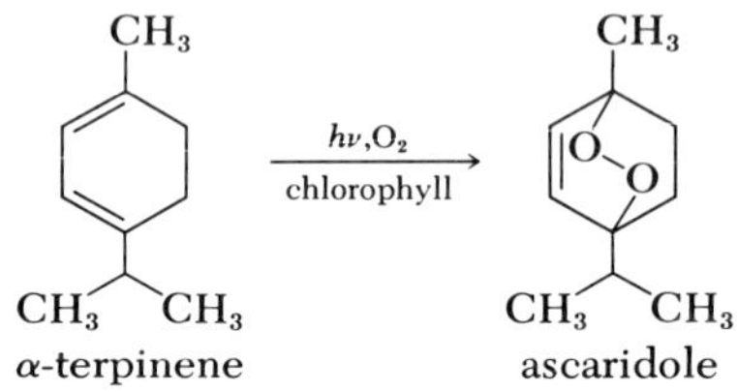

Other examples of 1,4-cycloadditions are listed below. The endoperoxides first formed are often unstable, and rearrange to other products.

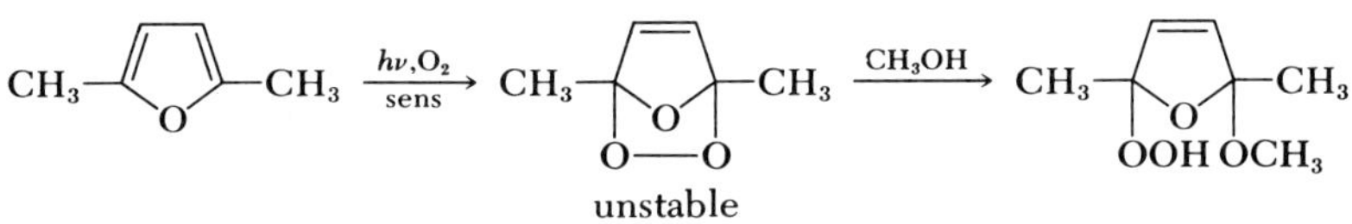

2,5-dimethylfuran (Foote and Wexler, 1964)

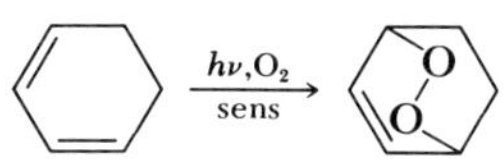

1,3-cyclohexadiene (Foote and Wexler, 1964; Foote *et al.*, 1968)

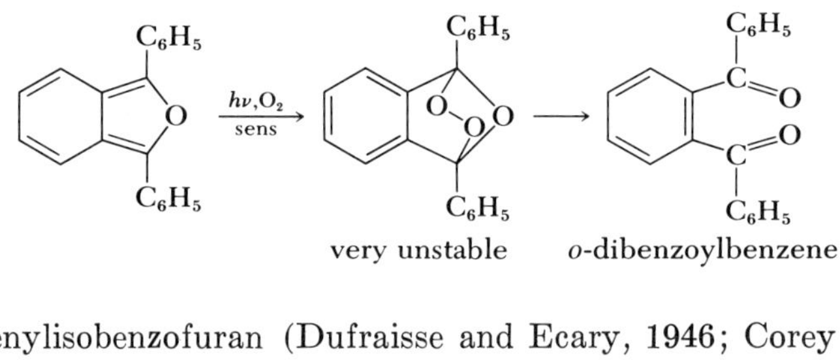

1,3-diphenylisobenzofuran (Dufraisse and Ecary, 1946; Corey and Taylor, 1964), 9,10-diphenylanthracene (Durfraisse and Etienne, 1935; Corey and Taylor, 1964; McKeown and Waters, 1966). This last reaction is reversible upon heating as discussed in Section 2.3.3. Moreover, the decomposition of these endoperoxides is accompanied by chemiluminescence, but the emission of light is probably the result of a side reaction (see Section 5.)

Many steroids form endoperoxides (see Arbuzov, 1965; Gollnick, 1968)

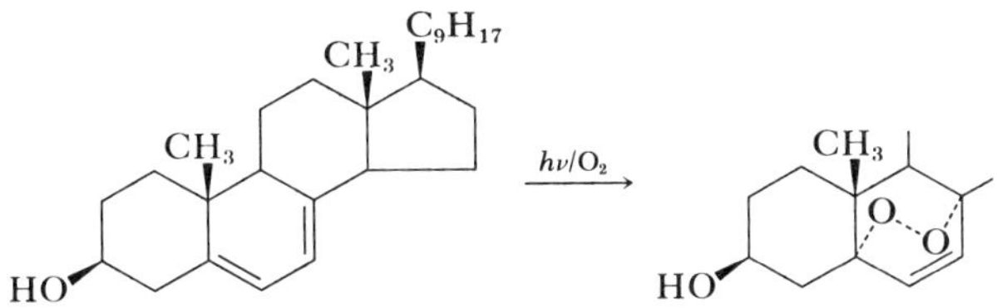

ergosterol (Fieser, 1936).

4.2 1,3-Cycloaddition ("ene" Reaction)

Singlet oxygen reacts with olefins to give the corresponding hydroperoxides with shift of the double-bond. The first step is probably a concerted 1,3-cycloaddition of singlet oxygen, where $^1O_2^*$ would here again behave like a dienophile in the "ene" reaction.

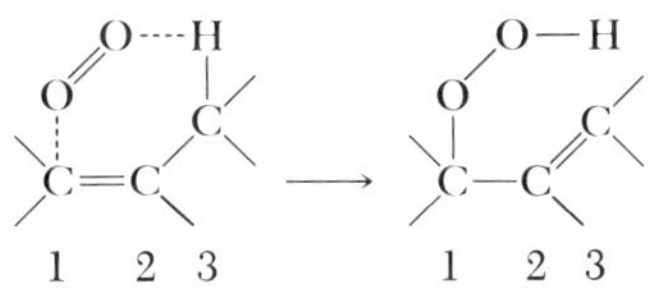

A cyclic intermediate explains the remarkable stereospecificity of the reaction in the cyclic systems: the transferred hydrogen atom always comes from the side of the double bond where the oxygen attack occurs: (e.g., the C—H bond broken with the C—O bond formed are on the same side).

Foote *et al.* (1968) have investigated the product distribution of the chemical oxygenation ($NaOCl + H_2O_2$) and photosensitized oxygenation of substrates giving more than one product. The oxygenation of limonene, for example, yields six different products; the remarkable agreement between the distributions of products obtained by these two modes of oxygenation is a striking argument in favor of the common intermediacy of singlet oxygen in both cases. Moreover, the distribution of the products of a singlet oxygen reaction and of a radical autoxidation are usually markedly different, as illustrated by the case of 1,2-dimethylcyclohexene (Table II).

Kopecky and Reich (1965) and Higgins *et al.* (1968) have determined, by a competition method, the relative reactivities of a large number of olefins towards singlet oxygen. They vary widely with the structures of the olefin, alkyl substituents increasing considerably its reactivity. For example, tetramethylethylene is about 10,000 times more reactive towards photooxidation than ethylene. Thus singlet oxygen appears to be a highly selective reagent.

An unusual type of product characterized by the presence of an

TABLE II

PRODUCTS FROM 1,2-DIMETHYLCYCLOHEXENE (FROM FOOTE, 1968) 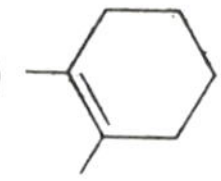

Product	Photooxygenation %	Chemical oxygenation (OCl^-/H_2O_2), %	Radical autoxidation %
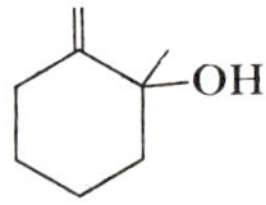	89	91	6
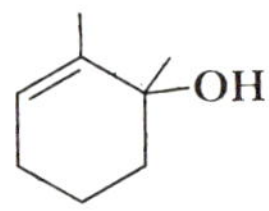	11	9	39
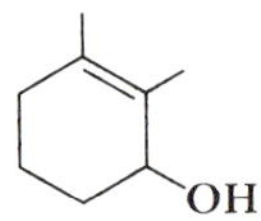	0	0	54

allenic group $\rangle C=C=C\langle$ has recently been reported among the products of reaction of singlet oxygen with model compounds of carotenoids (by photooxygenation or reaction with $NaOCl/H_2O_2$) (Mousseron-Canet *et al.*, 1969; Foote and Brenner, 1968).

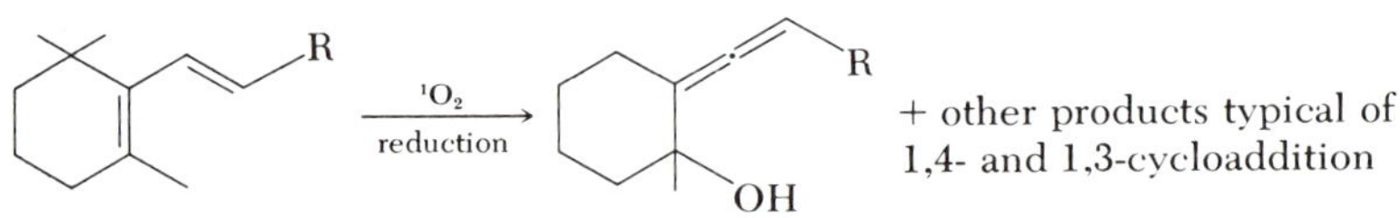

The allenic alcohol above can be visualized as resulting from an "ene" reaction on the cyclic double bond. The similitude between this type of compound and natural allenic carotenoids like fucoxanthin and neoxanthin is striking.

4.3 1,2-Cycloaddition

In some cases, it has been claimed that singlet oxygen may attack directly both sides of the double-bond, to form unstable 1,2-peroxides (*dioxetanes*), which subsequently could cleave into carbonyl fragments.

$$\rangle C=C\langle \xrightarrow{^1O_2^*} \rangle\overset{O-O}{C-C}\langle \rightarrow \rangle C=O + O=C\langle$$

According to Kearns (1969), molecular orbital and state correlation diagrams combined with empirical, spectroscopic and thermochemical data, support the prediction that 1,2-cycloaddition of $O_2(^1\Delta_g)$ to olefins may sometimes be allowed. For example, a 1,2-dixotane is considered as the probable intermediate in the photooxidation of indene, where an

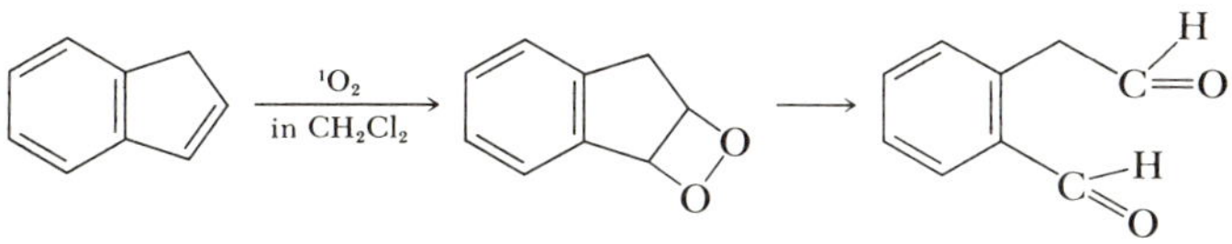

allylic hydroperoxide does not seem to be the reactive intermediate (Fenical *et al.*, 1969). Foote studied the photooxidation of enamines as models for the degradation of heterocyclic bases encountered in the "photodynamic action" (see Sections 1.3 and 6). In the examples below, the formation of an allylic hydroperoxide is precluded (Foote and Lin, 1968).

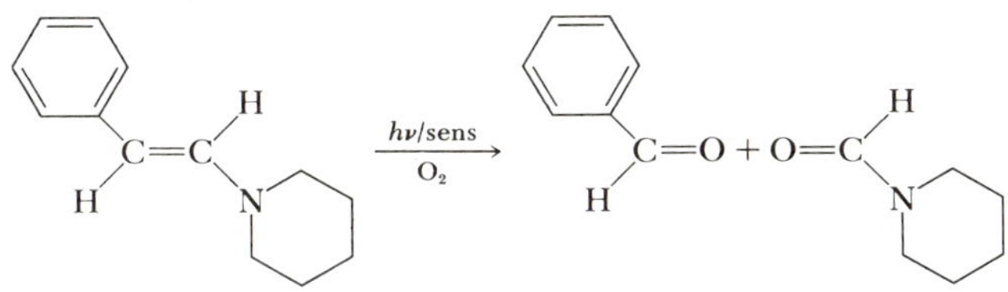

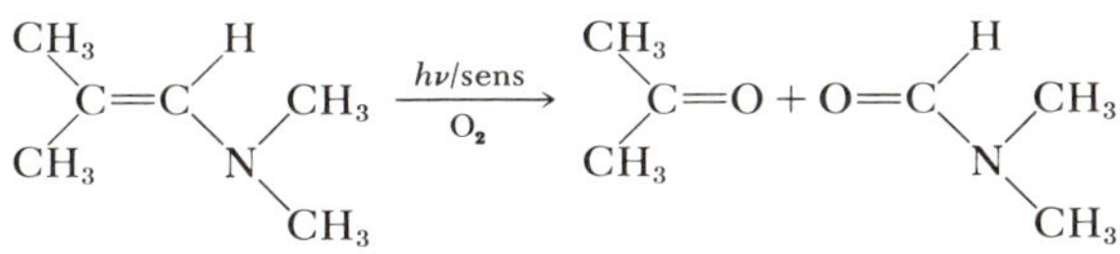

N-N-dimethylisobutenylamine

Foote and Lin carried out the photooxygenation of N,N-dimethyliso-butenylamine to completion at —60°, then allowed the solution to warm up while monitoring its nmr spectrum. An intermediate is clearly formed at low temperatures, which retains the C—C bond and breaks into the final carbonyl fragments at room temperature. However, the nmr spectrum of this intermediate is apparently not consistent with the expected dioxetane below:

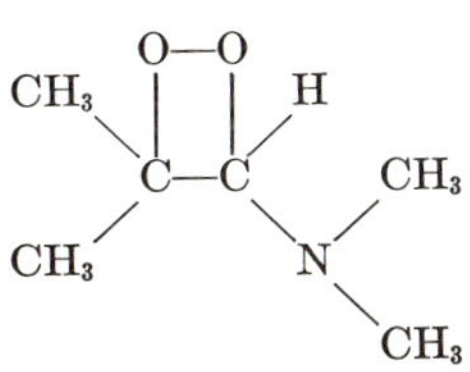

A dimeric or polymeric peroxide is a possibility.

Four-membered cyclic peroxides have often been suggested as reaction intermediates, but until now none has been isolated.

Kopecky and Mumford (1968) have claimed the development of a synthetic route to 1,2-peroxides of trisubstituted olefins, via their β-bromohydroperoxides (Kopecky *et al.*, 1968). Thus Kopecky reports that the first product of the reaction between base and 3-bromo-2-methyl-2-butylhydroperoxide I is the 3,3,4-trimethy-1,2-dioxetane II. This dioxetane intermediate thermally cleaves into acetone and acetaldehyde with a half-life of about 8 hours at 36°.

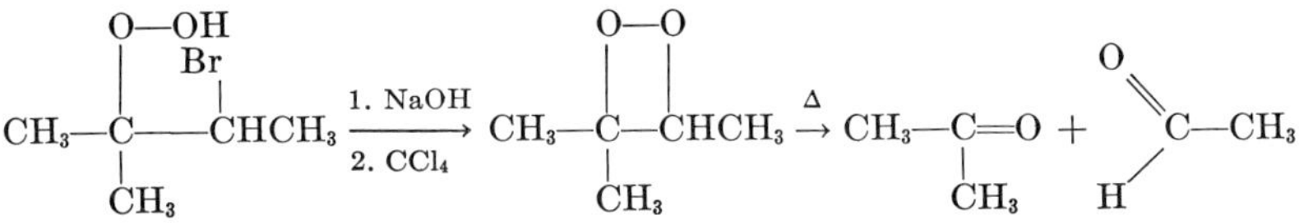

I II

The cleavage of this assumed dioxetane into carbonyl fragments has the additional interest of being accompanied by chemiluminescence (see Section 5) (Kopecky and Mumford, 1969).

It must be emphasized that the properties of the dioxetane II clearly rule it out as the intermediate in the dye-sensitized photooxidation of 2-methyl-2-butene.

4.4 Rearrangements of Peroxidic Products

As mentioned earlier, the first products of photooxygenation are often very unstable and rearrange sometimes into monomeric, sometimes into polymeric products. Rigaudy (1968) and Baldwin *et al.* (1968) have recently reported, independently, a novel type of rearrangement of transannular peroxides which occurs with cleavage of an aromatic ring, suggestive of some biochemical processes (Baldwin *et al.*, 1968). The 1,4-endoperoxide of 1,4-dimethoxy-9,10-diphenylanthracene III can rearrange under mild acid conditions into the aldehyde-ester IV and α-methoxyhydroperoxide VI, if water is present. Indeed, III is so sensitive to acid catalysis that traces of acid presumably formed during the photooxidation of the anthracene will lead to its rearrangement into IV.

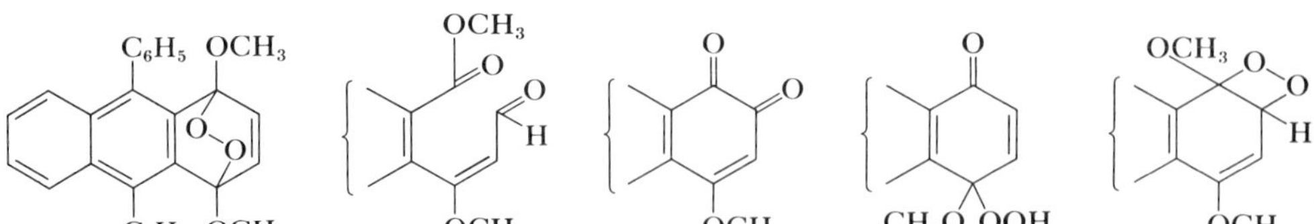

III IV V VI VII

It is tempting to suggest (with Baldwin *et al.*, 1968) that the first intermediate in this rearrangement may be the dioxetane VII; its subsequent cleavage into carbonyl fragments would be the cause of the observed chemiluminescence (see Section 5) (Wilson, 1969a).

4.5 Quenching of Singlet Oxygen

Some substances appear to have the property of "physically" quenching singlet oxygen, i.e., without themselves getting oxidized in the process. Foote and Denny (1968) report that β-carotene is a very effective quencher of photosensitized or chemical oxygenations probably via an energy transfer mechanism (see Section 6.1.3), whereas Ouannès and Wilson (1968) have observed that singlet oxygen in the gas phase or in solution is effectively quenched by tertiary amines. The role of a possible charge-transfer complex between $^1O_2^*$ and the amines was suggested to explain this effect.

4.6 Chemical Diagnosis of Singlet Oxygen

When none of the physical methods of identifying singlet oxygen in a reaction system (Section 3) are applicable, one has to resort to one or several of the following chemical tests:

(1) Addition of different "typical" singlet oxygen acceptors to the system presumed to proceed via singlet oxygen, in order to see if they undergo oxidation to their "typical" products. If they do indeed react,

(2) compare the product distribution so obtained from one of these acceptors, to the distributions of products of its photooxygenation or chemical oxygenation;

(3) determine the ratio of reactivies of two of these "typical" $^1O_2^*$ acceptors, and check if it matches the ratio of their reactivities toward photooxidation or chemical oxygenation.

(4) If a compound is presumed to get oxidized via singlet oxygen, one should compare the reaction products with the products of its photo- or chemical oxygenation, or with the product of its reaction with gaseous singlet oxygen electrically generated.

(5) The reaction system may be tested for the effect of adding one of the two singlet oxygen quenchers, β-carotene or, if the system is not base sensitive, a tertiary amine such as DABCO. Similarly, a typical free-radical autoxidation inhibitor such as a substituted phenol should be tried too. It should not alter in an appreciable way the rate and product distribution of an oxygenation *via* singlet oxygen.

If a very strong argument for the involvement of singlet oxygen can often be based on the results of these tests, it must be emphasized they can easily also be misleading.

Foolproof, highly reactive, typical singlet oxygen acceptors, easy to study, are rare. Thus tetramethylethylene reacts very efficiently with singlet oxygen, but the hydroperoxide formed (1,1,2-trimethylallyl hydroperoxide) is unfortunately identical to the product of its free-radical autooxidation (Howard and Ingold, 1967). The brightly fluorescent 1,3-diphenylisobenzofuran is extremely reactive toward singlet oxygen and the course of this reaction easy to monitor by the disappearance of the fluorescence, but this furan is also susceptible to autoxidation in the dark. 9,10-Diphenylanthracene reacts unambiguously with singlet oxygen to give the 9,10-endoperoxide, but it is important to ascertain that this peroxide is indeed formed, and not to base an argument for singlet oxygen simply on the disappearance of the starting material which could be the result of other reactions. Thus the need for caution in applying these chemical tests for singlet oxygen cannot be overemphasized.

5. Chemiluminescence

We have already briefly discussed (Sections 1.3, 2.1 and Fig. 2) the remarkable chemiluminescences directly due to singlet oxygen, which result from energy pooling processes. The reader is referred to Chapter 2 by E. A. Ogryzlo for a more thorough discussion. Reaction (1) represents

$$2O_2(^1\Delta_g) \rightarrow 2\,^3O_2 + h\nu_{634\text{nm}} \tag{1}$$

a double transition, i.e., two excited oxygen molecules undergoing simultaneous deactivation, whereas reaction (2)

$$\begin{array}{l} 2O_2(^1\Delta_g) \rightarrow O_2(^1\Sigma_g{}^+) + {}^3O_2 \\ \qquad\qquad\qquad \searrow \\ \qquad\qquad\qquad\quad {}^3O_2 + h\nu_{762\text{nm}} \end{array} \tag{2}$$

corresponds to a disproportionation, one of the $O_2(^1\Delta_g)$ being raised to its $^1\Sigma_g{}^+$ state from which it emits.

In other cases, energy pooling processes result in emission, not from an oxygen molecule itself, from other molecules capable of fluorescence. The following processes

$$3O_2(^1\Delta_g) + NO_2 \rightarrow 3\,^3O_2 + NO^*_2 \rightarrow NO_2 + h\nu \geq {}_{480\text{nm}}$$
$$3O_2(^1\Delta_g) + I_2 \rightarrow 3\,^3O_2 + I^*_2 \rightarrow I_2 + h\nu \geq {}_{500\text{nm}}$$
$$2O_2(^1\Delta_g) + \text{Violanthrone} \rightarrow 2\,^3O_2 + {}^1V^* + h\nu_{\text{max at 630 nm}}$$
$$2O_2(^1\Delta_g) + \text{Rubrene} \rightarrow 2\,^3O_2 + {}^1R^* \rightarrow R + h\nu_{\text{max at 580 nm}}$$

account for the excitation and fluorescence of nitric oxide and iodine (Arnold *et al.*, 1966), violanthrone (Ogryzlo and Pearson, 1968) and rubrene (Wilson, 1969b) in the presence of singlet oxygen.

In these examples, the emitter is a known singlet state of the "fluorescer." At least in the cases of violanthrone and rubrene, the excitation seems to be a stepwise process involving *successive* collisions with two molecules of singlet oxygen. The first collision results in the excitation of a metastable (triplet) state of the fluorescer. Energy transfer from a second molecule of singlet oxygen during the lifetime of that metastable state then raises the fluorescer to its emitting singlet state.

The details of these processes have not yet been fully worked out, but it appears that such a stepwise energy transfer can even occur when an activation energy of several kcal is required to reach the singlet state of the fluorescer (Wilson, 1969b).

What enables oxygen in its $^1\Delta_g$ state to bring about such luminescent processes is probably primarily its unusually long lifetime and relative immunity to deactivating collisions, but its high diffusion coefficient perhaps also plays a role. It is obvious, however, that such pooling processes can never be very efficient.

Among reactions involving organic molecules, it is striking that the large majority of known chemiluminescent processes involves molecular oxygen, hydrogen peroxide or other peroxides (see McCapra, 1966; Seliger and McElroy, 1965). The most intense and efficient chemiluminescences belong to these types of reactions. Examples are the chemiluminescences from luminol, lucigenin, oxalyl esters, lophine, acridinium salts and the bioluminescences.

When a luminescence is blue, the light quanta emitted correspond to 70 kcal or more; thus the energy involved is comparable to that of a fairly strong bond such as the C—O bond in organic molecules.

Until Ogryzlo's observations of energy pooling with singlet oxygen, no mechanism seemed available to account for large luminescence quanta by summation of the energies of two elementary processes. A triplet-triplet annihilation achieves just that, but is not observed in aerated solution. Thus to account chemically for the necessary exothermicity, the luminescing step needs to involve the formation of one or more strong bonds at the expense of only weak ones, preferably with the benefit of resonance stabilization in the new molecules formed. Evidently this energy requirement imposes rather stringent limitations on the possible mechanisms.

To obviate to the necessity of a "one-step" luminescent process, Khan and Kasha (1966) suggested the participation of singlet oxygen. They proposed a general mechanism whereby the energies of two singlet oxygen molecules would be pooled in a molecular pair, and then *simultaneously* transferred to an organic fluorescer. They pointed out that such pairs of oxygen molecules in the different excited singlet states, with or without additional vibrational energy, offer a stepladder of energy levels from which energy transfer could occur. Thus the singlet excited states of oxygen could provide at least a partial explanation for the abundance of chemiluminescent reactions involving oxygen. Sensitized chemiluminescence are indeed well known. But the simultaneous transfer in a triple collision of the energies of two singlet oxygens—without the intermediate excitation of a lower level of the fluorescer—could be expected to be very inefficient. Examples of sensitized luminescence involving $^1O_2^*$ pairs may well occur among the weaker luminescence known, but no cases have as yet been established.

Thus a direct physical mechanism involving singlet oxygen does not appear to be a likely cause of the many intense chemiluminescences involving oxygen. It may well be that the privileged property of oxygen here is simply a consequence of the relative weakness of the peroxidic link it tends to form when it combines with organic molecules. Not only does an average O—O bond require only about 35 kcal to be broken, but part of this energy expense can often be recuperated as new resonance energy of the carbonyl fragments formed. As singlet oxygen forms peroxides, in this indirect sense, it may play a role in chemiluminescence.

Starting from such energy consideration, McCapra (1968) and Rauhut (1969) have suggested the intermediacy of short-lived four-membered cyclic peroxides (dioxetanes, see Section 4.3) in the chemiluminescent step. They argued that orbital symmetry considerations predict that the concerted cleavage of such cyclic peroxides would lead to one carbonyl fragment in an excited electronic state. According to McCapra, many examples seem to substantiate this hypothesis, among which are some of the most intense cases of chemiluminescence known.

It is of course possible that singlet oxygen can in some molecules directly add to a double bond to form such 1,2-peroxides (Section 4.3) (McCapra and Hann, 1969). In other cases, rearrangement of a 1,4-peroxide (transannular peroxide) may lead to luminescence, perhaps via an unstable dioxetane. Examples of such rearrangements have been reported by Rigaudy (1968) and Baldwin *et al.* (1968) and discussed earlier (see Section 4.4). The chemiluminescence from the endoperoxide of 1,4-dimethoxy-9,10-diphenylanthracene seems to belong to the latter category (see Section 4.4).

6. Possible Biological Implications of Singlet Oxygen

Most authors concerned with the chemistry of systems involving photosensitized oxygenations and excited singlet oxygen have perceived and discussed biological implications of one kind or another. But apparently no concerted attempts have been made to evaluate the possible participation of singlet oxygen in biological systems, either theoretically or experimentally. For many of the biological systems it might seem a foregone conclusion that singlet oxygen is involved. For example, the biological systems involving dye-sensitized destructive effects of light where oxygen is required appear closely comparable to dye-sensitized photooxygenations in purely chemical systems, where the evidence for the involvement of singlet oxygen is strong (see Section 2). In other cases, such as bioluminescence, the possibilities of singlet oxygen involvement which we will discuss are far more speculative. Although the experimental evidence *with biological systems* is altogether inadequate, we do believe that excited oxygen may have important role(s) in biological systems, and that it merits more attention from workers in the field.

6.1 Photodynamic Action

6.1.1 Historical

Many biological systems, both *in vivo* and *in vitro*, suffer damage or destruction upon irradiation in the presence of an appropriate sensitizing dye and oxygen (Spikes, 1968). Modern experimental studies of this phenomenon are traced to Raab (1900) who found that *Paramecia* were rapidly killed in the light in the presence of oxygen and low concentrations of acridine or other dyes. Neither the dye alone nor light alone was damaging, and oxygen was absolutely required. Anticipating that this phenomenon might involve a mechanism common to all fundamental photobiological processes (chlorophyll was found to be an effective sensitizer), Tappeiner and Jodlbauer (1904) coined the term "photodynamic action" to refer to it. It also served to distinguish the phenomenon from the sensitization of photographic plates by dyes. By 1941, when Blum's comprehensive monograph was published, the phenomenon had been carefully studied in many systems and under a great variety of conditions. Although it appeared not to involve a mechanism common to all other photobiological processes (e.g., photosynthesis), Blum elected to retain the term in spite of its slightly misleading connotation.

Photodynamic action is now commonly understood to refer to biologically destructive photoreactions in which oxygen is consumed, i.e., involving dye-sensitized photochemical oxidations (Spikes, 1967). Such reactions are sometimes referred to as nonphysiological, but this may not necessarily be true in the cases in which the sensitizer molecule is an endogenous and normally occurring pigment, as for example in photosynthetic systems where destructive effects of light may be sensitized by chlorophyll, or in cells where cytochromes or other pigments play a similar role.

6.1.2 Exogenous Sensitizers in Photodynamic Action

Most experimental studies, however, have involved the addition of a dye to the system in order to achieve and study photodynamic effects. Many different kinds of dyes are known to be effective and have been used.[5] One of the earliest generalizations was that of the nonspecific role of the photosensitizing dye, in the sense that similar or identical photodynamic effects usually occurred with any of a variety of dyes. A similar situation pertains in purely chemical systems (see Section 2.2).

Using such exogenous sensitizers, all kinds of biological systems have been studied, ranging from the molecular to the cellular and organismic. Observations with whole organisms comparable to Raab's experiments with *Paramecia* have been reported repeatedly. If small mammals or other animals are injected with a photosensitizing dye (e.g., hematoporphyrin or eosin), they may be rapidly killed on exposure to light. Although the larger animals have not been used so frequently for experimental work, they are apparently equally susceptible. In fact, severe diseases of sheep, cattle and other grazing herbivores have been traced to the occurrence of photosensitizing pigments in plants on which the animals graze (Blum, 1941). The requirement for light (excluding u.v.) has long been recognized, and the protection which can be afforded by pigmented hair and/or epidermis is also clear. The black sheep thus comes into his own and emerges as superior in such circumstances.

External exposure of animals to photosensitizing dyes may likewise result in severe damage. In many, though surely not all of these cases, the oxygen dependency and/or utilization has been established, demonstrating that true photodynamic action is involved. The site of action, however, is not at all well defined in any of the cases, whether the organism be injected with or externally exposed to the sensitizing dye. Nor has the damage been found to have any general or common basis. It is attributed instead to multiple and diverse oxidations of key components, causing the loss of structural and functional integrity, and ultimately irreversible degradative changes.

It seems unnecessary to document further the fact that virtually all other kinds of organisms, though widely different physiologically, are similarly sensitive to photodynamic killing. The article by Spikes (1968) in an earlier volume of this series should be consulted. Sublethal damage may equally well occur if the treatment is less intense. In such cases any of a variety of processes may be impaired or inhibited, such as cell motility, cell division, membrane permeability and transport,

[5] Some of those which are most frequently used are rose bengal, methylene blue, eosin, neutral red, hematoporphyrin and chlorophyll.

glycolysis and respiration, etc. Moreover, subcellular materials isolated in an active form, ranging from organelles such as mitochondria to molecules, also suffer photodynamic damage. At the molecular level, biological substances such as amino acids, proteins, and nucleic acids have been studied in particular.

Numerous mechanisms have been proposed in these latter cases, where known compounds are involved, but the reaction schemes suggested have usually been based on the photooxidation of simpler model organic systems; their validity in the biological systems is still open to question. A thorough review of studies at the molecular biology level is that of Spikes and Livingston (1969). It is evident that up to the time of their review relatively little attention had been paid to the Kautsky mechanism and excited singlet oxygen. It seems of importance that this be done, keeping in mind that the dye-sensitized oxygenations, which occur under physiological conditions in biological systems, involve drastically different solvent and other conditions from those used in the model systems.

6.1.3 Endogenous Sensitizers in Photodynamic Action

A preoccupation with experiments involving the addition of sensitizing dyes to the system has tended to obscure and divert attention from the more "physiological" and probably more important type of photodynamic action—that which involves endogenous sensitizers.

It was noted many years ago (Swart-Füchtbauer, 1951; Swart-Füchtbauer and Rippel-Baldes, 1951) that the distribution of pigmented bacteria in fresh water may be positively correlated with locations where there is appreciable exposure to sunlight. Furthermore, it was demonstrated that sunlight has a bacteriocidal effect which is selective for the nonpigmented forms. All of these results can be interpreted in terms of killing by photodynamic action and the protection by those pigments, although the need for oxygen may not have been clearly proven.

In terms of indicating the intermediacy of singlet oxygen in biological systems of this sort, the most compelling and important experiments up to now are those which involve the carotenoids. It has been shown that, in several different biological systems, carotenoids can provide protection to organisms against the damaging effects of photodynamic action (Krinsky, 1968). More recently, it has been shown by Foote (1968, 1970) that in model organic systems, oxygenations involving singlet oxygen (produced either photochemically or chemically) are efficiently inhibited by β-carotene, suggesting that this consitutes the molecular basis for the *in vivo* protection.

A protective action by naturally occurring carotenoids against lethal photooxidations has been described in both photosynthetic and nonphotosynthetic organisms by Krinsky (1968) in his excellent review article. Since the porphyrins (both the Fe and the Mg derivatives) can act as photosensitizers, and since most cells contain such compounds, it follows that most cells are subject to photodynamic action via endogenous sensitizers.

The hypothesis that carotenoids serve to protect the photosynthetic organism against photodynamic action (sensitized by chlorophyll) was enunciated and developed by Stanier and his colleagues, based first on experiments with carotenoidless mutants of photosynthetic bacteria (Griffiths *et al.*, 1955; Sistrom *et al.*, 1956; Stanier, 1959). They discovered that these mutant bacteria are rapidly killed in the presence of light and oxygen, sensitized by endogenous bacterial chlorophyll, whereas the normal carotenoid-containing wild type bacteria are perfectly stable under these conditions.

The photosynthetic bacterium used by Stanier and his colleagues is uniquely suited to demonstrate the phenomenon. It is a nonsulphur purple bacterium (*Rhodopseudomonas spheroides*) which normally grows photosynthetically under anaerobic conditions using organic substrates as electron donors.[6] In the dark it can grow aerobically employing a respiratory metabolism of these same organic substrates. But oxygen inhibits the biosynthesis of pigments (both chlorophyll and carotenoids), so that cells grown aerobically ultimately loose their photosynthetic capacity[7] and switch from photosynthetic to respiratory metabolism. This oxygen inhibition of pigment biosynthesis is strictly that, and does not involve damage; as light is not involved it has no relation to photodynamic action.

The wild type and carotenoidless mutant have identical growth rates when grown photosynthetically in the absence of oxygen. But if oxygen is admitted the mutant stops growing and suffers massive killing, while the wild type continues normally.

The two strains also grow in a similar fashion aerobically in the dark, but if exposed to light (after only a relatively short time of aerobic growth) the wild type is unaffected whereas the mutant is quickly killed. However, if aerobic growth in the dark is continued long enough, the inhibition of pigment synthesis (which occurs in mutant and

[6] Since oxygen is not evolved in bacterial photosynthesis the system can be maintained truly anaerobic and the effects of light can be readily compared in the presence and absence of oxygen. Carotenoidless mutants of photosynthetic organisms which do evolve oxygen would be expected to be suicidal.

[7] As growth occurs the pigments are simply diluted out until cells have only a negligible quantity.

wild type alike) results in mutant cells lacking cholorophyll. Such cells are then no longer susceptible to photodynamic killing. From this and action spectra studies, it has been clearly demonstrated that bacteriochlorophyll acts as the sensitizer.

Additional studies of these and other systems have confirmed and extended these observations (Krinsky, 1968). Cells lacking carotenoids, obtained without genetic alteration by growth in the presence of diphenylamine, behave like the carotenoidless mutants above. Similar protective fuctions can be ascribed to the carotenoids in both higher and lower plants. Although less information is available for animals, it seems likely that such cases exist also. A very interesting report on the protective effect of β-carotene against lethal photodynamic action was reported by Mathews (1964). She injected mice intraperitoneally with haematoporphyrin both together with and without added β-carotene prior to exposure to light. The β-carotene clearly provided very significant protection against photodynamic killing.

In all these studies there had been much speculation concerning the molecular basis for the protective action of carotenoids. A mechanism involving a screening by absorption of photodynamic light has often been suggested, but was discarded in most cases where it was experimentally examined. A second and more attractive possibility is that the excited triplet of the sensitizer (e.g., chlorophyll) is quenched by carotenoids. A third hypothesis suggested by both Calvin (1955) and by Sistrom *et al.* (1956), is that carotenoids can serve as alternate and preferred substrates in photosensitized oxidations. Oxidized or oxygenated carotenoids should accumulate during the light; it was suggested that they should be reduced via subsequent dark reactions.

A fourth mechanism involving the quenching of excited oxygen has been proposed by Foote (1968). His recent experiments have extended the study of the protective action of carotenoids to model organic systems, where he has shown that methylene blue sensitized photooxygenation of 0.1 M 2-methyl-2-pentene to yield the hydroperoxides is 95% inhibited by 10^{-4}M β-carotene. The hypothesis that the inhibitor acts by reaction with singlet oxygen is supported by the fact that oxygenation of the same substrates using chemically generated singlet oxygen (using H_2O_2 and NaOCl; see Section 2.3) was similarly inhibited by β-carotene. In these cases no sensitizer or light was involved.

The first hypothesis, involving screening, is not compatible with these experiments since the absorption of methylene blue and β-carotene do not overlap.

The second hypothesis is essentially excluded by the fact that inhibition occurs in systems not involving sensitizers. Foote's analysis has shown in addition that substrate is competitive with β-carotene for an intermediate both in photooxygenation and in chemical (H_2O_2-NaOCl) oxygenation. He postulates that this intermediate is excited singlet oxygen.

Three possible mechanisms for quenching excited oxygen were considered by Foote. The first is essentially that mentioned above, namely that the carotenoid acts as an alternate substrate for photooxygenation. This may be ruled out as a major pathway in Foote's system; he estimates that about one thousand molecules of singlet oxygen are quenched for each β-carotene molecule which is consumed (oxygenated). The oxygenation which *does* occur may explain the frustrating ambiguity which had emerged over the years in experiments designed to study light dependent carotenoid oxidation *in vivo*. The possibility that specific (and physiologically significant) biochemical conversions might occur via such a pathway will be discussed later on.

The second and third mechanisms discussed by Foote involve a quenching of the excited state of oxygen. In one of these one might postulate the formation of some short-lived carotene-excited oxygen complex which somehow decomposes to carotene and ground state oxygen.

$$^1O_2^* + \text{car} \rightarrow \text{car}\text{—}^1O_2^* \rightarrow \text{car} + {}^3O_2 + \text{heat}$$

A mechanism which is more attractive involves energy transfer to populate the carotene triplet directly, the reverse of the reaction by which the O_2^* singlet is produced in the dye-sensitized pathway.

$$^1O_2^* + {}^1\text{car} \rightarrow {}^3O_2 + {}^3\text{car}^* \rightarrow {}^1\text{car} + \text{heat (or } h\nu)$$

Such a mechanism is possible only if the triplet energy level of β-carotene is near or below that of the $^1\Delta_g$ state of singlet oxygen (22.5 kcal). The exact value of the carotene triplet is not known at the present time (Foote, 1970), but there is evidence to indicate that it is smaller than 29 kcal, since β-carotene is known to quench chlorophyll triplets ($E_T = 29$ kcal). A scheme postulated by Foote to account for *in vivo* reaction of carotenoids is shown in Fig. 3.

6.2 Photobiological Systems

The possibility that singlet oxygen is involved in nondestructive biological phenomena does not seem improbable. As a chemical reagent it could conceivably be harnessed to participate in important cell chemistry, involving some kind of appropriate localization of the molecular species concerned. Several examples of biochemical reactions which might involve singlet oxygen have been suggested; it is equally possible that certain physiological responses to light might have steps with

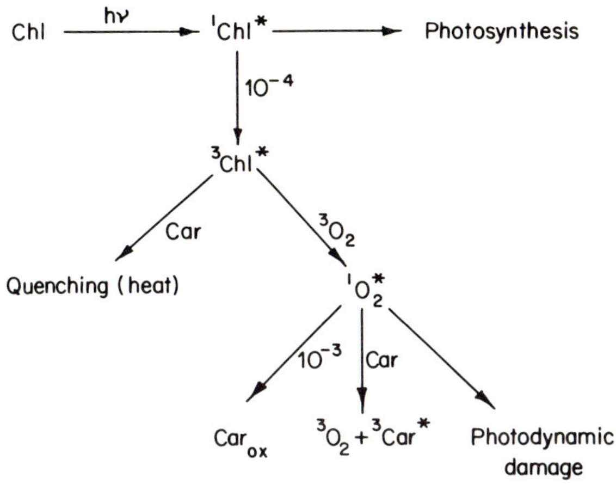

FIG. 3. In this scheme, one molecule of singlet chlorophyll in 10^4 escapes from the photosynthetic pathway (this number is roughly estimated from known lifetimes *in vivo*). The triplet chlorophyll which is produced is potentially lethal. Some of it may be quenched by interaction with carotene; the rest reacts with oxygen. The rate for reaction with carotene and with oxygen is the same, so that the proportion is simply governed by the local concentrations of carotene and oxygen. The resulting singlet oxygen can then react with chlorophyll or other acceptor molecules in the cell to produce photodynamic damage. However, it can also be quenched by carotene; of the quenching collisions, one in 10^3 would lead to oxygenated carotenes (from Foote, 1970).

singlet oxygen as an intermediate. Actually, many biologists might consider that singlet oxygen is eminently qualified as an intermediate in biological systems simply by virtue of its transitory—indeed elusive—nature.

There are numerous photobiological phenomena, such as growth or movement in response to light (phototropism and phototaxis), and physiological phenomena related to light-dark cycle (circadian rhythms and photoperiodism) for which the biochemical steps subsequent to light absorption remain unidentified, either in a general or specific way (Briggs, 1964; Hastings, 1964; Hendricks, 1964; Clayton, 1964; Page, 1968; Lees, 1968). In fact, it is probably true to say that not even the light absorbing pigment has been definitely identified in any of these phenomena.

If singlet oxygen were involved, its cellular production would presumably occur via energy transfer from the triplet level of a sensitizing pigment, this being appropriately localized so that the target chemistry would be possible and favored. The possibility that singlet oxygen could be produced chemically in the dark must also be recognized and kept in mind. Anbar (1966) has reported that singlet oxygen is a product in the breakdown of H_2O_2 with catalase, but no more recent studies on this system have been published and the mechanism seems obscure. However, the fact that singlet oxygen is produced upon the breakdown of certain endoperoxides seems clear (Wasserman and Scheffer, 1967; see Section 2.3.3).

In discussing photophysiological processes, it is most convenient to refer to phototropism, since this has been studied in far greater detail than any of the others (Thimann and Curry, 1960; Briggs, 1964). With regard to the light-absorbing pigment, flavins and carotenoids have been implicated, indeed briskly debated, as alternate candidates. But action spectrum studies have not allowed a clear selection between or even exclusion of either of these pigments. The three major peaks in the action spectrum (in the blue centered at about 445 nm and in the ultra violet at 360 and 290 nm) bear considerable resemblance to the absorption of flavins. But in the details of their relative heights and with regard to their precise wave-lengths there are substantial discrepancies.[8] While the blue peak of effectiveness is itself remarkably similar to the absorption of certain carotenoids, these pigments, except for the relatively uncommon cis-isomers, exhibit no absorption in the ultraviolet regions. Thimann (1967) has proposed that this might be accounted for by energy transfer from flavin to carotenoids, but this seems improbable in view of the apparent ineffectiveness of the flavin absorption at 260 nm which, if energy transfer occurs at all, should be equally as effective as the longer wave-length ultra violet. The details of the many other carefully executed experiments concerned with the identity of the light absorbing pigment need not be reviewed here. Suffice it to say that the hope that the identification of the pigment would provide definitive insights concerning the subsequent chemical steps in the phototropic response remains to be realized.

Attempts to specify the nature of those chemical events, which are presumably initiated by light absorption, have also been unsuccessful. In higher plants the discovery of the plant growth hormone, auxin (indole-3-acetic acid), was quickly followed by a theory of phototropism involving an asymmetric distribution of auxin caused by illumination (Cholodny, 1927). Went (1928) confirmed that the ratio of auxin in the two sides after unilateral illumination was indeed as predicted by the theory.

[8] A further observation of importance is the fact that red light, though in itself being inactive for phototropic responses, can significantly modify the quantitative aspects of the normal phototropic response. The red light effect is apparently phytochrome-mediated (Briggs, 1963; Briggs and Chon, 1966).

The hypothesis that this distribution might be due to the photochemical destruction of auxin on the illuminated side was attractive and received a strong impetus from Galston's (1949) early demonstration that a photooxidative destruction of auxin occurred *in vitro*, sensitized by riboflavin. But over the years a considerable amount of evidence has been obtained which argues convincingly against this hypothesis.[9]

The primary evidence may be summarized under three headings, as follows. (*a*) Careful studies with both endogenous and radioactive auxin indicate that asymmetric auxin distribution after unilateral irradiation involves auxin redistribution and little if any auxin destruction (Briggs, *et al.*, 1957; Briggs, 1963; Pickard and Thimann, 1964). Even prior to the tracer work, it had been shown that the insertion of a thin mica sheet in the tip of an intact seedling in a plane perpendicular to the incident light (acting as a mechanical barrier to the transport of auxin) inhibited phototropic curvature (Boysen-Jensen, 1928; Brauner, 1957; Briggs, 1960). In a control, irradiated parallel to the barrier, curvature occurred. (*b*) The measured quantum yield for the *in vitro* photodestruction of auxin is much lower than that which would appear compatible with the photodestruction hypothesis. From the measured differences in the auxin concentration after irradiation, together with estimates of the effective photon flux, it can be shown that a quantum yield of between 10^2 and 10^3 would have to pertain to account for photodestruction (Galston, 1959; Thimann and Curry, 1960). Although this has been called improbable (possibly because the *in vitro* quantum yield values with riboflavin as the sensitizer are only about 0.20), it should be remembered that photochemical reactions may have quantum yields greater than unity. This particular argument is therefore not applicable on a more general basis. (*c*) Although the phototropic response is evidently mediated by auxin and the differential translocation thereof, the auxin need not be present at the time of irradiation. In fact, the response can still be obtained if auxin application is delayed by as much as 8 hours (Von Guttenberg, 1959; Diemer, 1961). The experiment is as follows: decapitated seedlings fail to respond phototropically because the source of auxin has been removed. Decapitated plants can, however, be irradiated unilaterally, returned to the dark, and then treated uniformly with auxin applied at the cut tip. The auxin diffuses into the plant, and as a consequence of the earlier irradiation, a phototropic response occurs, presumably because of a translocation of auxin to the side remote from the earlier illumination.

[9] The fact that the specific hypothesis of flavin-sensitized auxin destruction has encountered difficulties should not be mistaken for evidence against the possible involvement of flavin as the light-absorbing pigment.

The hypothesis that phototropism is a consequence of the photochemical destruction of auxin is thus extremely difficult to reconcile with the experimental observations. No alternative hypothesis at the molecular level has been advanced. Thimann (1967) proposed that plastid movement might be the immediate result of photoreception and that the altered plastid location would then influence auxin transport. Irrespective of the merits of this particular idea, no biochemical mechanism or steps associated with the primary photochemistry were postulated.

None of the data concerned with phototropism—or indeed with any of the other photobiological processes mentioned—appear to be incompatible with a molecular mechanism in which singlet oxygen is involved as an intermediate. Free oxygen appears to be required for light perception in phototropism (Von Guttenberg, 1959), this feature being a necessary but not critically diagnostic property. The oxygen dependency appears not to have been widely studied; more data on the point would be very desirable.

A hypothesis involving singlet oxygen is attractive because options are available at both ends. On the one hand, a variety of different sensitizers would be equally capable in generating singlet oxygen via steps described in Section 2 and thus of initiating the same reaction. On the "target" side, singlet oxygen is an agent capable of reaction with individual members of a possibly diverse group of molecular species, assuming that each of those species participates as a regulatory molecule in the particular photophysiological system where it occurs. Thus a common photobiological mechanism could operate in connection with diverse phenomena.

The light-absorbing sensitizing pigment might occur in close association with the regulatory molecule, either by virtue of a complex, or a protein with sites for both, or by the association of both with a membrane. The association would presumably be such as to favor the reaction of the singlet oxygen produced upon irradiation with the substrate (regulatory) molecule. After reaction, the product might be free to diffuse elsewhere, or it might exert its effect *in situ* (e.g., in association with a membrane). No special or obligatory localization would be expected; in the case of membranes all cells would be individually and independently reactive.

On this theory the energy level of the excited (triplet) state of an effective pigment would only have to be 22.5 kcal or greater. The recent report (Poff, 1968) of a phototropic response in fungi influenced by near infrared light (900–1050 nm) is interesting and might be explained by this mechanism. Indeed, singlet oxygen would be an appropriate species to be generated by light ranging in energy all the way from the ultraviolet and blue to the near infrared.

An experiment reported many years ago is equally interesting to recall in connection with the singlet oxygen hypothesis. Roots, which are not normally phototropic, may exhibit such movements upon irradiation if they are previously treated with (and have absorbed) an appropriate dye (Blum, 1941). In those cases, the action spectrum for this apparent "phototropic" response corresponds to the dye absorption. The study of systems like this has apparently not been pursued. It would be interesting to know whether or not the "normal" physiological mechanisms in phototropism are called into play. If so, it would imply that the various tissues have similar phototropic potential and are differentiated only with regard to the localization of the pigment. Further studies of such an "artificial" system might be very instructive with regard to the mechanisms of phototropism.

6.3 Biochemical Systems

In addition to its role as a terminal electron acceptor, oxygen is required in biological systems for the biosynthesis of essential metabolites. Enzymatic systems involved in these syntheses were first demonstrated concurrently by two different laboratories. Mason *et al.* (1955) found that the oxidation of 3,4-dimethylphenol to dimethyl catechol catalyzed by a phenolase complex involved the incorporation of an atom of oxygen derived from molecular oxygen, not water. Hayashi *et al.* (1955) showed that in the oxidative cleavage of the aromatic nucleus of catechol both oxygen atoms in the product were derived exclusively from free oxygen.

Such oxygen-fixing reactions employ enzymes named oxygenases, and many more new reactions of this type have been reported in the years since their discovery (Hayashi, 1962; Mason, 1965; Bloch and Hayashi, 1966). Two subclasses have been recognized, mono-oxygenases and dioxygenases, corresponding to the two reaction types orginally described.

$$\text{Substrate} + \tfrac{1}{2}O_2 \rightarrow \text{Substrate-O}$$
$$\text{Substrate} + O_2 \rightarrow \text{Substrate-}O_2$$

As Hayashi and Nozaki (1969) point out in their recent review, reactions of this type occur ubiquitously in nature and exhibit versatile functions and unique properties, in both biosynthetic reactions and catabolic pathways (including the metabolic disposal of foreign compounds). With regard to reaction mechanism, these types of oxidative reactions are believed to involve some sort of "activation" of the molecular oxygen, so that it is able to react and be incorporated into the organic substrates. The present evidence with purified enzymatic systems is that in most or all of the dioxygenase systems some sort of iron-bound oxygen complex is formed, represented as

$$Fe^{++}\text{-}O_2 \rightarrow Fe^{+++}\text{-}O_2^-$$

In mono-oxygenases a reducing agent is commonly involved in the reaction and it seems possible that a species such as O_2^- or even O_2^{--} is the active intermediate.

But the possibility that excited singlet oxygen could function in the capacity of "active oxygen" in oxygenase type systems has often been mentioned, and a recent note by Baldwin *et al.* (1968) sets forth some experimental results which bear on the question. Briefly, they have found that the reaction of certain aromatic compounds with singlet oxygen results in ring cleavage giving products of the specific type found in certain aromatic dioxygenase enzyme systems (see Section 4.4).

In these speculations it has often been suggested that excited oxygen might be generated enzymatically. There is little evidence for this at the present time, aside from the report of Anbar (1966) mentioned previously, in which he found that the decomposition of hydrogen peroxide by catalase resulted in singlet oxygen production. In that study it was also found that molecular nitrogen can be oxidized nonenzymatically to nitrate, suggesting yet another type of biochemically important reaction of singlet oxygen.

The fact that excited oxygen is available in cells via the dye-sensitized pathway is evident, but whether or not there is any cellular utilization of the reagent so formed has not really been examined. One of the difficulties which can be envisioned in such a study is that of screening for reactions of this type. Assuming that they must involve some degree of localization, one can also imagine that *in vitro* systems would be difficult to isolate. Cells might utilize such pathways in a facultative fashion, possibly in detoxification reactions or in aromatic ring cleavage of compounds which could be used as metabolic energy sources. Such a mechanism would provide cells with the potential for dealing with compounds for which there were no constitutive enzymes and with compounds which occur only at relatively low concentrations, where there would be no possibility for specific enzyme induction.

The slow oxidation of carotenoids via a singlet oxygen mechanism has been noted by Foote (1970) to result in products which are structurally similar to compounds of the fucoxanthin and neoxanthin type (see Section 4.2). In fact it has been pointed out (Mousseron-Canet *et al.*, 1968) that synthetic pathways for many other naturally occurring oxygenated carotenoids can be rationalized on the basis of the decomposition of peroxidic intermediates of the type formed in the photooxygenations of the model compounds studied.

Still another suggestion for a way in which singlet oxygen might be formed in biological systems was put forward by one of us in a recent review (Hastings, 1968), namely that the species might be formed in bioluminescent reactions. The possibility that such reactions might represent the functional importance of bioluminescent systems in some organisms was suggested; there are indeed numerous cases where no particular use has been deduced for the light itself (Harvey, 1952). Several specific possibilities were suggested whereby excited oxygen might be produced in bioluminescent reactions. $^1O_2^*$ production might occur as a direct alternative to light emission, where the excited state in bioluminescence is represented by $^1L^*$.

$$^1L^* + {}^3O_2 \rightarrow {}^3L + {}^1O_2^*$$

Because of the short lifetime of the singlet state involved, this particular pathway appears unlikely. A second possible mechanism would involve the breakdown of an oxygen-containing (peroxide) intermediate to give excited oxygen as an alternative to the continued reaction to produce an excited species. Transannular peroxides are known to yield excited oxygen (see Section 2.3.3) as an alternate reaction pathway, in some cases, to a rearrangement yielding luminescence (Sections 4.4 and 5).

The possibility that excited oxygen might be an intermediate in the bioluminescent emission was also discussed, especially the possibility of the occurrence of emission from excited molecular pairs (enzyme-bound), which might be utilized as a method for energy summation. There is no evidence in support of any of these speculations at the present time.

References

Anbar, M. (1966). *J. Am. Chem. Soc.* **88**, 5924.

Arbuzov, Y. A. (1965). *Russ. Chem. Rev.* **34**, 558.

Arnold, S. J., Ogryzlo, E. A., and Witzke, H. (1964). *J. Chem. Phys.* **40**, 1769.

Arnold, S. J., Browne, R. J., and Ogryzlo, E. A. (1965). *Photochem. Photobiol.* **4**, 963.

Arnold, S. J., Finlayson, N., and Ogryzlo, E. A. (1966). *J. Chem. Phys.* **44**, 2529.

Arnold, S. J., and Ogryzlo, E. A. (1967). *Can. J. Phys.* **45**, 2053.

Arnold, S. J., Kubo, M., and Ogryzlo, E. A. (1968). *In* "Oxidation of Organic Compounds" III, *Advan. Chem. Series* **77**, 133.

Bader, L. W., and Ogryzlo, E. A. (1964). *Discussions Faraday Soc.* **37**, 46.

Baldwin, J. E., Basson, H. H., and Krauss, H. (1968). *Chem. Comm.* 984.

Bartlett, P. D., and Mendenhall, G. D. (1968). Paper presented at the 115th National Meeting of the Am. Chem. Soc., San Francisco, March 1968.

Bartlett, P. D., and Mendenhall, G. D. (1970a). *J. Am. Chem. Soc.* **92**, 210.

Bartlett, P. D., and Mendenhall, G. D. (1970b). (To be published.)

Bloch, K., and Hayashi, O. (eds.) (1966). "Biological and Chemical Aspects of Oxygenases," Maruzen, Japan.

Blum, H. F. (1941). "Photodynamic Action and Diseases Caused by Light," Reinholt, New York. (Reprinted in 1944 with an updated appendix by Hafner, N. Y.)

Bowen, E. J., and Steadman, F. (1934). *J. Chem. Soc.* 1098.

Bowen, E. J. (1963). *Advan. Photochem.* **1**, 23.
Boysen-Jensen, P. (1928). *Planta* **5**, 464.
Brauner, L. (1957). *Soc. Exptl. Biol. Symp.* **9**, 86–94.
Briggs, W. R., Tocher, R. D., and Wilson, J. F. (1957). *Science* **126**, 210.
Briggs, W. R. (1960). *Plant Physiol.* **35**, 951.
Briggs, W. R. (1963). *Am. J. Bot.* **50**, 196.
Briggs, W. R. (1963). *Plant Physiol.* **38**, 237.
Briggs, W. R. (1964). *In* "Photophysiology" (A. C. Giese, ed.), Vol. I, pp. 223–272. Academic Press, New York.
Briggs, W. R., and Chon, H. P. (1966). *Plant Physiol.* **41**, 1159.
Browne, R. J., and Ogryzlo, E. A. (1964). *Proc. Chem. Soc.* 117.
Cahill, A. E., and Taube, H. (1952). *J. Am. Chem. Soc.* **74**, 2312.
Calvert, J. G., and Pitts, J. N., Jr. (1966). "Photochemistry." Wiley, New York.
Calvin, M. (1955). *Nature* **176**, 1215.
Clark, I. D., and Wayne, R. P. (1969). *Chem. Phys. Letters* **3**, 405.
Clayton, R. K. (1964). *In* "Photophysiology" (A. C. Giese, ed.), Vol. II, pp. 51–78. Academic Press, New York.
Clyne, M. A. A., Thrush, B. A., and Wayne, R. P. (1965). *Photochem. Photobiol.* **4**, 957.
Cholodny, N. (1927). *Biol. Zentr.* **47**, 604.
Corey, E. J., and Taylor, W. C. (1964). *J. Am. Chem. Soc.* **86**, 3881.
Dianov-Klokov, V. I. (1959). *Opt. i. Spectroskopiya* **7**, 457.
Diemer, R., (1961). *Planta* **57**, 111.
Dijkgraaph, C., Sitters, R., and Hoytink, G. J. (1962). *Mol. Phys.* **5**, 643.
Dufraisse, C., and Etienne, A. (1935). *Compt. Rend.* **201**, 280.
Dufraisse, C., and Ecary, S. (1946). *Compt. Rend.* **223**, 735.
Elias, L., Ogryzlo, E. A., and Schiff, H. I. (1959). *Can. J. Chem.* **37**, 1680.
Ellis, J. W., and Kneser, H. O. (1933). *Z. Physik.* **86**, 583.
Evans, D. F. (1957). *J. Chem. Soc.* 1351, 3885.
Evans, D. F. (1959). *J. Chem. Soc.* 2753.
Evans, D. F. (1969). *Chem. Comm.* **367**.
Falick, A. M. (1967). "An EPR Study of the $O_2(^1\Delta_g)$ Molecule." Ph.D. Thesis, University of California, Berkeley, California.
Falick, A. M., Mahan, B. H., and Meyers, R. J. (1965). *J. Chem. Phys.* **42**, 1837.
Fenical, W., Kearns, D. R., and Radlick, P. (1969). *J. Am. Chem. Soc.* **91**, 3396.
Fieser, L. F. (1936). "The Chemistry of Natural Products Related to Phenanthrene." Reinholt, New York.
Foner, S. N., and Hudson, R. L. (1956). *J. Chem. Phys.* **25**, 601.
Fontijn, A., Meyer, C. B., and Schiff, H. I. (1964). *J. Chem. Phys.* **40**, 64.
Foote, C. S., and Wexler, S. (1964). *J. Am. Chem. Soc.* **86**, 3879, 3880.
Foote, C. S., Wexler, S., Ando, W., and Higgins, R. (1968). *J. Am. Chem. Soc.* **90**, 975.
Foote, C. S., and Brenner, M. (1968). *Tetrahedron Letters* **57**, 6041.
Foote, C. S., and Lin, J. (1968). *Tetrahedron Letters* **29**, 3267.
Foote, C. S., and Denny, R. W. (1968). *J. Am. Chem. Soc.* **90**, 6233.
Foote, C. S. (1968). *Accounts Chem. Res.* **1**, 104.
Foote, C. S., and Wong, S. Y. (1969). *Am. Chem. Soc.* Meeting, Minneapolis, Abstracts of Papers, April.
Foote, C. S. (1969). *Science* **162**, 963.
Foote, C. S. (1970). *In* "Symposium on Singlet Oxygen," New York Academy of Sciences. (In press.)

Gaffron, H. (1936). *Biochem. Z.* **287,** 130.

Galston, A. W. (1949). *Proc. Natl. Acad. Sci.* **35,** 10.

Galston, A. W. (1959). *In* "Encyclopedia of Plant Physiology, Vol. 17" (Ruhland, ed.), pp. 492-529. Springer-Verlag, Berlin.

Gollnick, K., and Schenck, G. O. (1964). *Pure Appl. Chem.* **9,** 507.

Gollnick, K., and Schenck, G. O. (1967). *In* "1,4-Cycloaddition Reaction" (J. Hamer, ed.), p. 255. Academic Press, New York.

Gollnick, K. (1968). *Advan. Photochem.* **6,** 1.

Griffiths, M., Sistrom, W. R., Cohen-Bazire, G., and Stanier, R. Y. (1955). *Nature* **176,** 1211.

Harvey, E. N. (1952). "Bioluminescence." Academic Press, New York.

Hastings, J. W. (1964). *In* "Photophysiology" (A. C. Giese, ed.), Vol. I, pp. 333–362. Academic Press, New York.

Hastings, J. W. (1968). *Ann. Rev. Biochem.* **37,** 597.

Hayashi, O., Katagiri, M., and Rothberg, S. (1955). *J. Am. Chem. Soc.* **77,** 5450.

Hayashi, O. (ed.) (1962). "Oxygenases." Academic Press, New York.

Hayashi, O., and Nozaki, M. (1969). *Science* **164,** 389.

Hendricks, S. B. (1964). *In* "Photophysiology" (A. C. Giese, ed.), Vol. I. pp. 305–332. Academic Press, New York.

Higgins, R., Foote, C. S., and Cheng, H. (1968). *In* "Oxidation of Organic Compounds" III, *Advan. Chem. Series* **77,** 102.

Howard, J. A., and Ingold, K. V. (1967). *Can. J. Chem.* **45,** 793.

Hoytink, G. J. (1969). *Accounts Chem. Res.* **2,** 114.

Kautsky, H., and de Bruyn, H. (1931). *Naturwissenschaften* **19,** 1043.

Kautsky, H. (1939). *Trans. Faraday Soc.* **35,** 216.

Kawaoka, K., Khan, A. U., and Kearns, D. R. (1967). *J. Chem. Phys.* **46,** 1842, 1883.

Kearns, D. R., Hollins, R. A., Khan, A. U., Chambers, R. W and., Radlick, P. (1967a). *J. Am. Chem. Soc.* **89,** 5455.

Kearns, D. R., Hollins, R. A., Khan, A. U., and Radlick, P. (1967b). *J. Am. Chem. Soc.* **89,** 5456.

Kearns, D. R., Khan, A. U., Duncan, C. K., and Maki, A. H. (1969). *J. Am. Chem. Soc.* **91,** 1039.

Kearns, D. R. (1969). *J. Am. Chem. Soc.* **91,** 6554.

Khan, A. U., and Kasha, M. (1963). *J. Chem. Phys.* **39,** 2105.

Khan, A. U., and Kasha, M. (1964). *Nature* **204,** 241.

Khan, A. U., and Kasha, M. (1966). *J. Am. Chem. Soc.* **88,** 1574.

Khan, A. U., and Kasha, M. (1970). *J. Am. Chem. Soc.* (In press.)

Kopecky, K. R., and Reich, H. J. (1965). *Can. J. Chem.* **43,** 2265.

Kopecky, K. R., van de Sande, J. H., and Mumford, C. (1968). *Can. J. Chem.* **46,** 25.

Kopecky, K. R., and Mumford, C. (1968). Abstract. 51st Annual Conference of the Chemical Institute of Canada, Vancouver, B. C., June 1968, p. 41.

Kopecky, K. R., and Mumford, C. (1969). *Can. J. Chem.* **47,** 709.

Krinsky, N. I. (1968). *In* "Photophysiology" (A. C. Giese, ed.), Vol. III, pp. 123–196. Academic Press, New York.

Lees, A. D. (1968). *In* "Photophysiology" (A. C. Giese, ed.), Vol. IV, pp. 47–138. Academic Press, New York.

McCapra, F. (1966). *Quart. Rev.* **20,** 485.

McCapra, F. (1968). *Chem. Comm.* 155.

McCapra, F., and Hann, R. A. (1969). *Chem. Comm.* 442.

McKeown, E., and Waters, W. A. (1966). *J. Chem. Soc.* 1040.

McNeal, R. J., and Cook, G. R. (1967). *J. Chem. Phys.* **47,** 5385.
Mallet, L. (1927). *Compt. Rend.* **185,** 352.
Mason, H. S., Fowlks, W. L., and Peterson, E. (1955). *J. Am. Chem. Soc.* **77,** 2914.
Mason, H. S. (1965). *Ann. Rev. Biochem.* **34,** 595.
Mathews, M. M. (1964). *Nature* **203,** 1092.
Mathias, A., and Schiff, H. I. (1964). *Discussions Faraday Soc.* **37,** 38.
Moureu, C., Dufraisse, C., and Dean, P. M. (1926). *Comp. Rend.* **182,** 1584.
Mousseron-Canet, M., Dalle, J. P., and Mani, J. C. (1968). *Tetrahedron Letters* No. 57, 6037.
Mousseron-Canet, M., Dalle, J. P., and Mani, J. C. (1969). *Photochem. Photobiol.* **9,** 91.
Murray, R. W., and Kaplan, M. L. (1968). *J. Am. Chem. Soc.* **90,** 537, 4161.
Nickon, A. and Mendelson, W. L. (1965). *J. Am. Chem. Soc.* **87,** 3921.
Ogryzlo, E. A., and Pearson, A. E. (1968). *J. Phys. Chem.* **72,** 2913.
Ouannès, C., and Wilson, T. (1968). *J. Am. Chem. Soc.* **90,** 6527.
Page, R. M. (1968). *In* "Photophysiology" (A. C. Giese, ed.), Vol. III, pp. 65–90. Academic Press, New York.
Pickard, B. G., and Thimann, K. V. (1964). *Plant Physiol.* **39,** 341.
Poff, K. H. (1968). Ph.D. Thesis, West Virginia University, Morgantown.
Raab, O. (1900). *Z. Biol.* **39,** 524.
Rauhut, M. M. (1969). *Accounts Chem. Res.* **2,** 80.
Rigaudy, J. (1968). *Pure Appl. Chem.* **16,** 169.
Rigaudy, J., Deletang, C., Sparfel, D., and Cuong, N. K. (1968). *Comp. Rend.* **267,** 1714.
Schenck, G. O., and Ziegler, K. (1944). *Naturwissenshaften* **32,** 157.
Schenck, G. O., Kintel, K. G., and Mertens, H. J. (1953). *Ann.* **584,** 125.
Schiff, H. I. (1969). *Can. J. Chem.* **47,** 1903.
Schnuriger, B., and Bourdon, J. (1968). *Photochem. Photobiol.* **8,** 361.
Schönberg, A. (1935). *Ann.* **518,** 299.
Seliger, H. H. (1960). *Analyt. Biochem.* **1,** 60.
Seliger, H. H., and McElroy, W. D. (1965). "Light: Physical and Biological Action." Academic Press, New York.
Sistrom, W. R., Griffiths, M., and Stanier, R. Y. (1956). *J. Cellular Comp. Physiol.* **48,** 473.
Snelling, R. D. (1968). *Chem. Phys. Letters* **2,** 346.
Spikes, J. D. (1967). *Ann. Rev. Phys. Chem.* **18,** 409.
Spikes, J. D. (1968). *In* "Photophysiology" (A. C. Giese, ed.), Vol. III, pp. 33–64. Academic Press, New York.
Spikes, J. D., and Livingston, R. (1969). (In press.)
Stanier, R. Y. (1959). *Brookhaven Symp. Biol.* **11,** 43.
Stevens, B., and Algar, B. E. (1967). *Chem. Phys. Letters* **1,** 58.
Swart-Füchtbauer, H. (1951). *Arch. für Mikrobiol.* **16,** 40.
Swart-Füchtbauer, H., and Rippel-Baldes, A. (1951). *Arch. für Mikrobiol.* **16,** 358.
Tappeiner, H. V., and Jodlbauer, A. (1904). *Deut. Arch. Klin. Med.* **80,** 427.
Thimann, K. V., and Curry, G. M. (1960). *In* "Comparative Biochemistry" (M. Florkin and H. Mason, eds.), Vol. 1, pp. 243–309. Academic Press, New York.
Thimann, K. V. (1967). *In* "Comprehensive Biochemistry" (M. Florkin and E. Stotz, eds), Vol. 27, pp. 1–20. Esevier, Amsterdam.
Thompson, Q. E. (1961). *J. Am. Chem. Soc.* **83,** 845.
Tsubomura, H., and Mulliken, R. S. (1960). *J. Am. Chem. Soc.* **82,** 15966.

Turro, N. J. (1965). "Molecular Photochemistry." Benjamin, New York.

von Guttenberg, H. (1959). *Planta,* **53,** 412.

Wasserman, E., Murray, R. W., Kaplan, M. L., and Yager, W. A. (1968). *J. Am. Chem. Soc.* **90,** 4160.

Wasserman, E., Kuck, V. J., Delavan, W. M., and Yager, W. A. (1969). *J. Am. Chem. Soc.* **91,** 1040.

Wasserman, H. H., and Scheffer, J. R. (1967). *J. Am. Chem. Soc.* **89,** 3073.

Whitlow, S. H., and Finlay, F. D. (1967). *Can. J. Chem.* **45,** 2087.

Wilson, T. (1966). *J. Am. Chem. Soc.* **88,** 2898.

Wilson, T. (1969a). *Photochem. Photobiol.* **10,** 441.

Wilson, T. (1969b). *J. Am. Chem. Soc.* **91,** 2387.

Went, F. W. (1928). *Rec. Trav. Botan. Néerl.* **25,** 1.

Young, R. A., and Sharpless, R. (1963). *J. Chem. Phys.* **39,** 1071.

Chapter 4

SPECTROPHOTOMETRIC STUDIES OF THE MECHANISM OF PHOTOSYNTHESIS

David C. Fork

Department of Plant Biology, Carnegie Institution of Washington, Stanford, California

Jan Amesz

Biophysical Laboratory, State University, Leiden, The Netherlands

1. Introduction

Although there are still important gaps, our knowledge about the mechanism of photosynthesis has advanced considerably during the last decades. Part of this advance results from modern biochemical research on cellular preparations of photosynthetic organisms; however, an important part is also due to the application of sensitive spectrophotometric methods, which have become an important tool of photosynthesis research.

Since fluorescence is one of the ways by which excited chlorophyll or bacteriochlorophyll dissipates excitation energy which is not used in photosynthesis, fluorescence measurements have been extensively applied to study energy transformation in photosynthetic pigments. Fluorescence studies have also been useful to obtain information about reactions of photosynthetic intermediates, either because the intermediate itself fluoresces, or because reactions of intermediates were found to be correlated with changes in fluorescence yield of chlorophyll or bacteriochlorophyll. A more widely used technique to study reactions of photosynthetic intermediates is that of absorption difference spectrophotometry. Many of the reactants of photosynthesis are colored and change their absorption spectrum upon oxidation or reduction. For this reason, spectral measurements of light-induced changes in absorption of photosynthetic material upon illumination and measurement of the action spectra for producing these absorption changes have provided much information about the identity and function of the reactants.

In the following section we shall briefly discuss some of the spectrophotometric instrumentation and methods used in photosynthesis studies. Some results of those studies will be discussed in the remaining sections.

2. Methodology

Various aspects of spectrophotometric instrumentation in photosynthesis and in photobiology in general have been treated in earlier Chapters in these Volumes (Allen, 1964; Seely, 1968). A clear discussion of the principles and of the experimental problems of optical techniques in photosynthesis is also given by Clayton (1965). Therefore, we will confine ourselves to only a brief discussion of this topic.

2.1 Absorption Difference Spectroscopy

Duysens (1952) was the first to use the technique of absorption difference spectrophotometry to study photosynthetic reactions. Since then, spectrophotometers of varying complexity have been constructed to measure changes in light absorption at various wavelengths induced by illumination. Because the absorbancy changes are small, optical compensation is often applied to enhance the stability of the apparatus. For the recording of relatively slow reactions this can be accomplished by use of a beam splitter and chopper. The measuring beam then passes alternatively through the sample vessel, which can be illuminated, or through a reference vessel with an adjustable light attenuator. The a.c. signal of the photomultiplier caused by transmittancy changes in

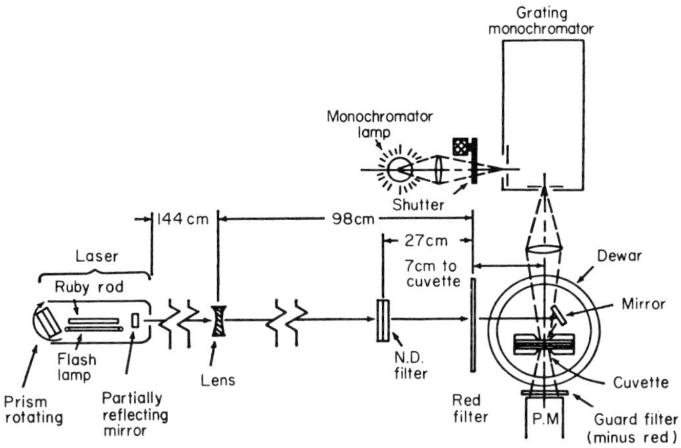

FIG. 1. Apparatus for measuring absorbancy changes induced by a laser flash. The negative lens and the neutral density filter serve to reduce the light intensity on the sample (DeVault and Chance, 1966). Reproduced by permission of the Rockefeller University Press and the authors.

the sample vessel is demodulated and recorded. An apparatus constructed by Chance (1951) employs two beams of different wavelengths which alternately pass through the same vessel. In an apparatus to measure fast absorption changes compensation is achieved electrically or by subtraction of the signal of a second photomultiplier (Witt *et al.*, 1959; Ke *et al.*, 1964). This is necessary for fast reactions because the chopping frequency in double-beam instruments limits the response time. Flash tubes or, more recently, a Q-switched ruby laser (DeVault and Chance, 1966; Parson, 1968) have been used for illumination to obtain the high light intensities and high time resolution needed to bring about and to measure fast reactions. An apparatus used for the study of the fast oxidation of cytochromes in purple bacteria and algae induced by a very short laser flash (about 10 nsec) is shown in Fig. 1.

Especially with small rapid transients the averaging of repeated signals may be necessary to obtain a better signal to noise ratio. This can be done either by means of a commercially available averager (Ke *et al.*, 1964) or by means of repeated sampling at various time intervals after the flash, as described by Rüppel *et al.* (1964) for very fast, and by Kok (1959) for somewhat slower phenomena.

2.2 Fluorescence

A general discussion of instrumentation for fluorescence measurement in biological research is given by Udenfriend (1962); more recent general information can be found in the books of Hercules (1966) and of Guilbault (1967). In the following we shall only mention some instruments especially constructed for photosynthesis studies.

Duysens and Sweep (1957) and Olson and Amesz (1960) described sensitive fluorescence spectrophotometers designed to measure small increases in fluorescence due to reduction of NAD and NADP in intact photosynthesizing organisms, and also for measuring the fluorescence emission and excitation spectra of these changes. The excitation beam was modulated; NAD(P) reduction was brought about by a nonmodulated beam of longer wavelength. Duysens and Sweers (1963) used a similar apparatus for studying changes in fluorescence yield of chlorophyll *a* induced by illumination. In this case both the (weak) modulated and the (stronger) nonmodulated beam excited fluorescence, but the fluorescence due to the latter beam, except for transient effects, was not recorded. Thus the effect of additional illumination on the yield of the fluorescence excited by the modulated beam could be studied. Instruments of the type discussed above, but without additional illumination have been used by various workers for studies on chlorophyll fluorescence. French (1955) has described an apparatus for the automatic recording of chlorophyll emission spectra.

Lavorel (1965) and Vredenberg and Duysens (1965) used a flow system to study light-induced changes in chlorophyll fluorescence. The advantage of the method is that for some types of experments a higher intensity of illumination can be applied, with resulting higher signal to noise ratio than with a stationary sample, because the individual cells or particles see only a brief "flash" of light when the suspension passes the illuminated area. For this reason the fluorescence properties of the material under "dark" conditions can be studied even with relatively intense illumination. A second illuminated area at a variable distance from the first one permits study of the effect of preillumination with varying dark time.

2.3 Experimental Analysis

Quantitative analysis of spectrophotometric measurements of photosynthetic material presents some special problems due to the complexity of the system and to optical inhomogeneity of the material.

The best material for studies on intact cells are suspensions of unicellular organisms and thin thalli, such as *Porphyra* and *Ulva*, but even with these the so-called flattening ("sieve") effect (Duysens, 1956) and light scattering (Amesz *et al.*, 1961; Butler, 1964) may give a distortion of absorption and of absorption difference spectra. The effect due to scattering can usually be largely eliminated by shortening the distance between sample and photocell or by use of an Ulbricht sphere; but the "sieve" effect, which is most severe in spectral regions of strong absorption, cannot be minimized in this way. Optical distortion may be even more serious in leaves of higher plants, where, in spectral regions of strong absorption by the chloroplasts, a large fraction of the light passing the leaf may have travelled through cell walls, veins and air chambers, so that the absorption spectrum of the pigments and especially the change of absorption spectrum due to components reacting in the chloroplast are "diluted." Virgin (1954) has shown the disturbing effect due to the scattering by air chambers on chlorophyll emission spectra. Apart from this, special precautions are usually needed to prevent scattered excitation light from reaching the photocell, and especially with weakly fluorescent material difficulties may arise due to the fact that the scattered light causes fluorescence by optical components, such as filters, between sample vessel and photocell.

Interpretations of light induced absorption changes and of the spectra of these changes may be impeded by various factors, apart from the ones mentioned above. Among these are the overlap of two or more difference spectra due to simultaneous occurrence of more than one reaction. The difficulty can sometimes be overcome by analysis of kinetics or by choosing more favorable experimental conditions. Fluorescence artifacts may occur in regions of chlorophyll absorption. Causes of error may also be scattering changes and poor penetration of the actinic light into a strongly absorbing sample. The latter factors are dependent among other things on the geometry of the optical arrangement. A discussion of these and other factors has been given by Duysens (1952), Amesz (1964a) and Clayton (1965).

A discussion of some of the problems involved in measuring action spectra of fluorescence and of absorption changes is given by Clayton (1965) and by Fork and Amesz (1969).

3. Photosynthetic Bacteria

3.1 The Primary Photochemical Reaction

There is considerable evidence that the primary photoreaction of bacterial photosynthesis consists of an oxidation-reduction reaction that takes place in the so-called reaction center. As will be discussed below, the reaction center probably contains one or a number of bacteriochlorophyll molecules, one or more cytochrome molecules, and possibly ubiquinone and other components. Most of the bacteriochlorophyll present in photosynthetic bacteria (bacteriochlorophyll *a* in most species and bacteriochlorophyll *b* in a few species of *Rhodopseudomonas*, Jensen *et al.*, 1964) is not located in the reaction center, but functions, together with carotenoids and the *Chlorobium* chlorophyll of green bacteria in collecting the light energy used for photosynthesis, as is indicated by fluorescence emission spectra and by action spectra of fluorescence and of photosynthetic reactions (Duysens, 1952; see also Duysens, 1964 and Fork and Amesz, 1969 for further references).

Upon illumination of intact bacteria or bacterial preparations, a decrease in absorption, which reverses again in the dark, occurs in the infrared. The maximum of the decrease in absorption is located at 840 (Sybesma and Vredenberg, 1963), 870–890 (Duysens, 1952; Arnold and Clayton, 1960), and 975 nm (Holt and Clayton, 1965) for green, purple and bacteriochlorophyll *b* containing bacteria respectively. According to the location of this maximum, the pigments which are bleached upon illumination have been called P840, P890 (P870) and P975 respectively. Apart from the location of its absorption band, there are various indications that the pigment is a specially bound bacteriochlorophyll, located in the reaction center. There is evidence that its bleaching represents its oxidation in a primary photochemical reaction.

"Titration" of the absorbancy change with mixtures of potassium ferro- and ferricyanide indicates that the standard oxidation reduction potential (at pH 7) of P890 is +0.44 or 0.48 V (Kuntz *et al.*, 1964; Cusanovitch *et al.*, 1968). Arnold and Clayton (1960) have shown that the light-induced oxidation proceeds at temperatures as low as a few degrees Kelvin. More recently, Parson (1968) reported that in *Chromatium* chromatophores the bleaching of P890 took place within 0.5 μsec upon excitation with a 30 nsec laser flash. Experiments of Vredenberg and Duysens (1963) and Clayton (1967) show a quantitative relation between bacteriochlorophyll fluorescence and P890 or P870 bleaching which indicates that excitation energy from the bulk bacteriochlorophyll is transferred to P890 (or P870). When this becomes bleached, the rate of transfer decreases, and the yield of bacteriochlorophyll fluorescence increases. P870 does not fluoresce unless its oxidation is inhibited by dithionite (Clayton, 1966b).

Except for the green bacteria examined so far, the bleaching is accompanied by the shift of a second bacteriochlorophyll band, located at 800 nm (P800) or 835 nm (P835), towards shorter wavelength. By treating chromatophores with oxidizing agents or detergents it is possible to destroy or remove all bacteriochlorophyll except P890 or 870 and P800 (Loach *et al.*, 1963; Clayton, 1966a; Beugeling, 1968; Reed and Clayton, 1968; Gingras and Jolchine, 1969; Thornber *et al.*, 1969). These "reaction center" preparations are still photochemically active (Fig. 2), and a quantum efficiency for the oxidation of P890 of close to one was

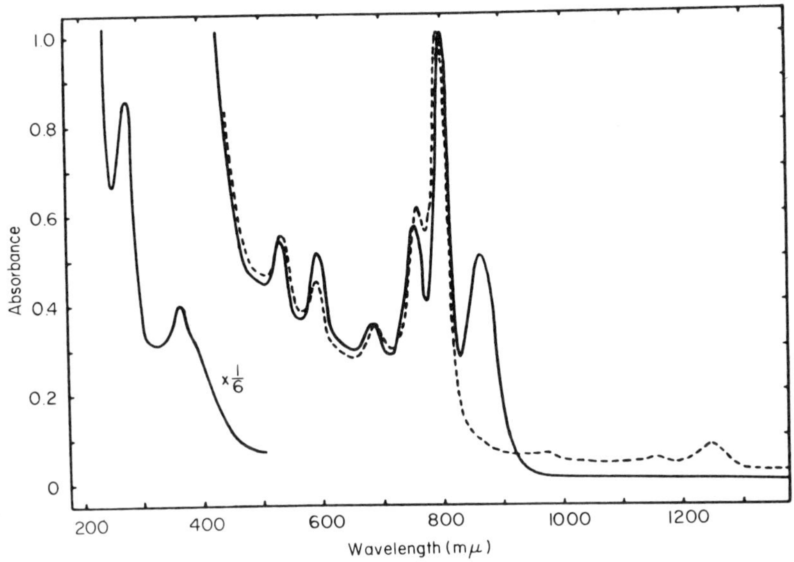

FIG. 2. Absorbancy spectra of a reaction center preparation prepared from *Rhodopsuedomonas spheroides* by detergent treatment, measured in the dark (solid line) and during illumination (dashed line) (Reed and Clayton, 1968). Note the disappearance of the band at 860 nm and the slight shift of the band at 800 nm. Reproduced by permission of the authors.

reported (Beugeling, 1968). Extraction experiments indicated that the preparations contained P870 (or P890) and P800 in a molar ratio of 1:2 (Clayton, 1966a). Active preparations without P800 or P830 have not been prepared. A close interaction between P800 and P890 or P870 is also indicated by optical rotatory dispersion and circular dichroism spectra of bleached and unbleached reaction center preparations (Sauer and Dratz, 1969).

The identity of the primary photooxidant, the "electron acceptor" for P890 that is reduced concomitantly with the oxidation of P890 in the primary photochemical act, is not yet established with reasonable certainty, although there is evidence that it may be a ubiquinone. Light-induced changes in ultraviolet absorption in dried chromatophore preparations of *Rhodopseudomonas spheroides* (Clayton, 1962), in chromatophore preparations from *Chromatium* (Ke *et al.*, 1967; Ke, 1969) and in reaction center preparations from *Rhodospirillum rubrum* (Beugeling, 1968), indicate a reduction of ubiquinone upon illumination (Fig. 3). The reduction can occur very rapidly (Ke, 1969) and at temperatures down to 77°K (Ke *et al.*, 1967) and has been reported to proceed with twice the quantum efficiency of P890 oxidation. This paradoxically high

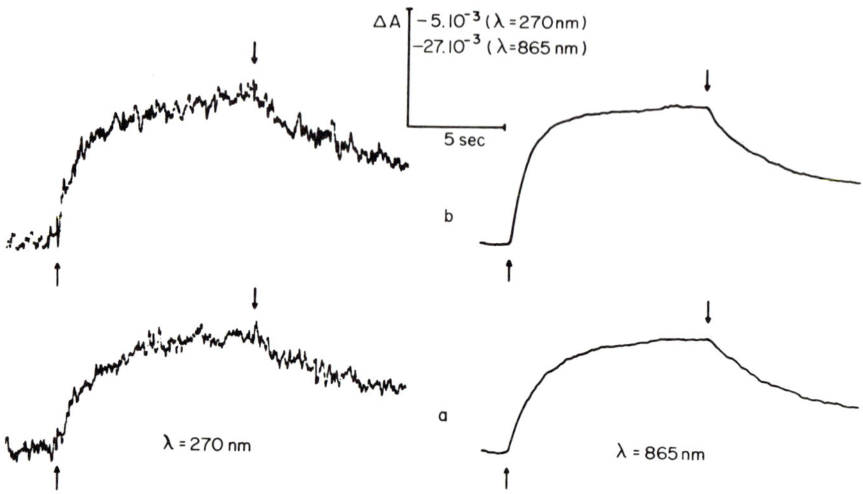

FIG. 3. Light-induced absorbancy changes in a reaction center preparation of *Rhodospirillum rubrum*, with identical kinetics at 270 and 865 nm. Upward and downward pointing arrows mark the beginning and the end of illumination. The intensity of illumination was twice as high for the upper as for the lower traces. An upward moving trace corresponds to a decrease in absorbancy. At 865 nm this decrease indicates the oxidation of P890; at 270 nm it suggests the reduction of ubiquinone (Beugeling, 1968). Reproduced by permission of Elsevier Publishing Co. and the author.

number might suggest that a second electron for the reduction of ubiquinone comes from the oxidation of an electron donor other than P890. Ke (1969) recently reported that the reduction of ubiquinone but not the oxidation of P890 was inhibited by a high concentration of ascorbate, which presumably reduced ubiquinone. This casts some doubt on the hypothesis that ubiquinone is the primary photooxidant.

Measurements of changes of $NADH_2$ fluorescence and of absorption changes in the near ultraviolet indicate that NAD is reduced at a high rate upon illumination of intact purple bacteria (Olson and Amesz, 1960; Amesz, 1963). Light-induced NAD reduction has also been observed in chromatophore preparations. The oxidation reduction potential of ubiquinone is higher than of NAD, which would imply that, if ubiquinone or a substance of about the same redox potential is the primary photooxidant, NAD cannot be reduced directly, but only by some form of "reversed" electron transport. Reduction of NAD by succinate by way of such reversed electron transport driven by the free energy of hydrolysis of ATP was observed by Keister and Yike (1967) and by Klemme (1969) in chromatophore preparations from *Rhodospiril-*

lum rubrum and *Rhodopseudomonas capsulata* respectively, but in both cases the rate was lower than that of light-driven reduction. HQNO (*2,-n-nonyl-hydroxyquinoline-N-oxide*), which inhibits "cyclic" light-induced ATP production, was found only partly to inhibit light-induced NAD reduction in *Rhodospirillum rubrum* (Amesz, 1963); recently, however, complete inhibition was reported in aerobic *Rhodopseudomonas spheroides* (Jackson and Crofts, 1968).

3.2 Function of Cytochromes

Starting with the work of Duysens (1954a) and of Olson and Chance (1960a,b) a large body of experimental data has accumulated concerning the role of cytochromes in bacterial photosynthesis. Space does not permit an elaborate treatment of the various evidence here; the reader may be referred to recent reviews (e.g., Duysens, 1964; Hind and Olson, 1968; Vernon, 1964, 1968).

Absorption difference spectra and kinetics of absorption changes have shown that in all species studied so far at least two different cytochromes are oxidized upon illumination. Although definite evidence is still lacking, it seems probable that one of the cytochromes functions as a redox carrier in a "cyclic" chain between oxidized P890 and the reduced form of the primary photooxidant, and that the other one reacts in a chain between P890 and the substrate (hydrogen donor).

The light induced oxidation of cytochromes proceeds with high rates and good efficiency at temperatures well below 0°C. However, the temperature sensitivity of the reaction is different for different cytochromes (Vredenberg and Duysens, 1964); the least sensitive reaction is the photooxidation of cytochrome 553 in *Chromatium* (DeVault and Chance, 1966; DeVault *et al.*, 1967) and of analogous cytochromes in other Thiorhodaceae (Chance *et al.*, 1969). This indicates that these cytochromes are situated close to P890 in the reaction center (see also Beugeling and Duysens, 1965 and Duysens, 1965) and that very little molecular movement is needed for the reaction. A clear indication that the oxidation of P890 rather than of cytochrome 553 is the primary photoreaction in *Chromatium* was given by the experiments of Parson (1968) mentioned in the previous section. After the very rapid oxidation of P890 brought about by a laser flash, a somewhat slower re-reduction was observed which took about 2 μsec, and which was accompanied by a simultaneous oxidation of cytochrome, indicating that the oxidation of cytochrome was brought about by oxidized P890. However, with preparations from *Rhodospirillum rubrum* no such simple relation between P890 and cytochrome kinetics was observed (Parson, 1967).

3.3 Other Absorption Changes

Light-induced changes in bacteriochlorophyll absorption not related to P890 or P870 have been observed in a number of species of purple bacteria (Clayton, 1963; Vredenberg *et al.*, 1965; Amesz and Vredenberg, 1966). The shape of the difference spectra of these changes indicates that they represent small shifts of the absorption bands of bacteriochlorophyll (Fig. 4). In some species these shifts are accompanied by shifts of the absorption bands of a carotenoid. Analysis of these phenomena in *Rhodopseudomonas spheroides* (Vredenberg and Amesz, 1966a) indicated that the absorption changes were not due to a chemical reaction of these pigments, but that they represent shifts of the absorption bands by only a few nm of a rather large fraction of the pigment present in the bacteria. The shifts take place with a high quantum efficiency (of the order of 15 molecules per absorbed light quantum). They are affected by uncouplers and may be caused by membrane effects related to phosphorylation (Fleischman and Clayton, 1968), in a similar way as the changes in pigment absorption in algae and higher plants discussed in Sections 4.4 and 4.5.

A very conspicuous light-induced increase in absorbancy occurs at

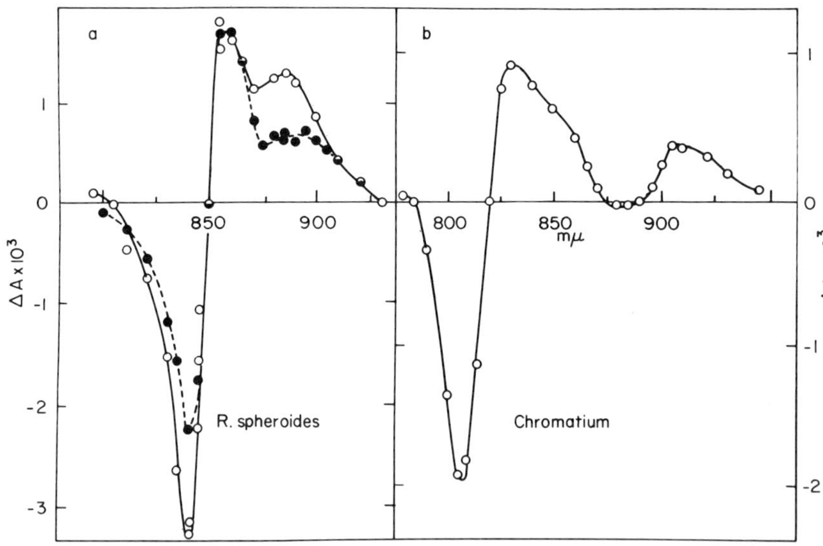

FIG. 4. Absorbancy difference spectra induced by illumination of a wild (solid circles) and of a mutant strain (open circles) of *Rhodopseudomonas spheroides* and of *Chromatium* (Vredenberg *et al.*, 1965). The spectra can be interpreted as shifts of bacteriochlorophyll bands centered at about 850 and 870 nm in *R. spheroides* and at about 820 and 890 nm in *Chromatium* (Vredenberg and Amesz, 1966b).

around 435 nm in purple bacteria under conditions where photosynthesis is inhibited or saturated (Duysens, 1957; Olson and Kok, 1959). The cause of this absorption change is not known. The change is very fast in strong light, (Parson, 1967, 1968) and occurs also at 100°K (Vredenberg, 1965) and in a reaction center preparation (Beugeling, 1968). The absorption increase is, at least under certain conditions, not correlated with bleaching of P890 (Olson and Kok, 1959; Vredenberg, 1965; Parson, 1968), and might represent a "waste" reaction caused by energy which is not used by the normal primary photochemical reaction.

3.4 One or Two Photosystems?

Until recently it has been generally assumed that photosynthetic bacteria, unlike oxygen evolving organisms, have only one photosystem, analogous to system 1 of algae. Recently, however, Morita (1968) and Sybesma and Fowler (1968) have measured different action spectra for the oxidation of different cytochromes in *Chromatium* and *Rhodospirillum rubrum*, and they interpreted these results as evidence in favor of the existence of more than one photosystem in bacteria. Sybesma (1969) also reported that the absorption changes around 800 nm due to the blue shift of P800 were smaller relative to the bleaching at 880 nm at low than at high intensities of illumination in *Rhodospirillum rubrum*. This might suggest the occurrence of two types of reaction center, e.g., one type with and one type without P800.

For a discussion of this and other evidence the reader is referred to Hind and Olson (1968) and to Fork and Amesz (1969). Here we shall confine ourselves to pointing out that there are various kinds of "older" evidence that fit better into a one photosystem than in a two photosystem scheme for bacterial photosynthesis (Fork and Amesz, 1969). Further experimentation will be needed to resolve this problem.

4. Algae and Higher Plants

Investigations on algae and higher plants over the last several years have shown that the photosynthesis of these plants involves the cooperation of two distinct light reactions each having its own characteristic pigments which serve to absorb and transfer energy to traps where the primary photochemical reactions take place. In essence, one light reaction, called system 1 (after Duysens and Amesz, 1962), is driven by chlorophyll forms absorbing maximally at relatively long wavelengths in the red region (the so-called "far-red" region). System 1 produces a strong reductant which leads, eventually, to CO_2 reduction. The other light reaction (system 2) is produced upon excitation of a chlorophyll form absorbing maximally at shorter wavelengths. This reaction leads to the production of a strong oxidant which, in turn eventually, brings about the oxidation of water. The other pigmens (carotenoids, chlorophylls *b* and *c*, and phycobilins) belong partly to system 1 and partly to system 2. The reader may be referred to the reviews by Allen, Whatley and Losada, Clayton and Blinks published in Vol. I of this series for a discussion of the origin and development of the concept of two light reactions in photosynthesis.

4.1 The Reaction Center of System 1

As with the photosynthetic bacteria, the reaction center of system 1 contains a small amount of a special chlorophyll, P700, (Kok, 1956, 1961; Kok and Hoch, 1961) which is the counterpart of bacterial P870 and P890. Energy transfer from the bulk chlorophyll to the reaction center apparently passes via another special, weakly fluorescent, long wavelength chlorophyll which is present in small amounts and located near reaction centers containing P700 (Kok, 1963; Butler, 1961). Changes in the redox state of P700 can be followed spectroscopically by observing absorption changes caused by this compound near 700 or 435 nm.

Witt and collaborators (Döring *et al.*, 1968a) have shown that P700 oxidation can occur in 0.4 μsec. At about 120°K the photooxidation is irreversible. Chance and Bonner (1963) also found irreversible photooxidation of P700 at 77°K in chloroplasts as well as in leaves. Vernon and Ke (1966) have reviewed results which would place cytochrome *f* rather than P700 as the primary photoreductant of system 1. As discussed in Section 3.1, Parson (1968) found in bacteria that the oxidation of P890 occurred before the oxidation of cytochrome 555 upon illumination of *Chromatium* with 30 nsec flashes from a Q-switched ruby laser.

If one observes that cytochrome oxidation takes place before oxidation of P700 this does not necessarily mean that cytochrome is the primary photoreductant of system 1 because a tight coupling between P700 and cytochrome *f* would permit rapid reduction by cytochrome of any photooxidized P700. Accumulation of oxidized P700 would then only occur after the cytochrome became oxidized. Such a response was seen in the red alga *Cryptopleura* (Fork and Amesz, 1967a, 1968a). Figure 5 shows that P700 is almost completely in the reduced state at light intensities where photosynthesis proceeds with maximum efficiency. P700 becomes oxidized only at light intensities where photosynthesis starts to depart from linearity. The behavior of the cytochrome is different, however. Photosynthesis still proceeds at maximum efficiency even when about a fifth of the cytochrome is oxidized. This result suggests that P700

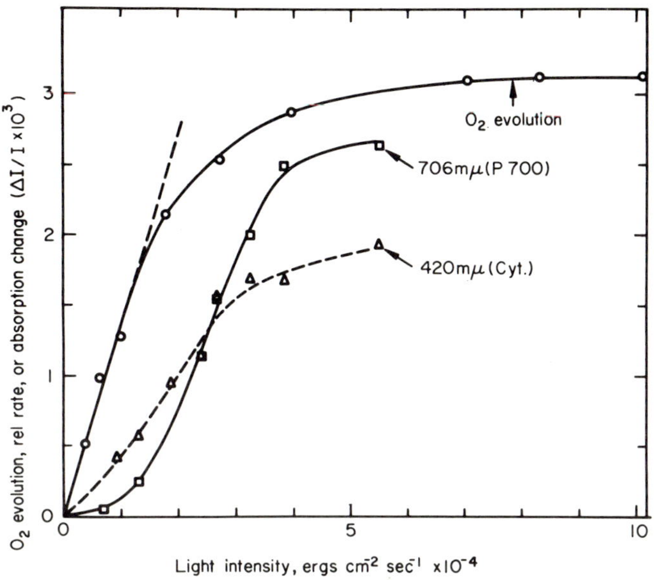

FIG. 5. Steady-state level of oxidation of P700, *f*-type cytochrome, and steady-state rate of O_2 evolution of *Cryptopleura violacea* as a function of the intensity of green light (540 nm max., 40 nm half-bandwidth, Fork and Amesz, 1968a).

and not cytochrome *f* behaves as the primary reactant of light reaction 1 because it would be difficult to envisage a mechanism whereby photosynthesis proceeds at maximum efficiency while a part of the primary photoreductant is in the oxidized form and thus unable to donate an electron in the primary reaction. Another intertidal red alga, *Iridaea*, proved particularly suitable for a study of the relationships between P700 and cytochrome *f*. A high quantum yield was found (Amesz and Fork, 1967b) for this alga (0.5 to 0.65) showing that the cytochrome functions in the main part of the electron transport chain. These high quantum yields were calculated when most of the cytochrome was allowed to become reduced in dark and when the initial net rate of P700 oxidation upon illumination was very low. However, illumination of this alga after a short dark time when most of the cytochrome was in the oxidized state gave an appreciable rate of P700 oxidation accompanied by a correspondingly lower rate of cytochrome oxidation. Extrapolation suggested a quantum yield for P700 of about 0.6 to 0.7 when all of the cytochrome is oxidized. The efficiency of photochemistry (measured as the sum of the quantum yields for P700 and cytochrome oxidation) was dependent on the oxidation level of P700 but not on the oxidation level of cytochrome. This is to be expected if it is assumed that a reaction center containing oxidized P700 is unable to do photochemistry. These results indicate that the reaction center in system 1 is active as long as P700 is reduced regardless of the oxidation level of cytochrome. They do not,however, directly demonstrate that P700 oxidizes cytochrome.

Evidence for energy transfer between reaction centers of system 1 was obtained for *Iridaea* because the sum of the quantum yields for P700 plus cytochrome oxidation declined less than did the level of reduced P700. This result suggested a probability of about 50% for transfer of energy to a second reaction center when the first one visited is inoperative. A similar result was obtained in a different way with *Cryptopleura* (Fork and Amesz, 1967b).

Absorption changes in the red and blue region in small particles (144,000 g sediment) prepared by digitonin fragmentation of spinach chloroplasts were shown to be produced by P700 and by cytochrome *f* (Anderson *et al.*, 1966). As expected for P700, the difference spectrum in the red had a main peak at 702 nm and, in addition, showed a second band at 683 nm. These small particles were enriched in P700 since the ratio P700/chlorophyll was about 1/200 in the small fragments and about 1/400 in chloroplast fragments before centrifugation. Similar results were reported by Vernon *et al.* (1966) for particles prepared by Triton treatment. Döring *et al.* (1968a) used a 50,000 g supernatant from centrifugation of digitonin-treated chloroplasts which was assumed to be enriched in system 1 activity. The absorption spectrum for this preparation also gave a double-peaked difference spectrum with maxima at 682 and 700 nm. Changes were also seen at 438 nm which were kinetically similar to those in the red. Döring *et al.* (1968a) suggested that two molecules could be combined in one dimer with an oblique structure which would cause a symmetrical splitting into one double band with peaks at 682 and 700 nm. Analysis of the shape of the light-induced difference spectrum of the blue-green alga *Schizothrix calcicola* (Amesz *et al.*, 1967) suggests that the band at about 680 and an asymmetry of the main band near 700 nm is due to a shift of a chlorophyll absorption band towards longer wavelength accompanying the bleaching of P700. The effect may be related to that causing the shifts of P800 and P835 in bacteria (see Section 3.1).

It is noteworthy that Kok already 12 years ago (Kok, 1957) noted bands near 680 and 700 nm in difference spectra of *Chlorella, Nitzchia, Porphyra, Nostoc*, and spinach chloroplasts. No evidence as to the nature of the primary photooxidant of system 1 has been obtained so far.[1]

[1] See, however, a recent paper by Yokum and San Pietro (1969).

4.2 The Reaction Center of System 2

Little is known about the reaction center of system 2 in terms of the nature of the chemical processes that take place upon illumination. Measurements of the yield of chlorophyll fluorescence (Duysens and Sweers, 1963) indicate that the photoreduction of the primary oxidant (called Q for quencher) is accompanied by an increase in the fluorescence of the chlorophyll of system 2. Measurements of O_2 evolution by short flashes (Joliot, 1968) indicate that the primary reductant is a two-electron donor which is oxidized stepwise and which moreover occurs in two different forms. However, direct evidence as to the chemical identity of the above-mentioned compounds based, e.g., upon absorption difference spectroscopy is lacking so far.

The observation of absorbance changes produced by a unique chlorophyll species that acts as a trap for system 2 analogous to the manner in which P700 functions in system 1 has proved elusive despite considerable effort by many workers over the past years. Recently, however, Döring *et al.*, (1967) observed with chloroplast fragments that absorbance changes measured at 690 nm exhibited a biphasic decay. The difference spectrum for the slowly-decaying component (half-life $= 2 \times 10^{-2}$ sec) was apparently caused by P700 because it had the expected maximum at 703 nm. The difference spectrum of the fast component (half-life $= 2 \times 10^{-4}$ sec) was not precisely determined but it had a maximum somewhere around 690 nm. This fast component was assumed to be produced by a chlorophyll; however, it was apparently related to system 2 because it was inhibited by DCMU, and could not be restored by addition of *N*-methylphenazonium methosulfate (PMS) and ascorbate as could P700, nor could it be produced by far-red exciting light.

By using the 10,000 g sediment from digitonin-fragmented chloroplasts Döring *et al.* (1968b) redetermined the maximum to be at 682 instead of 690 nm. Absorption changes were seen in the blue region, the kinetics of which were reported to match those at 682 nm. The difference spectrum for these changes showed a peak at 435 nm.

4.3 Plastocyanin and Cytochrome f

The role of plastocyanin as an electron carrier in photosynthesis has continued to be a matter of interest in recent years. This copper-protein was discovered by Katoh and co-workers (Katoh, 1960; Katoh *et al.*, 1961) and found to be localized in chloroplasts of green cells. It is apparently ubiquitous in higher plants and in green algae. It has also been found in a blue-green alga (Lightbody and Krogmann, 1967) but not in photosynthetic bacteria.

The blue (oxidized) form of plastocyanin has its main absorption band at 597 nm with shoulders at 460 and 770 nm. The half width of the 597-nm band is about 100 nm. The reduced form of plastocyanin is colorless and exhibits practically no absorption from 400 to 1000 nm (Katoh *et al.*, 1962). Spectrophotometric studies of changes in the redox state of plastocyanin are difficult because of its broad absorption in the visible mentioned above and also because the extinction coefficient of plastocyanin is low (about 10 mM^{-1} cm^{-1}), only about half that of the α-band of cytochrome *f*. Moreover, absorption changes caused by other, as yet unidentified, substances frequently obscure plastocyanin changes. Nevertheless, absorbance changes could be seen in the green alga *Ulva* (de Kouchkovsky and Fork, 1964) which were apparently produced by light-induced changes in the redox state of plastocyanin. The difference spectrum for these changes in *Ulva* in the region around 600 nm matched Katoh's oxidized-minus-reduced spectrum of purified plastocyanin. Since far-red light produced an increase of absorption (at 591 nm) it was concluded that system 1 oxidized plastocyanin. The decay of the change (reduction) upon turning off this light was faster in the presence of background light of 651 nm that is absorbed by system 2. An accelerated decay of the change could also be produced chemically by adding a strong reducing agent such as sodium hydrosulfite.

Observation of absorption changes attributable to plastocyanin in leaves and algae is made difficult because of overlapping by large unidentified change(s) of absorption with a maximum near 565 nm. Changes at this wavelength may be caused in part by a *b*-type cytochrome but most of the change must be caused by a different substance because a difference spectrum does not show the expected Soret band in the blue near 435 nm similar to what can be seen at 420 nm for cytochrome *f*. For a discussion of the role of the *b*-type cytochromes see the review by Hind and Olson (1968).

It seemed conceivable that absorbance changes measured in this region as well as in the region of plastocyanin absorption could be attributed to light-induced scattering changes and not to true light-induced absorbance changes at all. Therefore, the measurements have been repeated with the sample held inside an integrating sphere (Fork and de Kouchkovsky, 1968). With this arrangement scattered light eventually reaches the photomultiplier by multiple reflections within the sphere. One is thus able to distinguish between absorption changes and light-induced scattering changes because the latter will disappear or be diminished when the sample is in the sphere but true absorbance changes will be unaffected.

In leaves of wild cucumber, both the changes at 565 and 595 nm are apparently caused by absorbance changes because they were unaffected when measured in the sphere. However, it was possible to distinguish between the two changes because the decay of the 595-nm change was much slower than that at 565 nm, and a modified difference spectrum was determined in cucumber leaves by measuring this slowly-decaying component as a function of wavelength. The contribution of this slow component was maximal near 592 nm and the "spectrum" showed some correspondence to Katoh's plastocyanin curve both in cucumber leaves and in *Chlorella*.

In *Chorella* incubation with the copper chelator salicylaldoxime selectively inhibited the absorbance changes at 591 nm but did not inhibit those produced upon oxidation of cytochrome *f* (Fork and Urbach, 1965, 1966). This observation suggested that cytochrome *f* can still be oxidized by P700 even though plastocyanin is nonfunctional and that the cytochrome and not plastocyanin is the electron donor to P700.

A different electron transport sequence: system 2, cytochrome b_3, cytochrome *f*, plastocyanin, P700 was suggested by Gorman and Levine (1965, 1966) and Levine and Gorman (1966) partly on the basis of studies on a mutant of *Chlamydomonas* that lacks plastocyanin but contains the *b* and *f*-type cytochromes. Ferricyanide was added to chloroplast fragments from this mutant (Ac 208) to oxidize the cytochromes. After this treatment illumination with red light produced only increases of absorbance (reduction of cytochrome) at 553 and 559 nm. After reduction with sodium ascorbate no absorbance changes at 553 or 559 were induced by far-red actinic light. Addition of plastocyanin activated light-induced absorbance decreases at these wavelengths. It was concluded that system 2 reduced both the oxidized cytochrome *f* and b_3 but that system 1 was unable to oxidize these cytochromes after they were reduced by ascorbate since the missing plastocyanin broke the link between cytochrome *f* and P700. Unfortunately, no difference spectra for these effects were given to verify that the absorption changes at 553 and 559 were indeed due to cytochrome *f* and *b* respectively. A good check for the activity of cytochromes *f* and *b*, for example, would be to measure the difference spectra in the Soret region where absorbance changes caused by these substances are larger and farther apart from each other.

An experimental result which makes it unattractive to place plastocyanin as the electron donor for P700 was obtained already eight years ago by Müller and Witt (1961) who found that cytochrome *f* oxidation took place in spinach chloroplasts cooled to about 120°K. Also Chance and Bonner (1963) observed cytochrome *f* oxidation at 77°K in chloroplasts and leaves. Such a reaction is difficult to envisage if cytochrome *f* can only become oxidized by first reducing plastocyanin in a dark reaction. A charge-transfer complex involving both plastocyanin and cytochrome *f* would seem unlikely also.

4.4 Absorbance Changes of Chlorophyll b

Since its discovery some fifteen years ago by Duysens (1954b), the increase of absorbance seen at about 515 to 520 nm upon illumination of green algae and higher plants has continued to provide a challenge to all who have studied it in an attempt to understand its origin. During this time a number of substances have been put forward as being responsible for this change: a plastoquinone complex, chlorophyll *a*, a carotenoid, and chlorophyll *b*. It is apparent now that part of the difficulty experienced in the understanding of this change arises from the fact that different compounds can produce a positive change of absorbance in this region upon illumination. In green algae and higher plants, chlorophyll *b* apparently produces a large positive change near 515 nm which is associated with two negative bands near 475 and 650 nm (Rumberg, 1964; Fork *et al.*, 1966; cf., however, Hiller, 1969). Recent experiments with *Scenedesmus* (Amesz *et al.*, unpublished) indicate that the latter band is accompanied by about an equally large positive band at somewhat longer wavelength, suggesting a light-induced shift of the red band of chlorophyll *b*. The effect is most clearly seen in the presence of PMS, which stimulates the effect and suppresses P700 changes.

Carotenoids also produce a positive light-induced absorbance increase near 520 nm. This change has been observed in all plants so far examined that do not contain chlorophyll *b* (with the exception of blue-green algae) and also in a higher plant mutant lacking chlorophyll *b* (Fork and Amesz, 1967a, 1968b). It is reasonable to assume that carotenoids also contribute to some extent to absorbance changes near 520 nm in plants containing chlorophyll *b* as well.

Interpretations as to which photochemical system produces the 515-nm change have varied from system 2 alone (Witt *et al.*, 1965) to system 1 alone (Kok *et al.*, 1963; Rubinstein and Rabinowitch, 1964; Pratt and Bishop, 1968) to both systems 1 and 2 (Govindjee and Govindjee, 1965; Fork *et al.*, 1966; Schliephake *et al.*, 1968). No evidence has so far been found to show an antagonistic effect of red or far-red light on the 515-nm change (Fork and de Kouchkovsky, 1966). DCMU [3-(3,4-dichlorophenyl)-1,1-dimethylurea] produces only a partial inhibition of the 515-nm change (Fork and de Kouchkovsky, 1964, 1966; Govindjee and Govindjee, 1965; Fork and Urbach, 1966; Rubin, 1967). The difference spectrum in DCMU is like that seen before poisoning (Fork *et al.*, 1966).

At moderately high intensities of light this inhibitor eliminates the slow light-induced increase at 515 nm in intact algae without inhibiting the fast initial rise. The action spectrum for the latter showed system 1 to be active (Fork *et al.*, 1966). Pratt and Bishop (1968) reasoned that the slow phase of the change was produced by system 1 also and was a reflection of and dependent upon a flow of electrons provided from water by system 2. The partial regeneration in the presence of DCMU of the slow phase of the 515-nm change by the electron donor couple 2,3,5,6-tetramethyl-*p*-*phenylenediamine* (DAD) and ascorbate (Fork and Urbach, 1965) was seen by Pratt and Bishop as evidence that system 1 mediated the entire change including the slow phase. However, the action spectra of Pratt and Bishop (1968) were measured with dense suspensions and low actinic light intensity. The low-intensity incident actinic light is absorbed in the upper layers of the cell suspension and decreases rapidly in intensity as it traverses the suspension. Since it is known that the system-1 515-nm change is produced by low-intensity light while that caused by system 2 requires higher intensities, (Coleman and Rabinowitch, 1959; Govindjee and Govindjee, 1965; Rubinstein, 1965; Fork *et al.*, 1966) it is not unexpected that an action spectrum measured under these conditions shows only system 1 participation.

There is no doubt that under appropriate conditions a portion of the "slow" phase of the 515-nm change is brought about upon excitation of system 1, but system 2 is also responsible for the production of a component of this change. System 2 was shown to be responsible for the sensitization of the change in chloroplasts in the presence of ferricyanide. Ferricyanide oxidized P700 thereby rendering system 1 inoperative, and the action spectrum for the change clearly indicated system 2 participation (Müller *et al.*, 1963). This experiment was repeated "biologically" in the green alga *Ulva*. Here, the action spectrum for the slow change (which had a difference spectrum with a minimum at 475 and a maximum at 515 nm) showed system 2 participation when measured on a background of far-red (system 1) light (Fork *et al.*, 1966). A *Chlamydomonas* mutant without observable P700 showed an absorption increase at 515 nm in red (system 2) but not in far-red light (Chua and Levine, 1969).

De Kouchkovsky (1969) has recently made a parallel study of variable fluorescence (which originates in system 2 and is due to reduction of Q, see Section 4.2) and of the absorbance change at 515 nm. There was some correspondence between the 515-nm change and fluorescence induction. The slow fluorescence decrease in the light (P-S transition of Lavorel, 1959) was paralleled by similar variations of the 515-nm change in the light. Also, the dark regeneration curves for the slow increase of fluorescence in the light (Lavorel's P level) and the slow absorbance increase at 520 nm were parallel. However, there was not strict parallelism between the induction curves of fluorescence and the absorbance change. This is what would be expected if the 515-nm change is related to functioning of both systems 1 and 2.

Upon excitation by a short, intense laser flash, the change of absorbance at 515 nm is very fast (Hildreth *et al.*, 1966; Witt, 1967). Junge and Witt (1968) suggested that this very fast ($<2 \times 10^{-8}$ sec) light-induced absorbancy change indicates the formation of an electrical field across the thylakoid membrane, which somehow is supposed to cause a change in the absorption of chlorophyll *b*, followed by a translocation of H^+ ions. The dark decay of the change was said to be a reflection of the outward movement of hydrogen ions through the thylakoid membrane. This outward transfer of protons may or may not be coupled to phosphorylation.

Analysis showed that the biphasic decay of the 515-nm change after a brief flash of light was caused by a fast and a slow, first-order reaction (Junge and Witt, 1968). Rumberg and Siggel (1968) found that chloroplasts prepared according to the procedure of Winget *et al.* (1965) exhibited only the slow decay with a half-life from 20 to 500 msec, the fast component being practically absent. A linear relationship was found by Junge and Witt (1968) between the rate of phosphorylation and the amplitude of the slow absorbance change in spinach chloroplasts. The bigger the change the faster was the rate of phosphorylation. By providing phosphorylation conditions Rumberg and Siggel (1968) observed an acceleration of the decay of the 515-nm absorbance change up to two times from 90 to 40 msec accompanied by an acceleration of electron transport by a factor of up to two.

Another cause of increase of the dark decay was thought to be increased loss of H^+ through damaged or "leaky" membranes. This damage could be produced in a nonspecific way by subjecting chloroplast suspensions to hyper- or hypotonic sucrose solutions in which case an irreversible acceleration of the decay of the 515-nm change was produced. The thylakoid membranes were apparently affected more specifically by uncouplers such as methylamine hydrochloride and by the antibiotic gramicidin D. At an extremely low concentration (10^{-10} M) the latter substance already gave an acceleration of the decay in the presence of alkali ions. Gramicidin A has been shown by Chappell and Crofts (1966) to increase the permeability *in vivo* of the membranes of mitochondria and erythrocytes for alkali ions.

The 515-nm change (field formation), plastoquinone reduction and pH changes (proton transfer) were measured for chloroplasts in continuous far-red background light as a function of the time of illumination.

With flash times up to 20 msec (which reduced 80% of the active plastoquinone) a strict correspondence was observed between the size of the 515 and the pH changes and the amount of plastoquinone reduced (Reinwald *et al.*, 1968, see also Section 4.6). Schliephake *et al.* (1968) used the method of Chance and Mela (1966) to follow proton uptake in a chloroplast suspension by observing absorption changes of the pH indicator bromthymol blue at 605 nm. Measurements of the number of protons taken up (absorbance changes at 605 nm) in relation to electrons transferred (P700 changes) gave values of about 2 in agreement with the value reported by Izawa and Hind (1967) but not those of Karlish and Avron (1967, 1968) who found values ranging from 2 to 6 protons for 1 electron transferred.

On the basis of this and other evidence, Witt and co-workers (Schliephake *et al.*, 1968) suggested a chemiosmotic model of the type proposed by Mitchell (1966) for electron transport and proton uptake whereby an electrical field is generated by each light reaction upon rapid transfer of one electron from a donor molecule at the inner side of the thylakoid membrane to an acceptor at the outer side of the membrane. H^+ uptake from the outer compartment and H^+ production in the inner compartment is then brought about by subsequent "dark" electron transfer reactions: reduction of plastoquinone and of NADP and oxidation of reduced plastoquinone (by oxidized cytochrome *f*) and of water respectively. This creates a proton gradient across the membrane which drives phosphorylation after a sufficiently high gradient has been built up. For a more detailed description we may refer to the original publications.

4.5 Absorbance Changes of Carotenoids

As was mentioned in Section 4.4, carotenoids contribute to the light-induced absorption changes in many plants examined thus far (except in the blue-green algae). In purple bacteria, light-induced shifts in the absorption spectrum of carotenoids have already been known for a number of years (Smith and Ramírez, 1960; Nishimura and Chance, 1963; Olson and Morita, 1963; Vredenberg *et al.*, 1955; Amesz and Vredenberg, 1966).

The absorption spectra of typical carotenoids have three maxima separated by about 30 nm (Zechmeister, 1934; Goodwin, 1954). A shift of such a spectrum by a few nm to longer wavelengths yields a difference spectrum containing alternately three negative and three positive peaks. The positive maxima will be separated from each other by about 30 nm as will the negative peaks. On the long-wavelength side, the last positive peak of such a difference spectrum will show a gradually-decreasing portion reflecting the gradual decrease of absorption in the "tails" of the absorption bands of the original and shifted spectrum respectively (see Fig. 6).

Difference spectra which represent a shift to longer wavelength of the absorption bands of a carotenoid can be seen clearly in bacteria in the region from 450 to 550 nm because interference by other absorption changes is small. In higher plants and green algae most commonly used for study of photosynthesis, the chlorophyll *b* absorption changes at 515 and 475 nm obscure changes produced by carotenoids. Spectra which appear to represent a light-induced shift in the absorption spectrum of carotenoids have only recently been observed (Fork and Amesz, 1967a, 1968b; Fork, 1969a,b) in other types of photosynthetic organisms besides the bacteria.

The plants in which these carotenoid shifts were seen did not contain chlorophyll *b*. The shifts have so far been observed in red algae, brown algae, yellow-green algae (Xanthophyceae), and in a barley mutant lacking chlorophyll *b*. In red algae the carotenoid change produced by red light (absorbed largely by system 1) was qualitatively different from that produced by orange light (622 nm) absorbed by

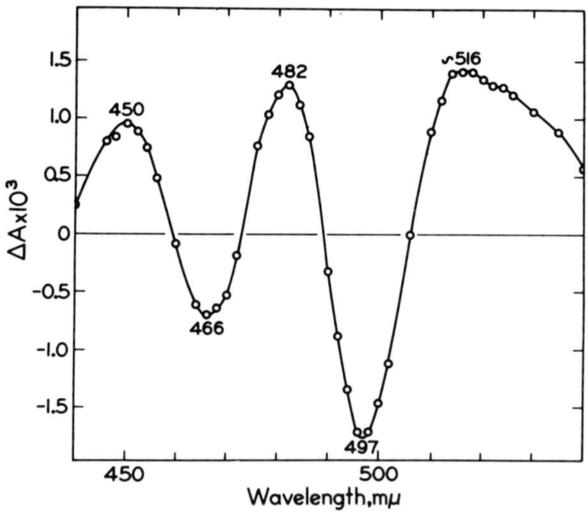

FIG. 6. Difference spectrum for the magnitude of the initial absorbance change produced in *Botrydiopsis alpina* in 10^{-5} M DCMU upon exposure to red light (Fork, 1969b). The number and position of the positive and negative bands indicate that the spectrum is produced by a shift towards longer wavelength of the absorption spectrum of a carotenoid with maxima near 445, 472, and 505 nm, as discussed in the text.

system 2 (phycocyanin). Excitation of system 1 in these algae produced a rapid change followed by a slower reversal; excitation of system 2 gave a sustained shift during illumination. Poisoning red algae with DCMU did not affect the rapid transient (red shift and reversal) of the carotenoid change. Measurements showed that the relative efficiencies of red and orange light in bringing about the carotenoid shift in the red alga *Iridaea splendens*, treated with DCMU, were the same as for cytochrome oxidation—a well-known system-1 reaction.

Similar results were obtained with *Botrydiopsis alpina* (Xanthophyceae). A difference spectrum for this alga is shown in Fig. 6. Action spectra measurements (Fork, 1969a,b) showed that both system 1 and 2 are responsible for the production of light-induced carotenoid shifts. A transient carotenoid shift persisted after addition of DCMU in *B. alpina* similar to what was observed with red algae. The action spectrum for this shift showed system 1 to be responsible. Addition of DAD and ascorbate to DCMU-treated *Botrydiopsis* regenerated a slow phase of the carotenoid change (similarly as for the chlorophyll *b* change described earlier) and the action spectrum showed that this slow change was also produced by system 1. A slow carotenoid change is produced by system 2 as well (similar to the situation observed for the chlorophyll *b* changes described in Section 4.4).

So far, there is no evidence to suggest that different carotenoids produce the shifts which occur when system 1 or system 2 is excited. Difference spectra for the carotenoid shift were quite similar in the red alga *Iridaea* and in *Botrydiopsis* and had maxima at or near 450, 482 and 515 nm and minima at 467 and 497 nm. But a barley mutant lacking chlorophyll *b* had a difference spectrum with minima at 457 and 493 and maxima at 447, 475 and 512 nm suggesting participation of a different carotenoid.

4.6 Plastoquinone

Measurements in the ultraviolet region are required in order to follow absorption changes produced by NADP and plastoquinones. Experimentally, such measurements are more difficult to make than those in the visible region.

Ultraviolet absorption changes were seen in spinach chloroplasts by Klingenberg *et al.* (1962). The difference spectrum for these changes showed some correspondence to the curve obtained upon reduction of plastoquinone *in vitro*. Moreover, Rumberg *et al.* (1964) demonstrated for *Chlorella* that red light led to a decrease of absorption at 254 nm (reduction of plastoquinone) while far-red light produced the opposite effect.

Amesz (1964b) measured absorbance changes in the blue-green alga *Anacystis nidulans* in the region from about 240 to 300 nm. The difference spectrum for these changes matched fairly closely that of oxidized-minus-reduced plastoquinone. Red light (system 1) was found to oxidize while system-2 light reduced this quinone-like substance. Both the reduction and oxidation proceeded with a high quantum yield (about 0.5). Kinetic studies (Amesz and Vredenberg, 1967) indicated a correlation between the oxidation-reduction level of the quinone and other components of the electron-transport chain. In weak light, rapid accumulation of oxidized P700 or cytochrome *f* was found to occur only when the *Anacystis* plastoquinone had been brought in the oxidized state by, e.g., prior illumination with system-1 light. Otherwise a lag was observed. The above observations indicate that the quinone functions in the main electron transport chain between the two light reactions. This quinone is apparently not acting as the primary oxidant of system 2 (Q, Section 4.2) because DCMU inhibited its reduction but not its oxidation. The quencher Q is reduced by system 2 but is not oxidized by system 1 in the presence of DCMU. Other evidence (Amesz and Fork, 1967a), however, does suggest that a different quinone-like substance may act as the primary reductant for system 2.

Difference spectra in the ultraviolet obtained by an averaging technique were recently reported by Stiehl and Witt (1968) for spinach chloroplasts and *Chlorella*. The "cleanest" spectra were obtained with chloroplasts. The spectrum for *Chlorella* was complicated by other changes as was seen earlier by Amesz (1964b). In chloroplasts a maximum was seen near 233 nm and a minimum near 263 nm. After extraction of the chloroplasts with petroleum ether only about 30% of the original quinone change persisted. The change was restored upon addition of plastoquinone A.

The ratio of the amount of plastoquinone to P700 was found by Schmidt-Mende and Rumberg (1968) to be 5 in spinach chloroplast fragments. Since each of these plastoquinones accepts 2 electrons upon reduction, the plastoquinone pool has a capacity of 10 electrons. The existence of a pool somewhere between the two light reactions had been earlier demonstrated in experiments with chloroplasts to which no electron acceptors had been added (de Kouchkovsky, 1963; Fork, 1963). Upon illumination an O_2 burst was seen which was sensitized by system 2 (Fork, 1963). Repeated illuminations gave successively smaller O_2 bursts. Far-red light was more effective in restoring the burst than was a dark period (Fork, 1963). A value of about 1 O_2/150 Chl or about 1 eq/40 Chl was found by de Kouchkovsky (1963). Since hexane-extracted chloroplasts no longer showed the O_2 burst upon illumination, de

Kouchkovsky and Briantais (1963) reasoned that the pool might be plastoquinone.

Schmidt-Mende and Witt (1968) observed in chloroplasts to which no electron acceptors were added that the plastoquinone reduced in a flash (about 0.1 sec) of red light (620 to 720 nm) remained reduced in the dark. The dark period in these experiments was of the order of 1 sec. Addition of benzylviologen caused a reoxidation of the plastoquinone reduced in the light of only about 50%. It was concluded that 5 of the 10 electrons from reduced plastoquinone were used to reduce those intermediates that became oxidized by system 1 upon reduction of benzylviologen. P700 and cytochrome *f* account for two of these electrons. The remaining three were thought to reduce another intermediate(s) of the electron-transport chain (plastocyanin ?). In the presence of both benzylviologen and continuous far-red background light the plastoquinone reduced in the light was reoxidized completely since excitation of system 1 kept all intermediates oxidized. About 1 msec was found to be needed for the light-induced transfer of one electron to plastoquinone. In agreement with this, difference spectra (Stiehl and Witt, 1968) showed that only a tenth as much plastoquinone was reduced in short flashes as in long flashes demonstrating in a different way the 10-electron capacity of the plastoquinone pool.

Quantitatively different conclusions were obtained from measurements of fluorescence increases brought about by short, saturating light flashes and of DCIP (2,6-dichlorophenol-indophenol) reduction (Forbush and Kok, 1968). The results indicated the existence of two pools of oxidant between system 2 and 1, with an electron-accepting capacity of about 6 and 12 respectively.

References

Allen, M. B. (1964). *In* "Photophysiology" (A. C. Giese, ed.), Vol. 1, p. 83. Academic Press, New York.

Amesz, J. (1963). *Biochim. Biophys. Acta* **66**, 22.

Amesz, J. (1964a). "Intracellular Reactions of Nicotinamide-adenine-dinucleotide in Photosynthetic Organisms." Thesis, University of Leiden.

Amesz, J. (1964b). *Biochim. Biophys. Acta* **79**, 257.

Amesz, J., and Fork, D. C. (1967a). *Biochim. Biophys. Acta* **143**, 97.

Amesz, J., and Fork, D. C. (1967b). *Photochem. Photobiol.* **6**, 903.

Amesz, J., and Vredenberg, W. J. (1966). *In* "Currents in Photosynthesis" (J. B. Thomas and J. C. Goedheer, eds.), p. 75. Ad. Donker, Rotterdam.

Amesz, J., and Vredenberg, W. J. (1967). *In* "Biochemistry of Chloroplasts" (T. W. Goodwin, ed.), Vol. II, p. 593. Academic Press, New York.

Amesz, J., Duysens, L. N. M., and Brandt, D. C. (1961). *J. Theoret. Biol.* **1**, 59.

Amesz, J., Fork, D. C., and Nooteboom, W. (1967). *Studia Biophysica* (Berlin) **5**, **175**.

Amesz, J., Nooteboom, W., Dirks, M. P., and Visser, J. W. M. (Unpublished).

Anderson, J. M., Fork, D. C., and Amesz, J. (1966). *Biochem. Biophys. Res. Commun.* **23,** 874.

Arnold, W., and Clayton, R. K. (1960). *Proc. Natl. Acad. Sci. U. S.* **46,** 796.

Beugeling, T. (1968). *Biochim. Biophys. Acta* **153,** 143.

Beugeling, T., and Duysens, L. N. M. (1965). *In* "Currents in Photosynthesis" (J. B. Thomas and J. C. Goedheer, eds.), p. 49. Ad. Donker, Rotterdam.

Butler, W. L. (1961). *Arch. Biochem. Biophys.* **93,** 413.

Butler, W. L. (1964). *Ann. Rev. Plant Physiol.* **15,** 451.

Chance, B. (1951). *Rev. Sci. Instr.* **22,** 619.

Chance, B., Kihara, T., DeVault, D., Hildreth, W., Nishimura, M., and Hiyama, T. (1969). *In* "Progress in Photosynthesis Research" (H. Metzner, ed.), **III,** 1321, *Proc. Intern. Congr. Photosynthesis Res.,* Freudenstadt, Germany.

Chance, B., and Bonner, W. D., Jr. (1963). *In* "Photosynthetic Mechanisms of Green Plants" *Natl. Acad. Sci. Natl. Res. Council Publ.* **1145,** p. 66. Washington, D. C.

Chance, B., and Mela, L. (1966). *Proc. Natl. Acad. Sci. U. S.* **55,** 1243.

Chappell, J. B., and Crofts, A. B. (1966). *In* "Regulation of Metabolic Processes in Mitochondria" (J. M. Tager, S. Papa, E. Quagliariello, and E. C. Slater, eds.), p. 293. Elsevier, Amsterdam.

Chua, N.-H., and Levine, R. P. (1969). *Plant Physiol.* **44,** 1.

Clayton, R. K. (1962). *Biochem. Biophys. Res. Commum.* **9,** 49.

Clayton, R. K. (1963). *Proc. Natl. Acad. Sci. U. S.* **46,** 769.

Clayton, R. K. (1965). "Molecular Physics in Photosynthesis." Random House (Blaisdell), New York.

Clayton, R. K. (1966a). *Photochem. Photobiol.* **5,** 669.

Clayton, R. K. (1966b). *Photochem. Photobiol.* **5,** 679.

Clayton, R. K. (1967). *J. Theoret. Biol.* **14,** 173.

Coleman, J., and Rabinowitch, E. (1959). *J. Phys. Chem.* **63,** 30.

Cusanovitch, M. A., Bartsch, L. G., and Kamen, M. D. (1968). *Biochim. Biophys. Acta* **153,** 397.

DeVault, D., and Chance, B. (1966). *Biophys. J.* **6,** 825.

DeVault, D., Parkes, J. H., and Chance, B. (1967). *Nature* **215,** 642.

Döring, G., Stiehl, H. H., and Witt, H. T. (1967). *Z. Naturforsch.* **22b,** 639.

Döring, G. Bailey, J. L., Kreutz, W., Weikard, J., and Witt, H. T. (1968a) *Naturwissenschaften* **5,** 219.

Döring, G., Bailey, J. L., Kreutz, W., and Witt, H. T. (1968b). *Naturwissenschaften* **5,** 220.

Duysens, L. N. M. (1952). "Transfer of Excitation Energy in Photosynthesis." Thesis. Univ. Utrecht.

Duysens, L. N. M. (1954a). *Nature* **173,** 692.

Duysens, L. N. M. (1954b). *Science* **120,** 353.

Duysens, L. N. M. (1956). *Biochim. Biophys. Acta* **19,** 1.

Duysens, L. N. M. (1957). *In* "Research in Photosynthesis" (H. Gaffron *et al.,* eds.), p. 164. Wiley (Interscience), New York.

Duysens, L. N. M. (1964). *Progress Biophys. Mol. Biol.* **14,** 1.

Duysens, L. N. M. (1965). *In* "Currents in Photosynthesis" (J. B. Thomas and J. C. Goedheer, eds.), p. 263. Ad. Donker, Rotterdam.

Duysens, L. N. M., and Amesz, J. (1962). *Biochim. Biophys. Acta* **64,** 243.

Duysens, L. N. M., and Sweep, G. (1957). *Biochim. Biophys. Acta* **25,** 13.

Duysens, L. N. M., and Sweers, H. E. (1963). *In* "Studies on Microalgae and

Photosynthetic Bacteria" (Japan. Soc. Plant Physiol., ed.). Univ. Tokyo Press, Tokyo.
Fleischman, D. E., and Clayton, R. K. (1968). *Photochem. Photobiol.* **8,** 287.
Forbush, B., and Kok, B. (1968). *Biochim. Biophys. Acta* **162,** 243.
Fork, D. C. (1963). *Plant Physiol.* **38,** 323.
Fork, D. C. (1969a). *In* "Progress in Photosynthesis Research" (H. Metzner, ed.), **II,** 800, *Proc. Intern. Congr. Photosynthesis Res.,* Freudenstadt, Germany.
Fork, D. C. (1969b). *Carnegie Inst. Yearbook* **67,** 496.
Fork, D. C., and Amesz, J. (1967a). *Photochem. Photobiol.* **6,** 913.
Fork, D. C., and Amesz, J. (1967b). *Biochim. Biophys. Acta* **143,** 266.
Fork, D. C., and Amesz, J. (1968a). *Carnegie Inst. Yearbook* **66,** 155.
Fork, D. C., and Amesz, J. (1968b). *Carnegie Inst. Yearbook* **66,** 160.
Fork, D. C., and Amesz, J. (1969). *Ann. Rev. Plant Physiol.* **20,** 305.
Fork, D. C., and de Kouchkovsky, Y. (1964). *Carnegie Inst. Yearbook* **63,** 441.
Fork, D. C., and de Kouchkovsky, Y. (1966). *Photochem. Photobiol.* **5,** 609.
Fork, D. C., and de Kouchkovsky, Y. (1968). *Biochim. Biophys. Acta* **153,** 891.
Fork, D. C., and Urbach, W. (1965). *Proc. Natl. Acad. Sci. U. S.* **53,** 1307.
Fork, D. C., and Urbach, W. (1966). *In* "Currents in Photosynthesis" (J. B. Thomas and J. C. Goedheer, eds.), p. 293. Ad. Donker, Rotterdam.
Fork, D. C., Amesz, J., and Anderson, J. M. (1966). *In* "Energy Conversion by the Photosynthetic Apparatus." *Brookhaven Symp. Biol.* **19,** 81.
French, C. S. (1955). *In* "The Luminescence of Biological Systems" (F. H. Johnson, ed.), p. 51. Am. Assoc. Advanc. Sci., Washington, D. C.
Gingras, G., and Jolchine, G. (1969). *In* "Progress in Photosynthesis Research" (H. Metzner, ed.), **I,** 209, *Proc. Intern. Congr. Photosynthesis Res.,* Freudenstadt, Germany.
Goodwin, T. W. (1954). "The Comparative Biochemistry of the Carotenoids." Chem. Publ. Co., New York.
Gorman, D. S., and Levine, R. P. (1965). *Proc. Natl. Acad. Sci. U. S.* **54,** 1665.
Gorman, D. S., and Levine, R. P. (1966). *Plant Physiol.* **41,** 1648.
Govindjee, and Govindjee, R. (1965). *Photochem. Photobiol.* **4,** 793.
Guilbault, G. G., ed. (1967). "Fluorescence; Theory, Instrumentation, and Practice," Marcel Dekker, New York.
Hercules, D. M., ed. (1966). "Fluorescence and Phosphorescence Analysis." Wiley (Interscience), New York.
Hildreth, W. W., Avron, M., and Chance, B. (1966). *Plant Physiol.* **41,** 983.
Hiller, R. G. (1969). *Biochim. Biophys. Acta* **172,** 546.
Hind, G., and Olson, J. M. (1968). *Ann. Rev. Plant Physiol.* **19,** 249.
Holt, A. S., and Clayton, R. K. (1965). *Photochem. Photobiol.* **4,** 829.
Izawa, S., and Hind, G. (1967). *Biochim. Biophys. Acta* **143,** 377.
Jackson, J. B., and Crofts, A. R. (1968). *Biochem. Biophys. Res. Commun.* **32,** 908.
Jensen, A., Aasmundrud, O., and Eimhjellen, K. E. (1964). *Biochim. Biophys. Acta* **88,** 466.
Joliot, P. (1968). *Photochem. Photobiol.* **8,** 451.
Junge, W., and Witt, H. T. (1968). *Z. Naturforsch.* **23b,** 244.
Karlish, S. J. D., and Avron, M. (1967). *Nature* **216,** 1107.
Karlish, S. J. D., and Avron, M. (1968). *Biochim. Biophys. Acta* **153,** 879.
Katoh, S. (1960). *Nature* **186,** 533.

Katoh, S., Suga, I., Shiratori, I., and Takamiya, A. (1961). *Arch. Biochem. Biophys.* **94**, 136.
Katoh, S., Shiratori, I., and Takamiya, A. (1962). *J. Biochem.* **51**, 32.
Ke, B. (1969). *Biochim. Biophys. Acta* **172**, 583.
Ke, B., Treharne, R. W., and McKibben, C. (1964). *Rev. Sci. Instr.* **35**, 296.
Ke, B., Vernon, L. P., Garcia, A. F., and Ngo, E. (1967). *Biochem.* **7**, 311.
Keister, D. L., and Yike, N. J. (1967). *Arch. Biochem. Biophys.* **121**, 415.
Klemme, J. H. (1969). *Z. Naturforsch.* **24b**, 67.
Klingenberg, M., Müller, A., Schmidt-Mende, P., and Witt, H. T. (1962). *Nature* **194**, 379.
Kok, B. (1956). *Biochim. Biophys. Acta* **22**, 399.
Kok, B. (1957). *Acta Botan. Neerl.* **6**, 316.
Kok, B. (1959). *Plant Physiol.* **34**, 184.
Kok, B. (1961). *Biochim. Biophys. Acta* **48**, 527.
Kok, B. (1963). *In* "Photosynthetic Mechanisms of Green Plants," *Natl. Acad. Sci. Natl. Res. Council, Publ.* **1145**, 45. Washington, D. C.
Kok, B., and Hoch, G. (1961). *In* "Light and Life" (W. D. McElroy and B. Glass, eds.), p. 379. Johns Hopkins Press, Baltimore, Maryland.
Kok, B., Cooper, B., and Yang, L. (1963). In "Studies on Microalgae and Photosynthetic Bacteria" (Japan. Soc. Plant Physiol., ed.) p. 373. Univ. Tokyo Press, Tokyo.
de Kouchkovsky, Y. (1963). "Induction Photosynthétique des Chloroplasts Isolés" Thesis, Univ. Paris (Orsay).
de Kouchkovsky, Y. (1969). *In* "Progress in Photosynthesis Research" (H. Metzner, ed.), **II**, 959, *Proc. Intern. Congr. Photosynthesis Res.*, Freudenstadt, Germany.
de Kouchkovsky, Y., and Briantais, J. M. (1963). *In* "Photosynthetic Mechanisms of Green Plants," *Natl. Acad. Sci. Natl. Res. Council Publ.* **1145**, p. 362. Washington, D. C.
de Kouchkovsky, Y., and Fork, D. C. (1964). *Proc. Natl. Acad. Sci. U. S.* **52**, 232.
Kuntz, I. D., Loach, P. A., and Calvin, M. (1964). *Biophys. J.* **4**, 227.
Lavorel, J. (1959). *Plant Physiol.* **34**, 204.
Lavorel, J. (1965). *Photochem. Photobiol.* **4**, 819.
Levine, R. P., and Gorman, D. S. (1966). *Plant Physiol.* **41**, 1293.
Lightbody, J. J., and Krogmann, D. W. (1967). *Biochim. Biophys. Acta* **131**, 508.
Loach, P. A., Androes, G. M., Maksim, A. F., and Calvin, M. (1963). *Photochem. Photobiol.* **2**, 443.
Mitchell, P. (1966). *Biol. Rev.* **41**, 445.
Morita, S. (1968). *Biochim. Biophys. Acta.* **153**, 241.
Müller, A., and Witt, H. T. (1961). *Nature* **189**, 944.
Müller, A., Fork, D. C., and Witt, H. T. (1963). *Z. Naturforsch.* **18b**, 142.
Nishimura, M., and Chance B. (1963). *Biochim. Biophys. Acta* **66**, 1.
Olson, J. M., and Amesz, J. (1960). *Biochim. Biophys. Acta* **37**, 14.
Olson, J. M., and Chance, B. (1960a). *Arch. Biochem. Biophys.* **88**, 26.
Olson, J. M., and Chance, B. (1960b). *Arch. Biochem. Biophys.* **88**, 40.
Olson, J. M., and Kok, B. (1959). *Biochem. Biophys. Acta* **32**, 278.
Olson, J. M., and Morita, S. (1963). *In* "Bacterial Photosynthesis" (H. Gest, A. San Pietro, and L. P. Vernon, eds.), p. 433. Antioch Press, Yellow Springs, Ohio.
Parson, W. W. (1967). *Biochim. Biophys. Acta* **131**, 154.
Parson, W. W. (1968). *Biochim. Biophys. Acta* **153**, 248.
Pratt, L. H., and Bishop, N. I. (1968). *Biochim. Biophys. Acta* **162**, 369.

Reed, D. W., and Clayton, R. K. (1968). *Biochem. Biophys. Res. Commun.* **30,** 471.

Reinwald, E., Stiehl, H. H., and Rumberg, B. (1968). *Z. Naturforsch.* **23b,** 1616.

Rubin, L. (1967). *Biol. Nauki* **3,** 60.

Rubinstein, D. (1965). *Biochim. Biophys. Acta* **109,** 41.

Rubinstein, D., and Rabinowitch, E. (1964). *Biophys. J.* **2,** 107.

Rumberg, B. (1964). *Nature* **204,** 860.

Rumberg, B., and Siggel, U. (1968). *Z. Naturforsch.* **23b,** 239.

Rumberg, B., Schmidt-Mende, P., and Witt, H. T. (1964). *Nature* **201,** 466.

Rüppel, H., Bültemann, V., and Witt, H. T. (1964). *Ber. Bunsengesellsch. Physik. Chem.* **68,** 340.

Sauer, K., and Dratz, E. A. (1969). *In* "Progress in Photosynthesis Research" (H. Metzner, ed.), **II,** 837, *Proc. Intern. Congr. Photosynthesis Res.,* Freudenstadt, Germany.

Schliephake, W., Junge, W., and Witt, H. T. (1968). *Z. Naturforsch.* **23b,** 1571.

Schmidt-Mende, P., and Rumberg, B. (1968). *Z. Naturforsch.* **23b,** 225.

Schmidt-Mende, P., and Witt, H. T. (1968). *Z. Naturforsch.* **23b,** 228.

Seely, G. R. (1968). *In* "Photophysiology" (A. C. Giese, ed.), Vol. III, p. 1. Academic Press, New York.

Smith, L., and Ramírez, J. (1960). *J. Biol. Chem.* **235,** 219.

Stiehl, H. H., and Witt, H. T. (1968). *Z. Naturforsch.* **23b,** 220.

Sybesma, C. (1969). *Biochim. Biophys. Acta* **172,** 177.

Sybesma, C., and Fowler, C. F. (1968). *Proc. Nat. Acad. Sci.* **61,** 1343.

Sybesma, C., and Vredenberg, W. (1963). *Biochim. Biophys. Acta* **75,** 439.

Thornber, J. P., Olson, J. M., Williams, D. M., and Clayton, M. C. (1969). *Biochim. Biophys. Acta* **172,** 351.

Udenfriend, S. (1962). "Fluorescence Assay in Biology and Medicine." Academic Press, New York.

Vernon, L. P. (1964). *Ann. Rev. Plant Physiol.* **15,** 73.

Vernon, L. P. (1968). *Bacteriol. Rev.* **32,** 243.

Vernon, L. P., and Ke, B. (1966). *In* "The Chlorophylls" (L. P. Vernon and G. R. Seely, eds.), p. 569. Academic Press, New York.

Vernon, L. P., Shaw, E. R., and Ke, B. (1966). *J. Biol. Chem.* **241,** 4101.

Virgin, H. I. (1954). *Physiol. Plantarum* **7,** 560.

Vredenberg, W. J. (1965). "Spectrophotometric Studies on Primary and Associated Reactions in Photosynthesis." Thesis, Univ. of Leiden.

Vredenberg, W. J., and Amesz, J. (1966a). *Biochim. Biophys. Acta* **126,** 244.

Vredenberg, W. J., and Amesz, J. (1966b). *In* "Energy Conversion by the Photosynthetic Apparatus." *Brookhaven Symp. Biol.* **19,** 49.

Vredenberg, W. J., and Duysens, L. N. M. (1963). *Nature* **197,** 355.

Vredenberg, W. J., and Duysens, L. N. M. (1964). *Biochim. Biophys. Acta* **79,** 456.

Vredenberg, W. J., and Duysens, L. N. M. (1965). *Biochim. Biophys. Acta* **94,** 355.

Vredenberg, W. J., Amesz, J., and Duysens, L. N. M. (1965). *Biochem. Biophys. Res. Commun.* **18,** 435.

Winget, G. D., Izawa, S., and Good, N. E. (1965). *Biochem. Biophys. Res. Commun.* **21,** 438.

Witt, H. T. (1967). *In* "Fast Reactions and Primary Processes in Chemical Kinetics" (Nobel Symposium V) (S. Claesson, ed.), p. 261. Almqvist and Wiksell, Stockholm; Wiley (Interscience), New York.

Witt, H. T., Moraw, R., and Müller, A. (1959). *Z. Physik. Chem.* (N. S.) **23**, 133.
Witt, H. T., Rumberg, B., Schmidt-Mende, P., Siggel, U., Skerra, B., Vater, J., and Weikard, J. (1965). *Angew. Chem.* (Intern. Ed.) **4**, 799.
Yokum, C. F., and San Pietro, A. (1969). *Biochem. Biophys. Res. Commun.* **36**, 614.
Zechmeister, L. (1934). "Carotinoide." Springer Verlag, Berlin.

Chapter 5

THE PHOTOPHYSIOLOGY OF VERTEBRATE COLOR VISION

H. Ripps

Department of Ophthalmology, New York University School of Medicine, New York, New York

R. A. Weale

Department of Physiological Optics, Institute of Ophthalmology, University of London, London, United Kingdom

1. Introduction

We should stress at the outset that we wish to consider some peripheral mechanisms of the vertebrate nervous system which may be utilized for the discrimination of different wavelengths of the electromagnetic spectrum. Color sensations, although a topic worthy of consideration, are mentioned only in so far as they relate to the mechanisms under discussion. This account, then, represents a review of some recent work which has helped to elucidate our understanding of the processes involved in signalling color: cognitive and associational functions lie outside the province of our survey.

1.1 The Absorption-Response Relationship

Only radiation which is absorbed can be physiologically active; hence, only radiation which is absorbed can be detected. The elements transducing electromagnetic energy into charge displacements which in turn initiate the visual response are to be found in the retina: the rod and cone outer limbs contain light-absorbing, photo-sensitive substances collectively referred to as the visual pigments.

Rods and cones, however, are not to be compared in their action with photographic emulsions. While it is true that every incident quantum has a finite probability of being absorbed by a visual pigment molecule, it is not to be inferred from this that the physiological response R stands in any simple relation to the number of quanta absorbed (however, see Section 2.4). Nevertheless, it is permissible to say that the response at any wavelength λ is a function of the extinction coefficient E_λ of the pigment responsible for evoking it, and the number of incident light quanta N_λ:

$$R_\lambda = f(E \cdot N)_\lambda \tag{1}$$

Now even though E_λ may depend upon the spectral composition of the light, wavelength discrimination is not possible because the responses elicited at different wavelengths (R_{λ_1} and R_{λ_2}) can be made equal by a suitable adjustment of the relative values of N_λ. On the other hand, wavelength discrimination is attainable if R is a function not of one extinction coefficient, but of two or more operating simultaneously:

$$R_\lambda = f(E', E'', N)_\lambda \tag{2}$$

For if N is shared between pigments having different absorption properties, it is no longer possible to juggle with its values at $\lambda 1$ and $\lambda 2$ so as to equate R_{λ_1} and R_{λ_2}. Clearly, the responses must arise from receptors which contain different pigments or different mixtures of pigment; otherwise, Eq. (2) degenerates to (1) with self-evident results. It does not matter what type of receptor the two or more pigments are contained in, but there is abundant evidence that the cones contain the photosensitive pigments subserving vertebrate color vision.

2. The Visual Pigments

2.1 Receptor Structure and Molecular Orientation

Electron microscopic studies (Sjöstrand, 1953) show that rods and cones are similar in that their outer limbs consist of membranous folds or lamellae, but the extent to which the folds themselves are in contact with the plasma membrane is not necessarily the same in the two structures (Cohen, 1968). In addition, recent studies on the optical activity of isolated receptors have shown that there are significant subelectron-microscopic differences between rod and cone membranes (Weale, 1968b) which may affect their relative permeabilities to some substances. The visual pigment molecules represent an integral part of the lamellae (Dewey *et al.*, 1969) forming, according to some authors (cf. Sjöstrand, 1961; de Robertis and Lasansky, 1961), lipid-protein bilayers. However, the molecules of visual pigment are not distributed randomly within the lamellae, but are so organized that their dipole moments lie in planes perpendicular to the axis of the receptor. This molecular arrangement within the outer limb gives rise to dichroism, for it maximizes the probability of interaction between the visual pigment and the electric vector of radiation travelling in a direction parallel to the axis (Fig. 1). It follows from this diagram that the ratio of the probabilities of absorption in the two cases is 2 and that, were the molecules pointing randomly in all directions as they do in solution, the probability would be only ⅔ of that existing in situation (a).

The question of the organization of the molecules within the lamellae has received attention from several authors (Hagins and Jennings, 1959;

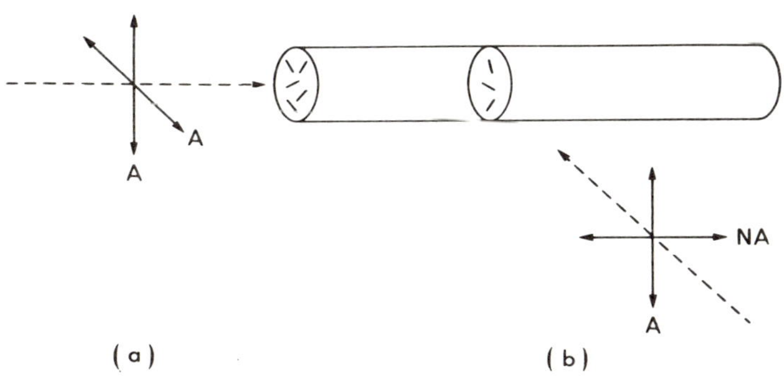

FIG. 1. (a) Light incident along the axis of a receptor is absorbed (A) independently of its plane of polarization. (b) When the direction of incidence is perpendicular to the axis, the component parallel to the axis is not absorbed (NA).

Liebman, 1962; Wald *et al.*, 1963; Pak and Helmrich, 1968) and it would appear that no dichroism can be conclusively demonstrated in their planes. However, as no one has so far studied individual lamellae, the existence of dichroism may be denied in lamellar planes for a receptor considered as a whole, but not yet for individual lamellae. The problem may appear to be only of academic interest in view of the fact that there is no evidence to show that vertebrate receptors can serve in the detection of polarized light (cf. Eakin, 1965; Waterman, 1954 for invertebrate structures). But it has recently been found (Weale, 1969a) that the presence of photo-sensitive pigment in a rod profoundly affects its birefringent properties (albeit in a direction perpendicular to the axis of the receptor). While this observation may be more relevant to the initiation of the visual response in general (Cohen *et al.*, 1968), it is premature to rule out any possible bearing it may have on wavelength discrimination if, as is likely, it also occurs in cones. Detailed consideration of this matter has to be deferred until cones have been studied, but it is clear that the method offers a mode of attack on pigment behavior essentially different from methods based on densitometry.

The latter have been applied during the last ten or fifteen years to the study of cone pigments in (*a*) the living eye, (*b*) isolated retinae, and (*c*) single cone receptors. The methods involved have been described fully (Rushton, 1965a,b,c; Carr and Ripps, 1967; Weale, 1965, 1968a; Ripps and Weale, 1963, 1964a; Brown and Wald, 1963, 1964; Liebman and Entine, 1964; Marks, 1965; Marks *et al.*, 1964) and there is little point in detailed repetition. However, in view of the claims made for them from time to time it may not be inapposite to consider some points which seem to have escaped attention, so that when these exciting results come to be incorporated into yet another theory of color vision, the basis on which this is done may not be unduly impugned.

2.2 Fundus Reflectometry of the Living Eye

The method of fundus reflectometry involves the principle of the ophthalmoscope with two important modifications: (*a*) the white light typically used for examining the eye is replaced by a series of monochromatic *test* lights, and (*b*) the examiner is replaced by a sensitive photo-electric device which measures accurately the fraction of the incident light that emerges from the subject's eye. Since the test beams traverse the retina both before and after reflection at the fundus oculi, the visual pigments, as well as other light-absorbing substances, leave their imprint on the spectral distribution of the emergent rays. But absorption by the visual pigments consitutes only a small fraction of the total light loss within the eye, and it would be impossible to extricate this information were it not for the fact that only the visual pigments are photo-labile; i.e., they alter their spectral properties when quanta are absorbed. Consequently, if a large number of visual pigment molecules are converted to relatively transparent photoproducts—the subjective connotation for this process is "bleaching"—a *measurable change* in retinal transmissivity will be effected. The intense bleaching exposure producing this effect is to be distinguished from the afore-mentioned test lights; the latter serve to indicate transmissivity changes across the spectrum, and are kept in reliable techniques as feeble as possible (Weale, 1962). Various methods for achieving satisfactory measurements have been evolved and can be studied in the above references. It should, nevertheless, be stressed again that the intensities required to achieve measurable changes during exposures lasting, say, 30 sec., are enormous (Fig. 2). The retinal irradiations are of the order of 10^6 photopic trolands,

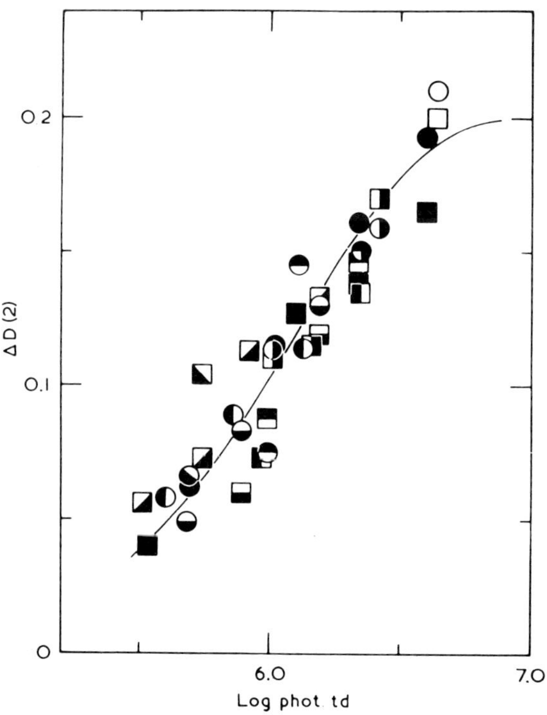

FIG. 2. Fundus reflectometry. Abscissa: retinal illumination in phot.td. Exposure = 30 sec. Ordinate: density change for double transit in the human fovea. The various symbols refer to equiluminant bleaching lights in various spectral ranges. Squares and circles are results for two different subjects after correction for photoreversal. The curve represents an exponential function. (Villermet and Weale, 1966.)

and correspond to those obtained from a bright summer sky when one's pupil is dilated.

The changes in retinal transmissivity (T) resulting from bleaching exposures are conveniently expressed in terms of density difference spectra, $\Delta D = f(\lambda)$:

$$\Delta D_\lambda = (D_D - D_L)_\lambda = \log_{10} (T_L/T_D)_\lambda$$

where subscripts $_D$ and $_L$ refer to dark- and light-adapted conditions, respectively, and λ is the wavelength of the test light. Although it has been shown (Weale, 1955a,b) that this approach may, in certain circumstances, represent an oversimplification, the relation applies with minimal error to measurements on the living human eye.

The fact of the matter is that the photocell may receive light other than that which has traversed the retinal receptors twice or even only once. Assume, for example, that it collects light reflected or scattered from one or more surfaces in front of the receptors, and let the fraction of the incident light I_{inc} so described be α_s. In addition, there are interstices between the retinal receptors, so that light may pass through the inter-receptor spaces; if the fractional cross section of retina covered by visual pigment is a, then that not covered by pigment is 1-a. Let the transmissivity of the dark-adapted retina be T_D, and let T_L and T_B represent the retinal transmissivities after partial and full bleaching exposures, respectively. The reflectivity of the fundus is R and the overall transmissivity of the ocular media, t. Then it follows that, for double transit through the eye, the measured intensity I of the emergent light is, for any test wavelength, given by:

$$I_D = I_{\text{inc}}[a(1 - \alpha_s)t^2T_D{}^2R + (1 - a)(1 - \alpha_s)t^2R + \alpha_s] \tag{3}$$

$$I_L = I_{\text{inc}}[a(1 - \alpha_s)t^2T_L{}^2R + (1 - a)(1 - \alpha_s)t^2R + \alpha_s] \tag{4}$$

$$I_B = I_{\text{inc}}[a(1 - \alpha_s)t^2T_B{}^2R + (1 - a)(1 - \alpha_s)t^2R + \alpha_s] \tag{5}$$

Dividing Eq. (3) by Eq. (4) we obtain the measured change in transmissivity (τ_m) produced by an exposure which only partially bleaches the visual pigments:

$$\frac{I_D}{I_L} = \tau_m = \frac{aT_D{}^2 + (1 - a) + \dfrac{\alpha_s}{(1 - \alpha_s)t^2R}}{aT_L{}^2 + (1 - a) + \dfrac{\alpha_s}{(1 - \alpha_s)t^2R}} \tag{6}$$

But $\Delta D_m = -\log_{10} \tau_m$.

Thus, setting $\dfrac{\alpha_s}{(1 - \alpha_s)t^2R} = \sigma$, we obtain:

$$\Delta D_m = \log_{10}\left[\frac{a(T_L^2 - 1) + \sigma + 1}{a(T_D^2 - 1) + \sigma + 1}\right] = \log_{10}\left[\frac{T_L^2 + \left(\frac{\sigma + 1}{a} - 1\right)}{T_D^2 + \left(\frac{\sigma + 1}{a} - 1\right)}\right] \tag{7}$$

Similar operations on Eqs. (3) and (5) give the density differences produced by a full bleach.

Since the actual double density change at the retinal level is log $[T_L^2/T_D^2]$ the measured density differences are affected by σ and a. Before discussing the extent to which these factors influence the measurements, let us consider another way in which the emergent rays I_D, I_L and I_B can be analyzed.

Rushton (1965a) has recently abandoned the density difference spectrum in favor of a "transmissivity difference spectrum" which, for a partial bleach, can be obtained by subtracting Eq. (4) — Eq. (3) and dividing by Eq. (5), the light emerging from the eye after the visual pigments have been bleached away *completely*. Thus,

$$\frac{I_L - I_D}{I_B} = \frac{a(T_L^2 - T_D^2)}{aT_B^2 + (1 - a) + \dfrac{\alpha_s}{(1 - \alpha_s)t^2R}} = \frac{T_L^2 - T_D^2}{(T_B^2 - 1) + \dfrac{\sigma + 1}{a}} \tag{8}$$

If it is assumed that photoproducts formed after a full bleach do not absorb in the visible part of the spectrum (i.e., $T_B \approx 1$), then Eq. (8) reduces to:

$$\frac{I_L - I_D}{I_B} = \frac{a(T_L^2 - T_D^2)}{\sigma + 1} \tag{9}$$

which is proportional to the change in absorptivity A, since

$$T_L^2 - T_D^2 = \Delta A$$

A comparison of Eqs. (7) and (9) is not without interest. It is readily seen that if $\sigma \to 0$ and $a \to 1$, the measured and receptoral density changes are equal, and that Eqs. (7) and (9) become essentially identical. For $a < 1$, and *in situ* densities ≤ 0.3, the measured values derived from both expressions will be proportionately reduced, although in practice a approaches unity. Both equations have the property of additivity—i.e., can operate on fractional bleaches—but Eq. (9) requires that values for a full bleach are also obtained. We shall have occasion to return to the topic of additivity in our discussion of the explanation of color defects in terms of visual pigments. However, the advantage of Eq. (9) over Eq. (7) is that if σ is independent of wavelength, and if stray light contributes a significant fraction of the light received by the photocell, the transmissivity spectrum given by Eq. (9) will not be distorted in shape because the denominator ($\sigma + 1$) acts merely as a scaling constant, whereas Eq. (7) may be significantly distorted by large values of σ. But it has to be stressed that what matters is the ratio

$$\sigma = \frac{\alpha_s}{(1 - \alpha_s)t^2R}, \text{ not } \alpha_s \text{ nor } R \text{ on its own,}$$

and σ can either be measured (Weale, 1955a,b; Rushton, 1965c) or minimized experimentally. However, σ is small when one deals with the human eye, which prompts one to ask why Rushton (1965a) felt impelled to *assume* that it was large, especially as he succeeded in showing by direct measurement that superficial stray light comprised only about 1 percent of the signal received from the eye (Rushton, 1965c). Thus, we may summarize our consideration of Eqs. (7) and (9) which have caused difficulties in the literature (cf. Alpern, 1968) with the observation that either function provides a reliable measure of the difference spectrum of human visual pigments.

2.2.1 Color Anomalies

The method of fundus reflectometry was varied in sundry ways in order to discover how many visual pigments are present in the human fovea, and there is little doubt that three can be detected there. One of them is maximally sensitive in a region ranging from 565 to 595 nm (Weale, 1959, 1968c; Rushton, 1958, 1965b; Ripps and Weale, 1963, 1965); as it is the most sensitive of the three to long wavelengths it is referred to as the red-sensitive pigment (R). The second, a green-sensitive substance (G), peaks at about 530–540 nm (Rushton, 1958, 1963); and the third (B) is believed to act mainly at 430–440 nm (Weale, 1968c) although the presence of products of bleaching which tend to absorb in this spectral region makes it difficult to be more specific as regards the $\lambda_{\max}$ of the blue-sensitive pigment. Rushton (1961), who claimed to have made—and then retracted (Rushton, 1962)—the discovery of (B), called it *cyanolabe* but Wald *et al.* (1953) had previously synthesized *cyanopsin* which absorbs largely at the other end of the spectrum. Neither curves nor pigments merit proper names, and in view of the confusion reigning over pigment nomenclature we shall use the symbols (R), (G), and (B) to designate the three cone pigments detectable in the human fovea.

The idea that the above "pigments" provide a basis for color vision rests on the following claims:

(1) Protanopes lack (R)—this explains their relative lack of sensitivity to light of long wavelengths.

(2) Deuteranopes lack (*G*)—this explains the shape of their foveal sensitivity curve and wavelength discrimination.

(3) The center of the normal fovea lacks (*B*)—this explains the relative hypo-sensitivity to blue light exhibited by this part of the retina.

Brown and Wald (1964) have criticized some of Rushton's evidence presented as a result of a classic series of experiments on the protanopic fovea. They correctly noted that, while the action spectrum of bleaching this fovea agrees with the subjective spectral sensitivity curve, Rushton's difference spectrum consists of six points, three of which provide a flat top, and the curve shows no sign of descent toward the short-wavelength end of the spectrum. Moreover, Rushton calculated that the density of (*G*), the only pigment he has been able to detect in the protanopic fovea, was 0.35. It has been pointed out (Weale, 1964) that if this were so, any agreement between the protanopic difference spectrum and the spectral sensitivity curve would not be expected on theoretical grounds. Nevertheless, Rushton may be said to have shown convincingly that following intense exposure, the protanopic density differences in the red part of the spectrum are much smaller than is true for normal foveae, and his data are not inconsistent with the notion that, at wavelengths longer than 550 nm, the protanopic fovea operates on essentially a single pigment, probably (*G*).

But the single pigment hypothesis is harder to buttress for this spectral range in the case of the deuteranopic fovea which Rushton (1965a,b) has studied. As the deuteranopic sensitivity function differs systematically from that of protanopes (Hsia and Graham, 1952), it is legitimate to expect a difference in the pigment contents of the two foveae. Originally Rushton (1958) had demonstrated in a preliminary paper that deuteranopic cones contain two cone pigments that were mixed. One of these was (*G*), the green-sensitive substance ($\lambda_{max} \approx 540$ nm) found in the protanope, and the other was (*R*), which Rushton then reported as absorbing maximally at 590 nm. The spectral position of the latter tended to oscillate in later preliminary publications, a circumstance which Ripps and Weale (1964b) attributed to variations in the intensity of the red bleaching light used. When the maximum finally settled down in Rushton's experiments (1965a) at 565 nm, he attributed its earlier oscillations to the effects of stray light; we have seen that these had been adumbrated ten years before (Weale, 1955a,b), and with reference to density difference spectra, Rushton (1958) found that "neither stray light nor absorption by photo-products enter as serious errors in the measurement of the difference spectrum." Nevertheless,

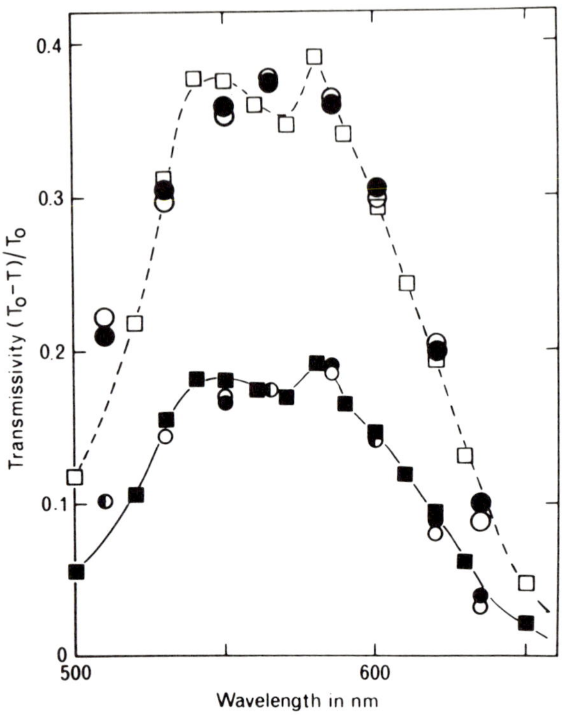

FIG. 3a. Small black circles: relative transmissivity differences when a deuteranopic fovea was bleached with red light; small white circles: ditto for blue-green light; large circles: corresponding data when the bleaches were completed with white light. Black and white squares: scaled values of Hsia and Graham's deuteranopic visibility function. (After Rushton, 1965a.)

Rushton believes that these difficulties are eliminated by plotting the data in terms of the transmissivity spectrum given by Eq. (9). We have already discussed the conditions wherein this is possible in theory, and Rushton's experimental data shall illustrate below how the story works out in practice.

There is clearly no need to measure actual intensities for a function such as that given by Eq. (9): intensity ratios can be expressed in terms of transmissivities for these represent ratios of intensities. Hence we can substitute in Eq. (9) by writing:

$$\frac{I_1 - I_2}{I_0} = \frac{T_1 - T_2}{T_0} \tag{10}$$

where T_0 represents a neutral wedge reading for the measuring beam when the fundus is fully bleached, and T_1 and T_2 are analogous readings obtained when the concentration of visual pigment assumes values designated by subscripts 1 and 2. Equation (10) is readily shown to be equal to the change in the fraction of light absorbed. As spectral sensitivity functions (S_λ) relate to pigment absorptivity, the success with which Rushton (1965a) matches Eq. (10) with S_λ provides ample support for the validity of this approach. In Fig. 3a the lower data represent the change in transmissivity following a partial bleach with deep red light, and, in a separate experiment, results for a partial blue-green bleach. The upper data show corresponding changes in transmissivity when the fractional chromatic bleaches were completed with white light. We have stressed that Eq. (9) and its derivatives are additive, and it is permissible, therefore, to substract the lower from the upper data. Rushton's claim is that since the changes obtained with red and blue-green lights, as well as those obtained after full bleaches, show similar spectral variations, the deuteranopic fovea can contain only

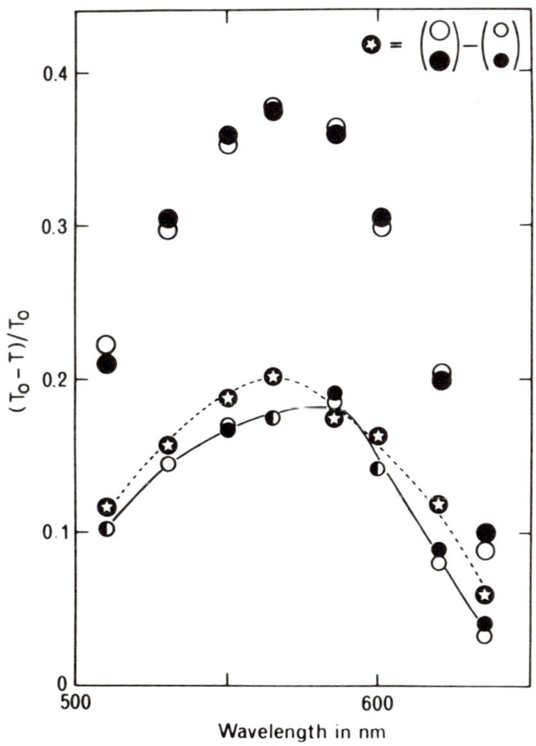

FIG. 3b. Circles: same as Fig. 3a; asterisks: the difference between the two sets of data. (Weale, 1969b.)

one pigment, namely the red-sensitive (apart from [*B*] which does not concern us in this context). However, it is plain from Fig. 3b that the differences between the upper and lower data follow a spectral function strikingly different from that due to the original chromatic bleaches (Weale, 1969b). Only two conclusions seem to us possible at this stage: either a white bleach following a fractional red or blue-green bleach of the deuteranopic fovea involves a pigment content remarkably different from that involved in the original bleach or, if the difference between the continuous and dotted curves in Fig. 3b should be deemed nonsignificant, the experimental error must be considered so large that the similarity between the "red" and "blue-green" difference spectra (lower data) is wholly fortuitous. The care which Rushton appears to have bestowed on the design of his apparatus and the apparent smallness of his experimental error in this study lend little support to the latter alternative. We are, of course, aware that Wald and Brown's (1965) studies and the recent studies of Alpern *et al.* (1968) stress the likelihood that deuteranopes have only one spectrally central pigment just as Pitt's (1935) data suggest that they have two: nevertheless the evidence obtained on this point by fundus reflectometry is so far not only inconclusive but also inconsistent.

2.3 Microspectrophotometry of Single Receptors

Although one of the obvious virtues of fundus reflectometry is that it affords a means for examining the visual pigments in their natural state, it suffers the disadvantage of being unable to provide information on the distribution of light-sensitive material within the receptors. It is true, of course, that the results of reflection measurements on the fovea are derived from the central cones, but the method cannot be used to determine whether the pigments detected therein are separated in individual receptors or whether two or more pigments are contained in each receptor. The spectrodensitometry of single receptors is not handicapped in this respect. The technique was introduced by Dobrowolski *et al.* (1955) in a study of the rods of Xenopus laevis, the clawed toad, and has been applied with notable success to the study of pigments in goldfish cones (Marks, 1965). In both investigations, the receptor outer segments were laid flat on the microscope stage, so that transmission measurements were made in a direction perpendicular to the axis of the receptor. When measured in this manner, the distribution of pigment throughout the receptor outer limb becomes an important factor in determining the *axial* density. The latter is of more than academic interest (cf. Weale, 1968a), and valid estimates of pigment density per unit length are badly needed in view of the discrepancies between some of the experimental data obtained by various methods (see below).

Different (but no less difficult) problems arise when receptors are studied along their axes, i.e., approximately in the manner in which they function under physiological conditions. The experimental skill required in these circumstances is even greater than in measurements done perpendicularly to the receptor axis, and has borne fruit only after the most modern resources have been mustered (Brown and Wald, 1964; Marks *et al.*, 1964). Since the preparations are aligned in the spectromicroscope in infrared light and single cone cross-sections are isolated with a focal stop in such light but examined with radiation of shorter (visible) wavelengths, the focusing problems encountered in an optical system not based on mirrors are formidable (Villermet and Weale, 1965). It is to be remembered, furthermore, that optical cross-sectioning defines the nature of the light bundle leaving the section but does not tell us much about its provenance; unless the numerical aperture of the system is kept relatively low, the acceptance angle may be such as to include rays having traversed receptors juxtaposed to the one selected by the focal stop. This may interfere with the production of absorption data relating to a single cone. Fortunately, most of these difficulties seem to have been overcome, although there is no guarantee that the results obtained by Brown and Wald (Fig. 4) or those due to Marks *et al.* (1964) are uncontaminated by stray light.

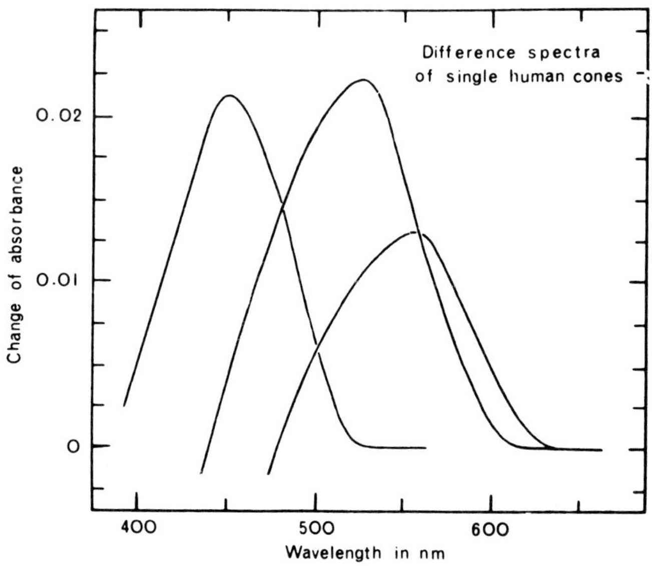

FIG. 4. Single receptor micro-densitometry. Difference spectra of single human cones. (After Brown and Wald, 1964.)

Ripps and Weale (1965) have already stressed that one of the problematical features of this earlier work on primate cones, contrary to that done on frog receptors (Liebman and Entine, 1968), is that the measured density values are unexpectedly low. This is true essentially of human receptors, no matter whether rods or cones. Thus, whereas fundus reflectometry suggests that the green-sensitive pigment absorbs (at its λ_{max}) some 30 percent of the light incident on it—and the red-sensitive pigment about 40 percent at its peak—the corresponding values obtained microdensitometrically are 5 percent and 3 percent, respectively. There is an analogous discrepancy connected with rods, but it amounts to a factor of only 3. The measuring light inevitably bleaches some of the pigment, as all single cell spectroscopists have demonstrated, and more sophisticated computation may be needed to allow for this than has, in fact, been used. Even so, it is hard to account for the factors of several hundred percent illustrated by the above comparisons, or the far greater disparity that exists between the single-cone measurements and some psychophysical data demanding extremely high pigment densities for their explanation (Brindley, 1953; Walraven and Bouman, 1960). However, it may be significant that the subjective and fundus reflectometric data were from the human fovea, whereas only parafoveal cones have so far been successfully measured by microdensitometry.

When all the reservations regarding the pioneering work in the examination of single receptors are made, the fact remains that this, and this alone, is in a position to decide such questions as whether anomalous color vision is due to pigments altogether different from those found in normal cones or whether, owing to some genetic quirk, more than one pigment was poured into one type of cone. Single receptor work is also needed to clinch the problems of dichromacy, which is only provisionally explained by fundus reflectometry done on fewer cases than one cares to admit. While color-defective human material may be hard to come by, this is not the case as regards other primate cones, and the relevant studies might help in providing an answer to the question of the species specificity of color defects.

2.4 The Early Receptor Potential

Some of the problems related to the study of cone pigments and defective color vision may possibly be tackled by a method no older than single-receptor densitometry, but one which, like fundus reflectometry, samples a fair number of receptors. We are referring to the early receptor potential (ERP), a response that can be elicited from all vertebrate retinae (Fig. 5) with intense flashes of light (for reviews see Brown, 1968; Arden, 1969). The ERP—which has two components of

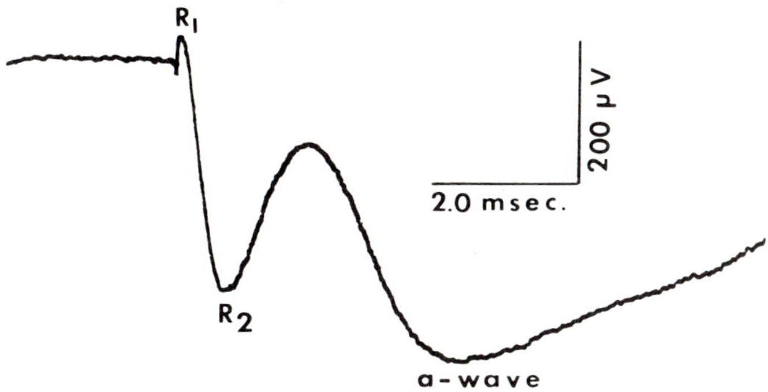

FIG. 5. The positive (R1) and negative (R2) components of the early receptor potential recorded from the living human eye. A small stimulus artifact preceding R1 marks the onset of an intense 300 μsec flash of "white" light from a Xenon discharge lamp. The a-wave of the electroretinogram develops before R2 has returned to the base-line. (Courtesy of Drs. R. E. Carr and I. M. Siegel.)

negligible latent period—changes its shape with temperature, but is singularly resistant to changes in its ionic environment: for these reasons it is thought to be linked directly to photochemical processes (Pak, 1968) although the intactness of the receptor membrane is probably essential for the potential to be recorded (Villermet and Weale, 1969).

There are two reasons why the ERP may be relevant to the study of color vision: (*a*) the amplitude of the response is directly related to the number of pigment molecules bleached (Cone, 1965), and (*b*) the action spectrum of the ERP obtained from the mixed retinae of some vertebrates indicates that the cones (although relatively few in number) are responsible for generating the major portion of the response. This has been observed so far in the frog (Goldstein, 1967, 1968), monkey (Brown *et al.*, 1965), and man (Carr and Siegel, personal communication). Thus Goldstein (1968) has shown in connection with frogs that the ERP due to their cones is much larger in amplitude than that of the rods even though the retinal concentration of the rod pigment exceeds that of the cones by a factor of 100. However, visual pigment concentration may be of less importance to the extracellularly recorded response than, for example, the ratio of the surface of the outer segment to its volume, which is up to 50 times greater in frog cones than in rods. Goldstein obtained an action spectrum for the (light-adapted) cone ERP in good agreement with the absorption spectrum of iodopsin. A residual action spectrum obtained after the retina had been adapted to red light could be due to the combined effect of pink and green rods.

It will be interesting to discover whether the ERP can be used to identify cone pigments when two or more are present in the same retina.

2.5 Spectral Selection

Although it might appear that the inevitable conclusion to be drawn from the results of retinal spectrophotometry is that three distinct cone pigments form the basis of vertebrate color vision, the evidence cited is not unequivocal in this regard. No one has yet managed to extract light-sensitive material from pieces of retina excised from the primate fovea, or for that matter, from the few pure-cone retinae examined so far (cf. Bridges, 1967). Even in those species from which presumptive cone pigments have been brought into solution (i.e., chicken, pigeon, turkey) only one photosensitive substance, in addition to rhodopsin, was detected (Wald, 1937; Bridges, 1962; Crescitelli *et al.*, 1964).

The failure to fill flasks with extracted cone pigments of various kinds has raised the possibility that color vision is subserved by a single photosensitive substance, rendered spectrally selective by adjuvant devices such as chromatic filters, or waveguides. The former are present in the receptors of several classes of vertebrate (Walls and Judd, 1933), and a possible anatomical substrate for the latter was proposed by Toraldo di Francia (1949) and Jean and O'Brien (1949). In the avian retina, for example, brightly colored oil globules are found in the inner limbs of cone receptors (cf. Strother, 1963; Meyer and Cooper, 1966). However, their effect on spectral sensitivity seems to be minimal (Armington and Thiede, 1956; Armington and Crampton, 1958; Ikeda, 1965), and the value of such filters for wavelength discrimination is doubtful (Meyer *et al.*, 1965). But whatever their function in birds, it is even less likely that prereceptoral filters are responsible for separating primate cones into different spectral classes since inclusions of this type have not been observed in mamalian receptors. Waveguide propagation, on the other hand, is probably an integral property of all vertebrate receptors, and more serious consideration has been given to its possible role in color selection.

2.5.1 The Retinal Receptors as Waveguides

The diameters of receptor outer segments approach the wavelengths of visible light, and the refractive indices of these elements are presumably greater than the index of the surrounding medium (Barer, 1957; Sidman, 1957). Thus, the propagation of photic energy down the length of the receptor can be considered analogous to that of the familiar "light pipe" or optical fiber (Snitzer and Osterberg, 1961).

One consequence of waveguide propagation is the nonuniform distribution of energy within the receptor. Viewed in cross-section the display of light and darkness takes the form of modal patterns, many of which have been demonstrated in the rods and cones of a variety of species including man (Enoch, 1961; 1963b). Although a number of factors influence modal distribution (e.g., orientation of the receptor with respect to the incident light, the index of refraction of the receptor as compared with that of the surrounding medium, etc.), the important fact is that the efficiency with which energy is transmitted depends on the wavelength. For a particular set of circumstances, therefore, some wavelengths may be almost entirely channeled outside the receptor and some allowed to enter and to course through the pigment container, while other wavelengths traverse only a portion of the full path length before leaking to the outside medium.

In the spectrophotometric study of receptors *in situ* all of the above components contribute to the test beam, and it becomes a relatively simple matter to demonstrate how the *measured* difference spectrum of a visual pigment can be significantly altered by such wavelength-selective transmission (cf. Murray, 1968). Indeed, by choosing propitiously the spectral characteristics of a light-sensitive substance and the physical parameters of the receptor, it is possible to simulate absorption spectra with λ_{max} spaced across the visible spectrum (Fig. 6). As only four such spectra are necessary (the color vision triad plus an additional one for achromatic rod vision), proponents of this view have suggested that modal coupling (i.e., interaction) results in strong excitation of three modes in cone receptors, whereas only one mode is strongly excited in rods (Biernson and Kinsley, 1965; Snyder, 1966).

Although the application of waveguide concepts to visual receptors can provide a suitable trichromatic analogue, there is strong evidence to show that such filter-like effects do not endow us with color vision but serve merely to modify the stimulus input to three cone systems which function by means of three different photolabile pigments. For example, the analysis of modal interactions, leading to differences in the spectral properties of rods and cones, demands receptoral dimensions which are at odds with measurements obtained from electron micrographs (Dowling, 1965). Even if such differential coupling were to occur, waveguide theorists would be hard-pressed to explain how, in dichromacy, the modes become uncoupled to yield the difference spectrum of only one photopigment, either red-sensitive or green-sensitive. Moreover, the three density spectra reported by Marks (1965) for the cone outer segments of goldfish, were obtained by measuring in a direction *perpendicular* to the axis of the receptor.

But perhaps the most telling blow to the one-pigment multiple-mode

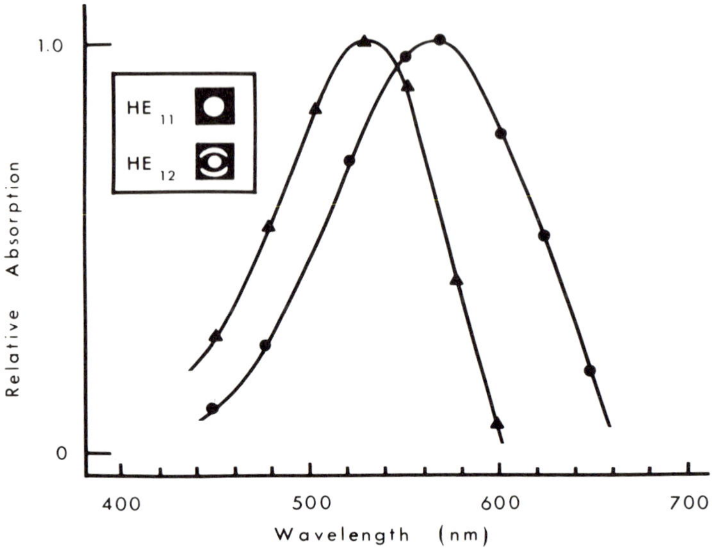

FIG. 6. Absorption spectra computed by Murray (1968) on the assumption that waveguide propagation modifies the spectral properties of a single light-sensitive pigment ($\lambda_{max} = 570$ nm). The effect of the HE_{11} mode (circles) is minimal, but the HE_{12} mode (triangles) causes the appearance of a "second" pigment maximal at 535 nm. Inset: The HE_{11} and HE_{12} modal patterns viewed microscopically near the distal ends of vertebrate receptors. (After Enoch, 1963b.)

hypothesis was dealt by the recent experiments of Murray (1968). Murray sonicated the fragments from a suspension of cone outer segments from the monkey fovea until their structural integrity and any significant waveguide properties were lost. In spite of this harsh treatment, difference spectra recorded after successive chromatic bleaches revealed the presence of at least two photolabile substances, both of which were spectroscopically similar to pigments detected by axial measurements on single cones of the same species.

It should not be assumed, however, that just because we strongly believe in the evidence for the existence of light-sensitive material in the outer limbs of cones, this is true of all writers. Bülow (1968), for example, thinks that color vision is based on the cone outer limbs acting as resonators, and Lau (1962) holds that the photo-receptors are basically mechanoreceptors. Either of these nonpigmentary hypotheses may prove to be correct and our faith in the validity of the pigmentary foundation of color vision ephemeral. But it seems to us that the notion of mechanoreception is too weakly rooted in its consideration of receptors being derived from cilia, and it is hard to see how a system which does not involve quantum counting—as would appear to be true of resonators—can act as a governor in that most important of all secondary sensory properties, namely adaptation.

3. Receptor Physiology and Information Flow

3.1 Electrical Responses from Single Cones

If each of the three cone pigments of vertebrate receptors is confined to its own cell, as we are led to believe from the microspectrophotometric data, then the electrical activity of a given cell should reflect the absorption characteristics of its photopigment. Until recently, however, attempts to introduce a microelectrode into an individual receptor have proved disastrous to either the cell, or the electrode, or both. Tomita (1965) has shown us the reason for these early failures, as well as the prerequisites for success. Apparently the receptor does not take kindly to gentle probing; to impale a single receptor it is necessary to mount the retina on a stage, set the latter into rapid oscillatory motion, and essentially 'jolt' the retina against a 0.1 μ pipette until a unit is pierced. Even with this elegant technique, damage ensues, for the responses from relatively large cones of the carp could not be maintained much longer than a minute, as evidenced by a loss of the 30–40 mV negative resting potential.

Measurements carried out by the above technique showed that the single cone responded to light with a graded, sustained, hyperpolarizing potential, unaffected by an increase in the size of the stimulus spot (Tomita *et al.*, 1967). Further evidence that the electrode had penetrated the cone inner segment was obtained by measuring the depth to which the micropipette had penetrated and by histological localization of a dye electrophoretically injected into the recording site (Kaneko and Hashimoto, 1967). Unfortunately, the brief time available for recording did not permit the determination of action spectra, but it was clear from the response amplitudes that different spectral regions provided more effective stimuli for some cells than for others. By scanning the spectrum with a series of monochromatic light flashes, responses such as those shown in Fig. 7 were obtained. There is no entirely satisfactory way of deriving spectral sensitivity functions from such data (particularly as the stimulus may have filtered through adjacent receptors), but analysis of 142 cells which gave acceptable records showed that the wavelengths eliciting the largest responses were grouped in the regions of 462, 529, and 611 nm. These values are close to the absorption maxima of goldfish cone pigments (i.e., 455, 530, and 625 nm) obtained by Marks (1965).

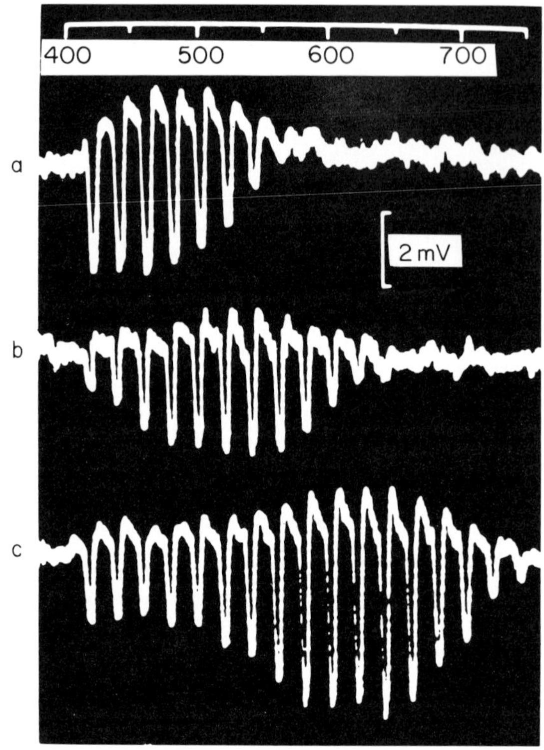

FIG. 7. Graded potentials recorded from single cones in the carp retina. The responses are from (a) blue-, (b) green-, and (c) red-sensitive elements. A downward deflection indicates negativity. (Tomita *et al.*, 1967. By kind permission of the authors and the publishers of *Vision Research*.)

3.1.1 Distribution of Cone Types

There appears to be a propensity for estimating the population of cone types within the retina based on the identification of those elements successfully measured electrophysiologically or by microdensitometry. Thus, out of a total of 113 goldfish cones examined by Marks, fewer than half were single cones (i.e., not members of joined pairs), and these occurred in the ratio of 2:4:1 for the red-, green-, and blue-sensitive elements, respectively. However, in the experiments of Tomita, *et al.* (1967) red-sensitive cones were the most numerous, the blue-sensitive were next in line, and green-sensitive units comprised only 10 percent of the total. To attribute these discrepancies to methodological or species differences seems quite futile. While there is clearly some value to be derived from a correlation of visual function with the distribution of receptor types, little faith can be placed in small-sample counting procedures employing such hit-or-miss methods.

One of the main obstacles to determining the population densities of the various cone types is that morphological differences (if they exist) are not detectable even at the level of observation provided by the electron microscope. Perhaps the most promising avenue to explore in this connection is the application of a histochemical technique developed by Enoch (1963a) for staining 'active' receptors. It has been known for some time that the ellipsoids of rods and cones (as well as other retinal regions) which are rich in mitochondria exhibit strong oxidative enzyme activity, which can be demonstrated by the reduction of colorless tetrazolium salts to highly colored insoluble formazan pigments (Kuwabara and Cogan, 1959; Cogan and Kuwabara, 1959; Pearse, 1961). What Enoch has shown is that this reaction (at least for succinate dehydrogenase) depends upon quantal absorption by visual pigment molecules, for its action spectrum closely resembles the absorption spectrum of the retinal photopigment (Enoch, 1964a). Since the density of the stain increases with light intensity (Enoch, 1964b, 1966), it may well be possible to bleach selectively the different classes of cone receptor and to observe on a large scale their distribution throughout the retina.

3.2 The Anatomical Basis for Retinal Interaction

The results obtained at the receptoral level support the trichromatic principle which forms the basis of the Young-Helmholtz theory of color vision: i.e., at least one stage in the processing of chromatic information is limited to three independent variables. But a linear transformation of three variables cannot mimic *all* messages sent to the brain, and there is ample evidence that the signals generated by the cones undergo considerable modification by intervening mechanisms (cf. Judd, 1949, 1966; Hurvich and Jameson, 1957). Since electrophysiological studies indicate that a significant degree of "color-coding" occurs at the level of the retina (cf. Section 4), it behooves us to take at least a cursory look at the neuronal organization of this tissue.

However, a few words of caution seem appropriate at this juncture. Although the anatomy of the vertebrate retina affords considerable latitude for hypothesizing on the nature and possible sites of information processing, the reader should be warned that the conduction pathways within the retina are only now coming to be understood. Nevertheless, it may prove well nigh impossible to predict *a priori* how a neuron will respond even when its presynaptic element and the nature of its transmitter substance have been identified; e.g., a single interneuron in the abdominal ganglion of *Aplysia* is capable of mediating, via the same chemical transmitter, both excitation and inhibition to its post-synaptic element (Wachtel and Kandel, 1967). Thus, while it is clearly desirable to know the terrain into which one is probing, attempts to correlate structure with function must be considered, for the most part, speculative.

It is well known that in the primate retina approximately 120 million rods and more than 6 million cones ultimately channel their signals into a million or so optic nerve fibers. In addition to this convergence, there are other avenues for intercellular communication, the details of which have been examined recently with the aid of the electron microscope (cf. Cohen, 1963; Dowling and Boycott, 1966; Kidd, 1962; Missotten, 1965; Sjöstrand, 1961). As shown in Fig. 8, lateral expansions

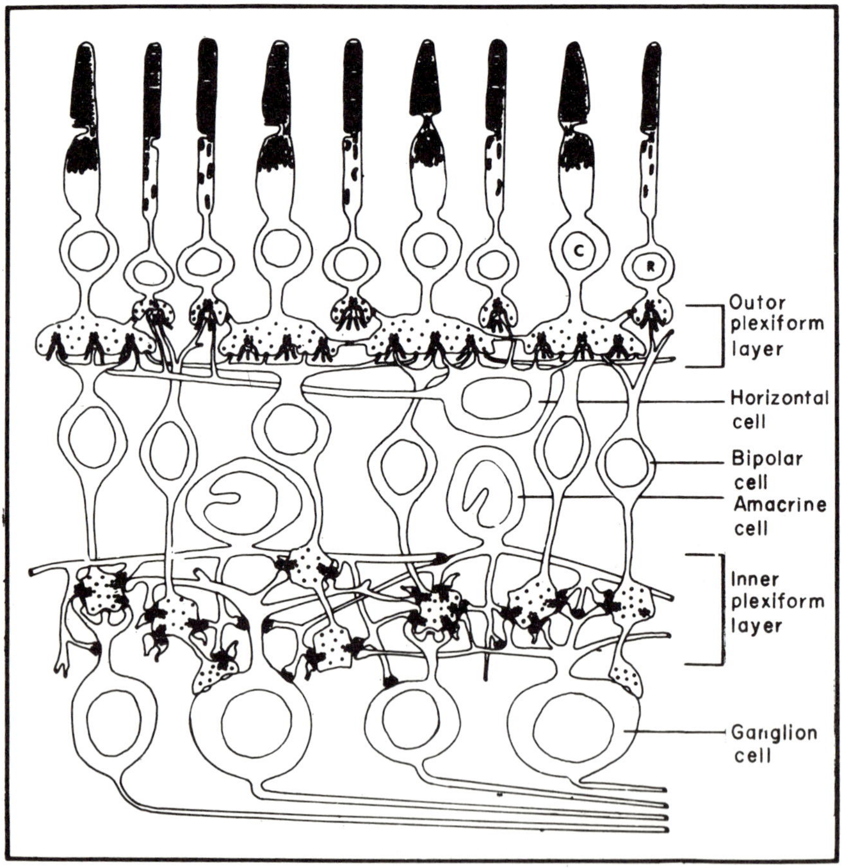

FIG. 8. Schematic representation of the synaptic organization of the primate retina. (After Dowling and Boycott, 1966.)

of each cone pedicle abut similar projections from adjacent cones, and may also contact the surface of a rod terminal (Cohen, 1965; Missotten, 1965). Interest in these contacts stems from their superficial similarity to so-called "tight" junctions, which are the sites of electrical synapses elsewhere in the nervous system (Pappas and Bennett, 1966). It seems unlikely, however, that these contacts provide a pathway for electrotonic spread of excitation, since they form an extensive network at the terminal ends of foveal cones (Cohen, 1965) and would probably interfere with the excellent resolving power and color discrimination mediated by the foveal receptors. A more likely alternative is that these junctions assist the retinal glia in providing mechanical support for elements which depend heavily on the maintence of their orientation for efficient function.

In the outer plexiform layer, a more elaborate system of interconnections can be observed. The bipolar dendrites and the processes of the horizontal cell contact the base of the pyramid-shaped cone pedicle, and join neighboring receptors; but the nature and significance of these connections are still uncertain. However, both anatomical (Dowling *et al.*, 1966) and electrophysiological (Werblin and Dowling, 1969) studies suggest that the horizontal cell is a true interneuron, and the "strategic" location of its terminals within the receptor base is suggestive of a functional role, possibly in the regulation of transmission from the receptor to the bipolar cell (Dowling and Boycott, 1966).

The bipolar cells transmit information directly to the inner plexiform layer where their axons form synapses with amacrine cell processes and the dendrites of ganglion cells. The amacrine processes, on the other hand, spread laterally through the inner plexiform layer, in the course of which they synapse with a number of bipolar cells, other amacrine processes, and the dendrites and somata of ganglion cells (Dowling and Boycott, 1966). By virtue of these contacts, the amacrine cells can exert an influence over a considerable distance. It seems likely, therefore, that these cells are involved in mediating some of the spatial and chromatic receptive-field properties which have been observed in electrical recordings obtained from the retinal ganglion cells and single units in the lateral geniculate nucleus (see Sections 4.3 and 4.4),

4. Electrical Signs of Spectral Mechanisms

4.1 The Electroretinogram

The most frequently recorded electro-retinal response is probably the electroretinogram (ERG). But while it has served nobly for the detection of photopic and scotopic mechanisms, and has proven of great value in clinical applications, the sources of its components have still not been satisfactorily identified. With remote recording electrodes, the principal features of the ERG are a cornea-negative a-wave, followed by a larger positive b-wave. The a-wave, however, attains its wave-like appearance owing to the intervention of the opposing b-wave, for when the latter is eliminated (e.g., by anoxia) a sustained negative potential, known as the PIII component (Granit, 1933), is recorded (Fig. 9). The depth at which an intraretinal pipette picks up the maximum negative response indicates that, in the mammalian retina, PIII arises from the receptors (cf. Brown, 1968). In cold-blooded vertebrates, on the other hand, Tomita (1963) found that the main source of PIII is the region of the inner nuclear layer. However, his suggestion that the a-wave may consist of more than one component has been borne out by recent studies. Murakami and Kaneko (1966) have shown that there are, in fact, short-latency (distal) and long-latency (proximal) subcomponents in the PIII response of cold-blooded retinae, and similar findings have since been obtained in mammals (Pautler *et al.*, 1968). Their results also help to resolve the discrepancy between the results of Tomita and those of Brown and his co-workers, since the distal (receptoral) PIII component of the cat retina is much larger than the proximal component, whereas the reverse is true, for example, in frog and turtle.

The origins of the b-wave are difficult to ascertain, but like the a-

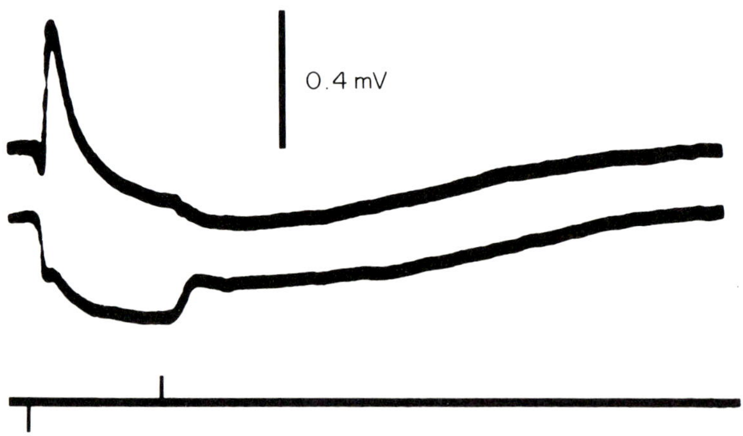

FIG. 9. The electroretinogram recorded across the retina of a cynomolgus monkey before (upper trace) and after (lower trace) clamping the retinal circulation. Ischaemia of the inner retinal layers eliminates the b-wave, revealing the time course of the negative a-wave. Stimulus duration: 320 msec. (After Brown *et al.*, 1965.)

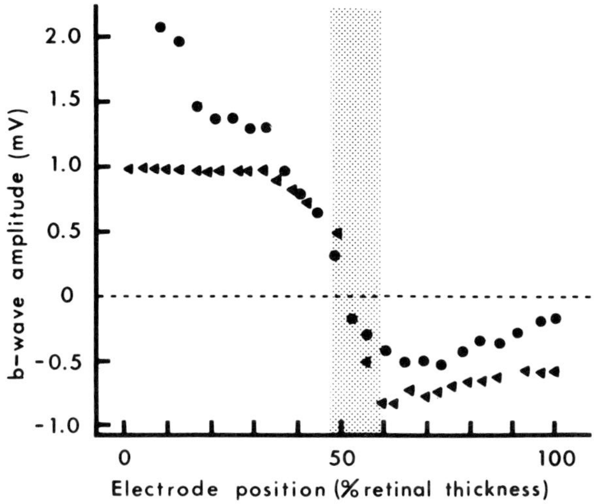

FIG. 10. The amplitude of the b-wave as a function of electrode depth in the cat retina. Zero on the scale of abscissae represents the inner (vitreal) surface of the retina. The potentials reverse polarity when the electrode reaches the inner nuclear layer (shaded area). Data from two experiments are shown. (After Arden and Brown, 1965.)

wave, it represents the synchronous activity of large numbers of cells. When a pipette is advanced through the cat retina large amplitude b-waves of the local ERG are obtained from the outer margin of the inner nuclear layer and from the inner margin of the inner plexiform layer (Arden and Brown, 1965). But the polarities are opposite at these sites, with the inversion occurring at an intermediate depth (Fig. 10). This shows that the b-wave is generated by radially oriented elements. Recent studies (Miller and Dowling, personal communication) indicate that both the Müller cells and bipolar cells contribute to the response.

4.1.1 SPECTRAL MECHANISMS

Given the probable origins of the ERG, and the ease with which it can be recorded in the living eye, it might be supposed that spectral processes, particularly those which depend on the activity of different classes of cone, can be readily distinguished electroretinographically. In fact, unequivocal evidence that mechanisms for wave-length discrimination can be detected was obtained by Forbes and his co-workers (1955, 1958) in the retinae of terrapins which are dominated by cones. These investigators employed a technique in which two beams of different wavelengths were rapidly and smoothly alternated; i.e., the two were strictly out of phase. Since the stimuli were first adjusted in intensity so as to equate the amplitudes of the ERGs, it was possible to effect a "silent" substitution when the two sources were of the same, or nearly the same, wavelength. When $\Delta\lambda$ was increased, however, substitution produced a response which could not be suppressed by adjusting the relative intensities of the paired wavelengths. Clearly, receptors with different spectral sensitivities were at work, and in the turtle eye, three photopic mechanisms were detected (Forbes *et al.*, 1958).

But the identification of the cone mechanisms of the human retina by means of electroretinography has proved a more formidable task. It has been known for some time that rods and cones contribute to both major components in the primate ERG (Adrian, 1944, 1946; Motokawa and Mita, 1942; Armington *et al.*, 1952), and the large number of rods strongly favors the scotopic system. Nevertheless, the use of a variety of techniques which exploit the differential responsiveness of rods and cones (e.g., to wavelength, level of adaptation, rate of intermittent stimulation) has made it possible to obtain responses which are derived primarily from one system or the other, and in some circumstances to show that photopic activity is subserved by more than one spectral mechanism (Armington, 1952; Biersdorf and Armington, 1957).

Recently, however, some measure of success has been achieved in separating the cone systems of human color vision, and in demonstrating the manner in which they contribute to the photopic ERG. Riggs and his co-workers (1964, 1966) devised a wavelength-substitution method similar in some respects to that used by Forbes *et al.* (1958), although it incorporates several new features. The test pattern consisted of alternate stripes of two different wavelengths (Fig. 11) which could be rapidly interchanged, i.e., λ_a was substituted for λ_b, and vice versa. Since the stimulus field undergoes no change in either energy content or spectral composition, neither the *effective* level of retinal adaptation nor the stray-light distribution alters during wavelength substitution. Furthermore, by using a closely-spaced grating, and working in the photopic luminance range, they eliminated any significant contribution from the scotopic system. Now when the ERG effectivity of all wavelengths was equated (based on previous spectral sensitivity studies [Johnson *et al.*, 1966]), responses to phase alternation were elicited only when the two wavelengths were of sufficient spectral separation, and increased in amplitude as the wavelengths became more divergent (Fig. 11). Making the assumption that the responses are proportional to the differences between the spectral sensitivities of the receptors to each wavelength pair, Riggs *et al.* (1964, 1966) were able to derive a putative set of trichromatic coefficients representing the relative responsiveness of three cone mechanisms. While a number of questions con-

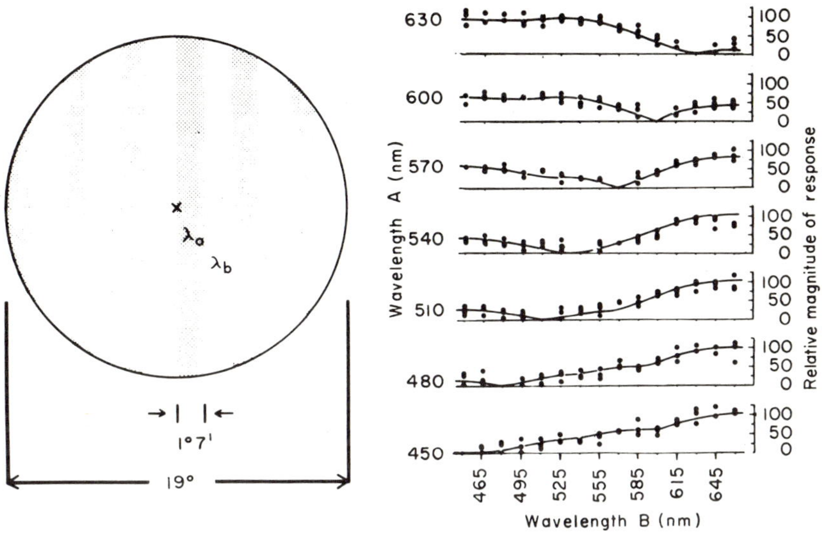

FIG. 11. Left: the appearance of the striped pattern used to obtain phase alternation between two wavelengths. The circular edge of the field and the fixation point (×) remain stationary as wavelengths A and B alternate position. Right: relative amplitudes of response elicited by alternation between various paired wavelengths. (After Riggs *et al.*, 1966.)

cerning the character of the potentials evoked with this technique await answers, additivity implies that the responses originate peripherally to the interactive processes which are evident in proximal regions of the retina. The difference between the chromaticity diagram computed by Riggs *et al.* (1964, 1966) and that based on subjective color matches (CIE) needs accounting for in this context.

Electroretinographic trichromacy, however, does not seem to hold for all species. In the carp, for example, Witkovsky (1968) found that the a- and b-waves exhibit marked differences in spectral sensitivity, and that neither function could be mimicked by the simple addition of three cone spectra (derived from Tomita's data) irrespective of the proportions in which they were added (but compare our caveat regarding Tomita's results, Section 3.1). In fact, b-wave sensitivity closely approximated the "action" spectrum of the red-sensitive cone, although chromatic adaptation clearly demonstrated the presence of more than one spectral component. Similar difficulties were encountered by Burkhardt (1968) who attempted to match the ERG sensitivity of goldfish to Marks' microdensitometry results on single cones. In addition to finding wavelength-dependent differences in the ERG waveform, he observed that adaptation to a weak, short-wavelength background *enhanced* the response evoked by long-wavelength stimulation. Thus, either the ERG potentials in fish reflect the responses of different retinal elements as compared with mammals (see Section 4.1), or alternatively, perhaps the trichromatic message is acted upon earlier in fish retinae which contain an extensive, organized system of interconnecting horizontal cells (Yamada and Ishikawa, 1965).

4.2 S-potentials

From the standpoint of color processes, there is little doubt that one of the most exciting (and puzzling) avenues being explored con-

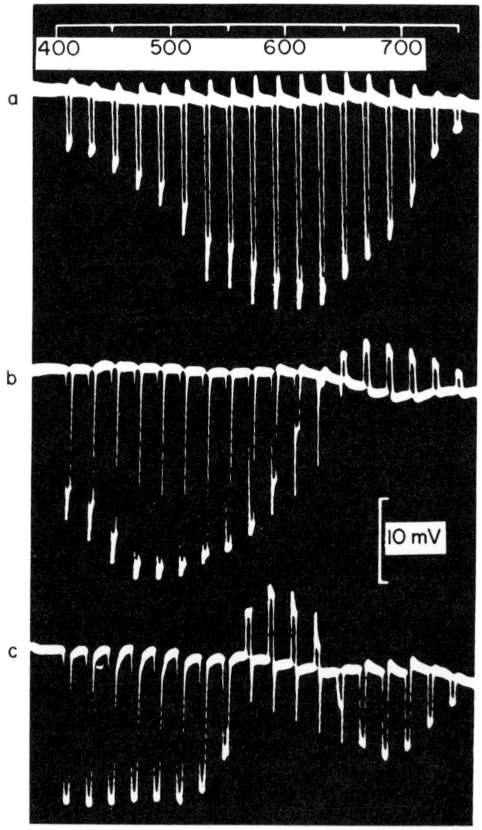

FIG. 12. Three types of S-potential recorded from the carp retina in response to chromatic stimuli (wavelength scale at top). (a) Luminosity unit; (b) biphasic chromaticity unit; and (c) triphasic C-unit. A downward deflection indicates negativity (Tomita, 1965. By kind permission of T. Tomita and the editors of the *Cold Spring Harbor Symposia on Quantitative Biology*.)

cerns the role of some unusual intraretinal potentials first recorded by Svaetichin (1956) from isolated fish retinae, and now referred to as S-potentials. Some of these responses (chromatic or C-potentials) exhibit the remarkable property of reversing their polarity as the stimulus is derived from different parts of the spectrum, occasionally undergoing two reversals in a spectral scan; others (luminosity or L-potentials) are negative (hyperpolarizing) for all wavelengths (Fig. 12).

In spite of extensive research, no clear picture has emerged of the relation between S-potentials and the message carried by the fibers of the optic nerve. The peculiar electrical properties of these responses, in addition to constituting a major obstacle to interpretation, have aroused a good deal of controversy with regard to their origins and mechanisms of generation. For example, if a triphasic unit (Fig. 12) is subjected to a short-wavelength adapting light, the hyperpolarizing response to red light ($\lambda = 710$ nm) is enhanced irrespective of the change in resting level induced by the adapting source. As shown in Fig. 13, light of $\lambda =$ 580 nm depolarizes the unit, whereas adaptation to $\lambda = 400$ nm causes

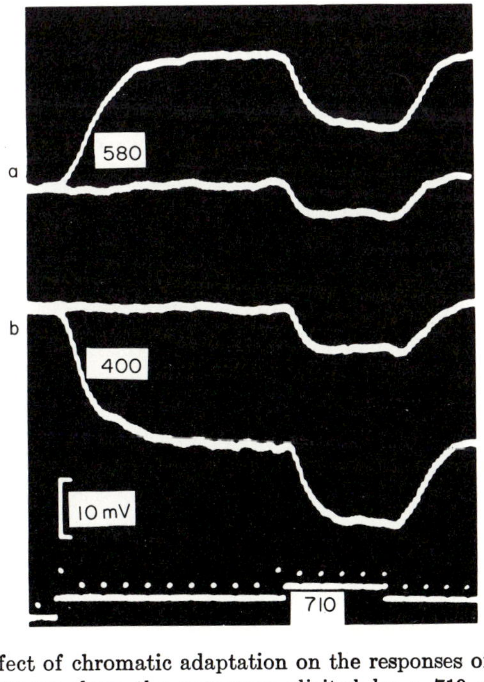

FIG. 13. The effect of chromatic adaptation on the responses of a triphasic C-unit. The two central traces show the responses elicited by a 710 nm stimulus before adaptation. A 580 nm adapting field (a) depolarizes the unit, whereas a 400 nm adapting field (b) causes hyperpolarization. However, in both cases the responses to the original (710 nm) stimulus are enhanced. Time marks at 0.1 sec intervals. (After Tomita, 1965.)

further hyperpolarization; yet in both cases the response to a 710 nm stimulus is increased (Tomita, 1965). Furthermore, the application of large currents of either polarity negligibly affect the responses to light (MacNichol and Svaetichin, 1958; Watanabe *et al.*, 1960; Negishi *et al.*, 1968). These and other characteristics, inconsistent with the behavior of typical neurons (cf. Tomita, 1965), have led to the suggestion that the elements repsonsible for generating S-potentials are a specialized form of glia which affect the neuronal transmission of excitation from photoreceptors to ganglion cells (Svaetichin *et al.*, 1961). Although there is no evidence that the activity of glial cells can alter neuronal responses, some elaborate hypotheses have been advanced suggesting that there is a metabolic linkage between neuron and glia which leads to reciprocal alterations in their membrane potentials (Lipetz, 1963; Negishi *et al.*, 1968). However, other authors (notably Dowling and Boycott, 1966) are of the opinion that information processing within the retina is handled by horizontal, bipolar and amacrine cells which, by virtue of their synaptic contacts, should be considered true interneurons. Many of the questions concerning the functional significance of these cell types may soon be resolved now that it is possible to record the electrical activity of these elements (Bortoff, 1964; Werblin and Dowling, 1969; Kaneko and Hashimoto, 1969).

Although the cellular origins of S-potentials are still uncertain, these units seem capable of integrating activity from large areas of the retina, and even its full extent (Norton *et al.*, 1968). Electrode-tracking and electrophoretic dye-injection techniques (Tomita, 1957; MacNichol and Svaetichin, 1958; Tomita *et al.*, 1959) have implicated primarily the inner nuclear layer, where the horizontal, bipolar, Mueller, and amacrine cells are to be found; L-potentials are obtained from distal, and C-potentials from proximal parts of this region. It should be noted that whereas S-potentials have been recorded from nearly all classes of vertebrates, the chromatic variety have yet to be demonstrated in primates, and little has been done by way of spectral analysis on the responses from mammalian retinae. Consequently, we must confine our remarks to results obtained on fish retinae, where the relatively large internal cells are easily penetrated and their activity can be monitored over extended periods of time.

4.2.1 Luminosity Units

It has been suggested that L-potentials, which probably arise from the external horizontal cells of the teleost retina (MacNichol and Svaetichin, 1958; Fatechand *et al.*, 1966), are concerned not with spectral processes, but with signalling luminosity. This may well be so in some species (e.g., mullet), where it has been shown that L-potentials are affected independently of the spectral nature of the adapting field (Svaetichin and MacNichol, 1958). But if the L-potential is, in fact, a product of horizontal cell activity, this result suggests that "luminosity" is derived from a single class of cone, while the signals from all others bypass the horizontal cells. Recently, however, Witkovsky (1967) found that the photopic L-potentials of carp are influenced by two different photopic mechanisms, and Naka and Rushton (1966) state that three, or possibly four, cone types contribute to the hyperpolarizing L-units of tench.

The latter authors report interaction phenomena in these units akin to those seen in color-coded ganglion cells of goldfish (Wagner *et al.*, 1960a); i.e., responses elicited by a red stimulus were enhanced by adaptation to a green background, and vice versa. But drawing analogies between results from tench and goldfish is a risky business, for the findings on carp (another close relative of the tench) are strikingly different. Comparing the response properties of ganglion cells and L-units in carp, Witkovsky (1967) was unable to detect either blue- or green-sensitive components in the L-potentials, although both were evident at the ganglion cell. Furthermore, the behavior of L-potentials during light- and dark-adaptation (Witkovsky, 1967; Naka and Rushton, 1968) seems to bear no relation to the sensitivity and receptive field changes characteristic of the ganglion-cell responses (Witkovsky, 1967; Barlow *et al.*, 1957a,b). On the available evidence, therefore, it is difficult to entertain the notion that L-units influence ganglion cell discharge.

4.2.2 Chromatic Units

The functional significance of the chromatic S-potentials is no less perplexing. C-potentials have long been regarded as representing the most peripheral signs of an "opponent-color" mechanism, whereby, e.g., the responses of one class of cone receptor serve ultimately to excite some optic nerve fibers and inhibit others, while the opposite effect can be induced by cones of a different class (cf. MacNichol, 1964, 1966; Svaetichin *et al.*, 1965). Indeed, the antagonistic action evoked by different wavelengths on the C-potentials has been compared with the color-coded discharge pattern of ganglion cells, where a shift in wavelength can convert the output of a single element from an "on" to an "off" response (Wagner *et al.*, 1960b). For example, a biphasic C-potential may depolarize in response to green light, and hyperpolarize to red light; ganglion cells, in turn, may respond with a burst of activity when a green light is flashed onto the retina, but discharge only on termination of a long-wavelength stimulus. At intermediate wavelengths, the interplay of excitatory and inhibitory effects can lead to a cancellation of activity whereby both elements fail to respond.

While these parallels serve to strengthen the view that C-units perhaps control ganglion-cell responses, the correspondence shown above is somewhat contrived, for such concordant behavior is not typical of the results obtained from a single species. In this connection, it is instructive to recall again Witkovsky's (1967) quantitative analysis of ganglion cell response and S-potentials in the carp retina. He found that whereas ganglion cells could be either "on" to short-wavelength and "off" to long-wavelength stimuli, or vice versa, the biphasic C-potentials invariably hyperpolarized to short wavelengths and depolarized to long ones. Furthermore, not only were there marked differences in times of recovery after light adaptation, but spectral sensitivity determinations on these units were also different; wavelength-dependent "on-off" ganglion cells were maximally sensitive at 480 nm and 594–620 nm, while the C-potential sensitivity functions peaked at 520–550 nm and 665 nm. Thus, Witkovsky could find no evidence that S-potentials control, or are influenced by, neuronal firing. However, the lack of correspondence between S-potentials and ganglion-cell responses may be a consequence of intervening processes or of properties inherent in the ganglion cell itself. Until we have a better understanding of the origins of these potentials and their conduction mechanism, it is premature to dismiss them as mere epiphenomena which play no part in the color vision code.

4.3 Color-coded Ganglion Cells

We have already alluded to one aspect of ganglion cell discharge which suggests that, in some species, the firing pattern represents a color-coded response. In the goldfish retina, Wagner *et al.* (1960b) found the majority of ganglion cells to be of the "on-off" variety, responding transiently to the onset and termination of stimulation. With monochromatic lights, however, the "on" response of a unit was elicited by short-wavelength stimuli, and the "off" discharge by longer wavelengths (Fig. 14), or vice versa. Spectral sensitivity determinations (Wagner *et al.*, 1963) showed that red- and green-sensitive mechanisms contribute to such opponent-color units, and an extension of these studies (Bicking, 1965) has demonstrated that all of the possible pairs of antagonistic responses which can be derived from the red-, green-, and blue-sensitive mechanisms of goldfish are detectable (although red-green antagonism is far and away the most prevalent).

Perhaps the most interesting feature of color-coded units is their receptive-field organization. When explored with small monochromatic

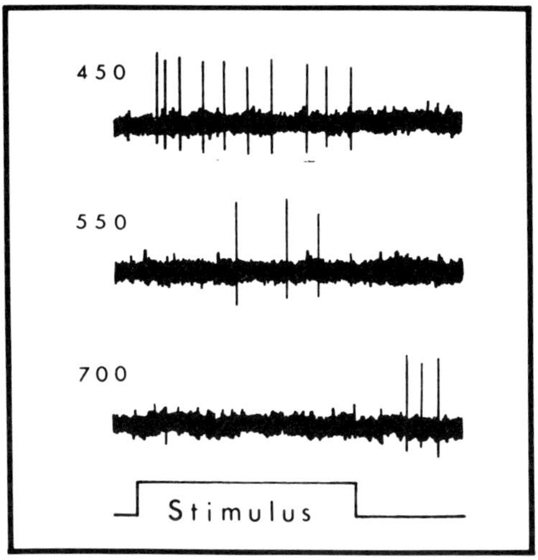

FIG. 14. Chromatic responses of an "on-off" ganglion cell in the goldfish. At 450 and 550 nm, the cell discharges following the onset of illumination; at the longer wavelength, the cell fires only when the light is turned off. Stimulus duration is 0.5 sec. (After Wagner *et al.*, 1960.)

spots (Wagner *et al.*, 1963), a cell which fired, for example, when a red light was turned on in the center of its receptive field, gave the same type of response to red in the periphery of its receptive field; green stimulus spots would, in this case, evoke an "off" response from center to periphery. Thus, the receptive fields for both red and green spots were superimposed and more or less concentric, except that the receptive field for green was invariably larger than that for red irrespective of whether the responses were excitatory or inhibitory. This type of opponent-color organization seems, therefore, to be well suited for effecting successive color contrast: if a given ganglion cell signals "red," it will respond to the onset of a red stimulus, but not to green; turning off a green light, however, elicits the "red" response.

Recently, however, Daw (1967, 1968) has shown that the majority of color-coded ganglion cells in the goldfish retina have much larger receptive fields than was revealed by spot stimulation, and that the center-surround organization of these fields maximizes the response to contrasting colors presented simultaneously. Through the use of contiguous stimulus fields consisting of central spots and annular surrounds (with independent wavelength controls), Daw was able to demonstrate that the surround evoked the same type of response when illuminated with red light, as that elicited by a green stimulus in the center of the receptive field. Thus, when center and surround were delivered simultaneously, the units gave maximal "on" responses to a red spot in a green surround, and maximal "off" responses to a green spot in a red surround, or vice versa. For contrasting colors, therefore, center-surround stimulation *enhanced* the ganglion cell response—precisely the mechanism required for mediating simultaneous color contrast.

Although the class of unit described by Daw has not been observed in other vertebrate retinae, it is true that few investigators have examined receptive-field properties with similar stimulus configurations. This does not apply, however, to studies (Michael, 1968) on the pure-cone retina of the ground squirrel, whose optic nerve fibers also exhibit opponent-color responses (albeit to green and blue). Michael was unable to find evidence of units which could subserve simultaneous color contrast, although he reports having located such cells in the lateral geniculate nucleus (LGN) of the ground squirrel. It is noteworthy that a few color-contrast detectors were identified in the monkey striate cortex (Hubel and Wiesel, 1968), but not at the level of the LGN (Wiesel and Hubel, 1966). It will be interesting to see how these findings are eventually reconciled.

Considering that some genera of monkey have color vision similar to that of man (Grether, 1939; de Valois, 1965; Sperling *et al.*, 1968), there is a surprising paucity of data on the chromatic responses of primate ganglion cells. Color-coded "on-off" units have been demonstrated in a few ganglion cells of the spider monkey (Hubel and Wiesel, 1960) but, until recently, this area of study has not been vigorously pursued. Indeed, the examination of ganglion cell activity might seem superfluous in view of the splendid researches on the lateral geniculate nucleus of primates by Wiesel and Hubel (1966) and by de Valois and his coworkers (see Section 4.4). But when it is recalled that LGN recordings reflect typically postsynaptic events, the significance of analyzing the presynaptic message is clear: only by comparing signals from ganglion cells with those from the LGN is it possible to establish the degree to which each participates in the encoding of visual information.

To this end, Gouras has undertaken the study of Rhesus ganglion cells, recording *in vivo* from the retinae of anesthetized animals. The initial phases of this work (Gouras, 1965; Gouras and Link, 1966) dealt primarily with the manner in which ganglion cells distinguish between, and transmit, rod and cone signals, both of which are carried by the same optic nerve fibers. In some respects, the results suggest a relatively simple sorting process; the ganglion cell transmits those signals which reach it first. As cone signals have the shorter latency, they invariably prevail when the stimulus intensity exceeds cone threshold. What is puzzling, however, is how the same optic nerve fiber can code for achromatic signals in the dark-adapted state, and for chromatic signals when light-adapted.

On examining the receptive-field organization of Rhesus ganglion cells with chromatic stimuli, Gouras (1968) recorded from two types of cell—*phasic* and *tonic*—both of which were excited by stimulation of the center and inhibited by stimulation of the periphery of their receptive fields. Phasic units, which discharged transiently to maintained stimuli of all wavelengths, received signals from both green- and red-sensitive cone mechanisms, but these mechanisms were not spectrally opponent; i.e., center-surround antagonism was independent of wavelength. Tonic units, on the other hand, respond continuously when activated, and display the antagonistic center-surround arrangement found in a majority of cells of the LGN (Wiesel and Hubel, 1966); i.e., center and surround have different spectral sensitivities. This type of cell receives excitatory signals from only one cone mechanism (either blue-, green-, or red-sensitive) in the center of its receptive field, and inhibitory signals from another cone mechanism in the surrounding field. Since the spectrally antagonistic mechanisms have different spatial distributions, Gouras suggests that these cells may function in form as well as hue discrimination—a property attributed also to opponent-color units of the LGN (Wiesel and Hubel, 1966). It is apparent, therefore, that some (and perhaps all) types of response seen at the lateral geniculate nucleus already exist at the level of the retina.

4.4 The Lateral Geniculate Nucleus

What influence, if any, the LGN exerts on in-coming signals from the optic nerve fibers has yet to be determined, but there is little doubt that it differs from species to species, and that a major part of the color-coding process is completed before the message is transmitted to the visual cortex. Evidence for this view stems primarily from the work of de Valois and his associates (cf. de Valois *et al.*, 1958, 1959; de Valois, 1965) on the electrical and behavioral correlates of primate color vision. Recording from single units of the monkey LGN, they have shown that, with diffuse retinal illumination (15° stimulus fields), two types of cell are distinguishable on the basis of their responses to chromatic stimuli: (*a*) "nonopponent" units which respond uniformly to all wavelengths by either increasing (excitation) or decreasing (inhibition) their rates of spontaneous discharge, and (*b*) spectrally "opponent" cells which are excited by some wavelengths and inhibited by others.

The opponent-color cells can be further subdivided (de Valois *et al.*, 1966) into four groups since for each opponent pair (*R*-*G*, *B*-*Y*) there exists a mirror image that responds in opposite fashion. For example, some units (*R-G*) are excitatory in the red and inhibitory in the green; others (*G-R*) show the reverse response. But in both instances the change from one type of activity to the other occurs in the region of 600 nm (the "cross-point") where the excitatory and inhibitory inputs from two different cone mechanisms (presumably red- and green-sensitive) are balanced; stimuli of this wavelength elicit little or no response. An analogous situation obtains for the *Y-B* and *B-Y* cells (which receive antagonistic inputs from blue- and red-sensitive cone mechanisms), with the cross-point at about 490 nm.

As a consequence of these interacting processes, the opponent-color cells are remarkably sensitive to wavelength shifts (de Valois *et al.*, 1964) that upset the existing relationship between excitation and inhibition. Thus, if all wavelengths are equated for luminance, it is possible to determine the smallest shift in wavelength that produces some criterion change in the cell's discharge rate. This procedure enabled de Valois *et al.* (1967) to derive a wavelength-discrimination function which clearly reflects the principal features of the animal's performance on behavioral tests (Fig. 15). As expected, wavelength-discrimination is best in those spectral regions where the cell's responses cross from excitation to inhibition, i.e., at about 600 nm for *R-G* units, and in the region of 490 nm for the *Y-B* units. Thus, when *R-G* and *Y-B* units act in concert, the cell which is most sensitive to a particular $\Delta\lambda$ probably determines the animal's perceptual response.

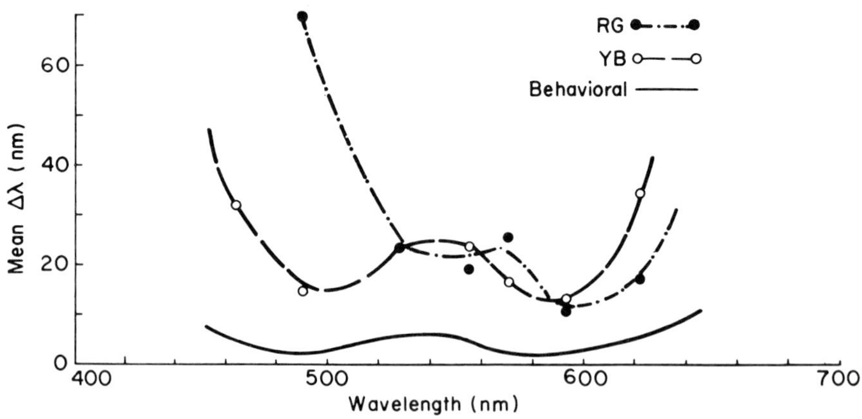

FIG. 15. Wavelength discrimination data derived from recordings from spectrally-opponent cells of the LGN of the macaque compared with the results obtained by a behavioral method. (De Valois *et al.*, 1967. By kind permission of R. L. de Valois and the publishers of the *Journal of Neurophysiology*.)

Whereas the spectrally-opponent cells appear to be capable of signalling color, the nonopponent units may be considered brightness indicators, since their responses at all wavelengths can be equated by an appropriate adjustment of stimulus energy. When this is done under light-adapted conditions, the spectral sensitivity of such cells resembles the photopic luminosity function determined behaviorally (de Valois and Jacobs, 1968). Further evidence that the nonopponent units carry brightness information was provided by analyzing the relative saturation of various spectral lights (de Valois *et al.*, 1966; de Valois and Jacobs, 1968). The latter affect opponent cells as well, but their signals presumably relay information as to chromaticity. It was assumed, therefore, that saturation is determined by the ratio of the sensitivities of opponent to nonopponent cells at each stimulus wavelength. Since the nonopponent units are maximally sensitive in the region of 560–580 nm, it follows that chromatic stimuli should be less saturated in this region than at the extremes of the spectrum where the activity of opponent cells is dominant. Figure 16 shows that this is indeed the case, both behaviorally and electrophysiologically.

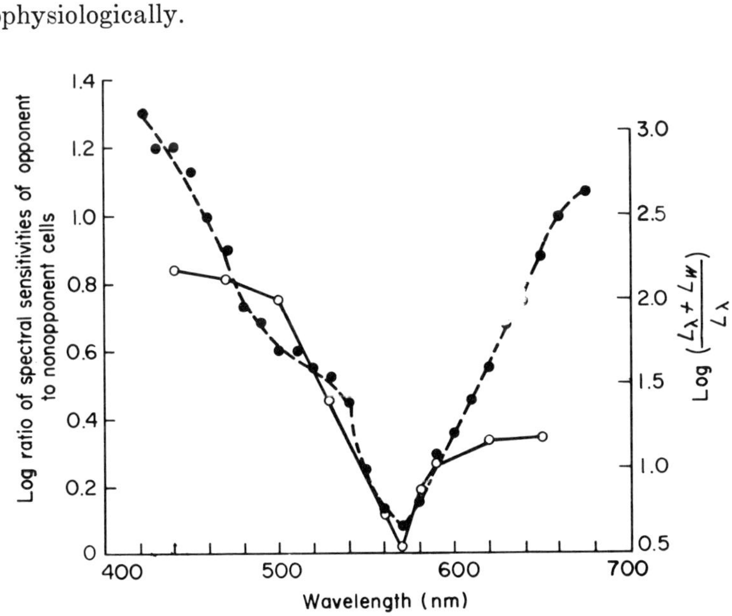

FIG. 16. The relative saturation of various spectral lights determined psychophysically (open circles and right-hand scale of ordinates) and from single unit recordings in the LGN (filled circles and scale at left). Both sets of data are for the macaque. (De Valois and Jacobs, 1968. By kind permission of R. L. de Valois and the publishers of *Science*. Copyright 1968 by the *American Association for the Advancement of Science*.)

ACKNOWLEDGMENTS

We wish to thank Miss Gabrielle Villermet and Miss Jane Zakevicius for assistance in the preparation of the manuscript and the illustrations. Partial support for this work was provided by Grants 5K3-EY-18,766 and EY-00285 from the National Institutes of Neurological Diseases and Blindness, U. S. Public Health Service.

REFERENCES

Adrian, E. D. (1944). *Nature* **154**, 361–362.

Adrian, E. D. (1946). *J. Physiol.* **105**, 24–37.

Alpern, M. (1968). *Ann. Rev. Physiol.* **30**, 279–318.

Alpern, M., Mindel, J., and Torii, S. (1968). *J. Physiol.* **199**, 433–456.

Arden, G. B. (1969). *Progr. Biophys. Mol. Biol.* **19**, 387–423.

Arden, G. B., and Brown, K. T. (1965). *J. Physiol.* **176**, 429–461.

Armington, J. C. (1952). *J. Opt. Soc. Am.* **42**, 393–401.

Armington, J. C., and Crampton, G. H. (1958). *Am. J. Ophthalmol.* **46** (Part II), 72–87.

Armington, J. C., and Thiede, F. C. (1956). *Am. J. Physiol.* **186**, 258–262.

Armington, J. C., Johnson, E. P., and Riggs, L. A. (1952). *J. Physiol.* **118**, 289–298.

Barer, R. (1957). *J. Opt. Soc. Am.* **47**, 545–556.

Barlow, H. B., Fitzhugh, R., and Kuffler, S. W. (1957a). *J. Physiol.* **137**, 338–354.

Barlow, H. B., Fitzhugh, R., and Kuffler, S. W. (1957b). *J. Physiol.* **137**, 327–337.

Bicking, L. A. (1965). Cited in MacNichol, E. F., Jr. (1966). *Proc. Natl. Acad. Sci. U. S.* **55**, 1331–1344.

Biernson, G., and Kinsley, D. J. (1965). *IEEE Trans. Microwave Theory Tech.* **MTT-13**, 345–356.

Biersdorf, W. R., and Armington, J. C. (1957). *J. Opt. Soc. Am.* **47**, 208–215.

Bortoff, A. (1964). *Vision Res.* **4**, 627–635.

Bridges, C. D. B. (1962). *Vision Res.* **2**, 125–137.

Bridges, C. D. B. (1967). *In* "Comprehensive Biochemistry" (M. Florkin and H. Stotz, eds.), **27**, pp. 31–78. Elsevier, London.

Brindley, G. S. (1953). *J. Physiol.* **122**, 332–350.

Brown, K. T. (1968). *Vision Res.* **8**, 633–677.

Brown, K. T., Watanabe, K., and Murakami, M. (1965). *Cold Spring Harbor Symp. Quant. Biol.* **30**, 457–482.

Brown, P. R., and Wald, G. (1963). *Nature* **200**, 37–43.

Brown, P. R., and Wald, G. (1964). *Science* **144**, 45–52.

Bülow, N. (1968). *Acta Ophthalmol.* **46**, 1054–1066.

Burkhardt, D. A. (1968). *Vision Res.* **8**, 839–853.

Carr, R. E., and Ripps, H. (1967). *Invest. Ophthalmol.* **6**, 426–436.

Cogan, D., and Kuwabara, T. (1959). *J. Histochem. Cytochem.* **8**, 334–341.

Cohen, A. I. (1963). *Biol. Rev.* **38**, 427–459.

Cohen, A. I. (1965). *J. Anat.* **99**, 595–610.

Cohen, A. I. (1968). *J. Cellular Biol.* **37**, 424–444.

Cohen, L. B., Keynes, R. D., and Hille, B. (1968). *Nature* **218**, 438–441.

Cone, R. A. (1965). *Cold Spring Harbor Symp. Quant. Biol.* **30**, 483–491.

Crescitelli, F., Wilson, B. W., and Lilyblade, A. L. (1964). *Vision Res.* **4**, 275–280.

Daw, N. W. (1967). *Science* **158**, 942–944.

Daw, N. W. (1968). *J. Physiol.* **197**, 567–592.

de Robertis, E., and Lasansky, A. (1961). *In* "The Structure of the Eye" (G. K. Smelser, ed.), pp. 29–49. Academic Press, New York.

de Valois, R. L. (1965). *Cold Spring Harbor Symp. Quant. Biol.* **30,** 567–579.

de Valois, R. L., and Jacobs, G. H. (1968). *Science* **162,** 533–540.

de Valois, R. L., Smith, C. J., Kitai, S. T., and Karoly, A. J. (1958). *Science* **127,** 238–239.

de Valois, R. L., Smith, C. J., and Kitai, S. T. (1959). *J. Comp. Physiol. Psychol.* **52,** 635–641.

de Valois, R. L., Jacobs, G. H., and Abramov, I. (1964). *Science* **146,** 1184–1186.

de Valois, R. L., Abramov, I., and Jacobs, G. H. (1966). *J. Opt. Soc. Am.* **56,** 966–977.

de Valois, R. L., Abramov, I., and Mead, W. R. (1967). *J. Neurophysiol.* **30,** 415–433.

Dewey, M. N., Davis, P. K., Blasie, J. K., and Barr, L. (1969). *J. Mol. Biol.* **39,** 395–405.

Dobrowolski, J. A., Johnson, B. K., and Tansley, K. (1955). *J. Physiol.* **130,** 533–542.

Dowling, J. E. (1965). *Science* **147,** 57–59.

Dowling, J. E., and Boycott, B. B. (1966). *Proc. Roy. Soc. B* **166,** 80–111.

Dowling, J. E., Brown, J. E., and Major, D. (1966). *Science* **153,** 1639–1641.

Eakin, R. M. (1965). *Cold Spring Harbor Symp. Quant. Biol.* **30,** 363–370.

Enoch, J. M. (1961). *Am. J. Ophthalmol.* **51** (Part II), 1107–1118.

Enoch, J. M. (1963a). *Invest. Ophthalmol.* **2,** 16–23.

Enoch, J. M. (1963b). *J. Opt. Soc. Am.* **53,** 71–85.

Enoch, J. M. (1964a). *J. Opt. Soc. Am.* **54,** 368–374.

Enoch, J. M. (1964b). *J. Opt. Soc. Am.* **54,** 1027–1030.

Enoch, J. M. (1966). *J. Opt. Soc. Am.* **56,** 116–123.

Fatechand, R., Svaetichin, G., Negishi, K., and Drujan, B. (1966). *Vision Res.* **6,** 271–283.

Forbes, A., Burleigh, S., and Neyland, M. (1955). *J. Neurophysiol.* **18,** 517–535.

Forbes, A., Deane, H. W., Neyland, M., and Gongaware, M. S. (1958). *J. Neurophysiol.* **21,** 247–262.

Goldstein, E. B. (1967). *Vision Res.* **7,** 837–845.

Goldstein, E. B. (1968). *Vision Res.* **8,** 953–956.

Gouras, P. (1965). *Science* **147,** 1584–1593.

Gouras, P. (1968). *J. Physiol.* **199,** 533–547.

Gouras, P., and Link, K. (1966). *J. Physiol.* **184,** 499–510.

Granit, R. (1933). *J. Physiol.* **77,** 207–240.

Grether, W. F. (1939). *Comp. Psychol. Monogr.* **15,** 1–38.

Hagins, W. A., and Jennings, W. H. (1959). *Disc. Faraday Soc.* **27,** 180–190.

Hsia, Y., and Graham, C. H. (1952). *Proc. Natl. Acad. Sci. U. S.* **38,** 80–85.

Hubel, D. H., and Wiesel, T. N. (1960). *J. Physiol.* **154,** 572–580.

Hubel, D. H., and Wiesel, T. N. (1968). *J. Physiol.* **195,** 215–243.

Hurvich, L. M., and Jameson, D. (1957). *Psychol. Rev.* **64,** 384–404.

Ikeda, H. (1965). *Vision Res.* **5,** 19–36.

Jean, J., and O'Brien, B. (1949). *J. Opt. Soc. Am.* **39,** 1057.

Johnson, E. P., Riggs, L. A., and Schick, A. M. L. (1966). *In* "Clinical Electroretinography" (H. M. Burian and J. H. Jacobson, eds.), pp. 75–91. Pergamon Press, Oxford.

Judd, D. B. (1949). *J. Res. Natl. Bur. Std.* (Washington, D. C.) **42,** 1–16.

Judd, D. B. (1966). *Proc. Natl. Acad. Sci. U. S.* **55,** 1313–1330.

Kaneko, A., and Hashimoto, H. (1967). *Vision Res.* **7,** 847–851.

Kaneko, A., and Hashimoto, H. (1969). *Vision Res.* **9,** 37–55.

Kidd, M. (1962). *J. Anat.* **96,** 179–188.
Kuwabara, T., and Cogan, D. G. (1959). *J. Histochem. Cytochem.* **7,** 329–333.
Lau, E. (1962). *Proc. 6th Intern. Congr. Opt.* Munich, 145.
Liebman, P. A. (1962). *Biophys. J.* **2,** 161–178.
Liebman, P. A., and Entine, G. (1964). *J. Opt. Soc. Am.* **54,** 1451–1459.
Liebman, P. A., and Entine, G. (1968). *Vision Res.* **8,** 761–775.
Lipetz, L. (1963). *IEEE Trans. Mil. Elect.* MIL-**7,** 144–155.
MacNichol, E. F., Jr. (1964). *Vision Res.* **4,** 119–133.
MacNichol, E. F., Jr. (1966). *Proc. Natl. Acad. Sci. U. S.* **55,** 1331–1344.
MacNichol, E. F., Jr., and Svaetichin, G. (1958). *Am. J. Ophthal.* **46** (Part 2), 26–40.
Marks, W. B. (1965). *J. Physiol.* **178,** 14–32.
Marks, W. B., Dobelle, W. H., and MacNichol, E. F., Jr. (1964). *Science* **143,** 1181–1183.
Meyer, D. B., and Cooper, T. G. (1966). *Amer. J. Anat.* **118,** 723–734.
Meyer, D. B., Cooper, T. G., and Gerney, C. (1965). *In* "The Structure of the Eye," II Symposium, Eighth Intern. Congr. Anatomists (J. W. Rohen), pp. 521–533. Schattauer, Stuttgart.
Michael, C. R. (1968). *J. Neurophysiol.* **31,** 249–282.
Missotten, L. (1965). "The Ultrastructure of the Human Retina." Editions Arscia S. A., Brussels.
Motokawa, K., and Mita, T. (1942). *Tohoku J. Exptl. Med.* **42,** 114–133.
Murakami, M., and Kaneko, A. (1966). *Nature* **210,** 103–104.
Murray, G. C. (1968). Doctoral dissertation. Johns Hopkins Univ., Baltimore. Maryland.
Naka, K. I., and Rushton, W. A. H. (1966). *J. Physiol.* **185,** 587–599.
Naka, K. I., and Rushton, W. A. H. (1968). *J. Physiol.* **194,** 259–269.
Negishi, K., Laufer, M., and Svaetichin, G. (1968). *Nature* **218,** 39–40.
Norton, A. L., Spekreijse, H., Wolbarsht, M. L., and Wagner, H. G. (1968). *Science* **160,** 1021–1022.
Pak, W. L. (1968). *Photochem. Photobiol.* **8,** 495–503.
Pak, W. L., and Helmrich, H. G. (1968). *Vision Res.* **8,** 585–589.
Pappas, G. D., and Bennett, M. V. L. (1966). *Ann. N. Y. Acad. Sci.* **137,** 495–508.
Pautler, E. L., Murakami, M., and Nosaki, H. (1968). *Vision Res.* **8,** 489–491.
Pearse, A. G. E. (1961). *In* "The Structure of the Eye" (G. Smelser, ed.), pp. 53–72. Academic Press, New York.
Pitt, F. G. J. (1935). *Med. Res. Council Spec. Rept. Ser.* 200.
Riggs, L. A., Johnson, E. P., and Schick, A. M. L. (1964). *Science* **144,** 567.
Riggs, L. A., Johnson, E. P., and Schick, A. M. L. (1966). *J. Opt. Soc. Am.* **56,** 1621–1627.
Ripps, H., and Weale, R. A. (1963). *Vision Res.* **3,** 531–543.
Ripps, H., and Weale, R. A. (1964a). *J. Physiol.* **173,** 57–64.
Ripps, H., and Weale, R. A. (1964b). *J. Opt. Soc. Am.* **54,** 272–273.
Ripps, H., and Weale, R. A. (1965). *Nature* **205,** 52–56.
Rushton, W. A. H. (1958). *N. P. L. Symposium No. 8 on Visual Problems of Colour* **1,** 77–101, H. M. S. O. London.
Rushton, W. A. H. (1961). *The New Scientist* **10,** 374–377.
Rushton, W. A. H. (1962). *Sci. Am.* **207,** 120–132.
Rushton, W. A. H. (1963). *J. Physiol.* **168,** 360–373.
Rushton, W. A. H. (1965a). *J. Physiol.* **176,** 24–37.
Rushton, W. A. H. (1965b). *J. Physiol.* **176,** 38–45.

Rushton, W. A. H. (1965c). *J. Physiol.* **176,** 46–55.
Sidman, R. L. (1957). *J. Biophys. Biochem. Cytol.* **3,** 15–30.
Sjöstrand, F. (1953). *J. Cellular Comp. Physiol.* **42,** 15–44.
Sjöstrand, F. S. (1961). *In* "The Structure of the Eye" (G. K. Smelser, ed.), pp. 1–28. Academic Press, New York.
Snitzer, E., and Osterberg, H. (1961). *J. Opt. Soc. Am.* **51,** 499–505.
Snyder, A. W. (1966). *J. Opt. Soc. Am.* **56,** 705–706.
Sperling, H. G., Sidley, N. A., Dockens, W. S., and Jolliffe, C. L. (1968). *J. Opt. Soc. Am.* **58,** 263–268.
Strother, G. K. (1963). *Exp. Cellular Res.* **29,** 349–355.
Svaetichin, G. (1956). *Acta Physiol. Scand.* **39,** Suppl. 134, 17–46.
Svaetichin, G., and MacNichol, E. F., Jr. (1958). *Ann. N. Y. Acad. Sci.* **74,** 385–404.
Svaetichin, G., Laufer, M., Mitarai, G., Fatechand, R., Vallecalle, E., and Villegas, J. (1961). *In* "The Visual System: Neurophysiology and Psychophysics" (R. Jung and H. Kornhuber, eds.), pp. 445–456. Springer, Berlin.
Svaetichin, G., Negishi, K., and Fatechand, R. (1965). *In* "Colour Vision: Physiology and Experimental Psychology" (A. V. S. de Reuck and J. Knight, eds.), 178–205. Little, Brown, Boston, Massachusetts.
Tomita, T. (1957). *Jap. J. Physiol.* **7,** 80–85.
Tomita, T. (1963). *J. Opt. Soc. Am.* **53,** 49–57.
Tomita, T. (1965). *Cold Spring Harbor Symp. Quant. Biol.* **30,** 559–566.
Tomita, T., Murakami, M., Sato, Y., and Hashimoto, Y. (1959). *Jap. J. Physiol.* **9,** 63–68.
Tomita, T., Kaneko, A., Murakami, M., and Pautler, E. L. (1967). *Vision Res.* **7,** 519–531.
Toraldo di Francia, G. (1949). *J. Opt. Soc. Am.* **39,** 324.
Villermet, G. M., and Weale, R. A. (1965). *J. Roy. Microscop. Soc.* **84,** 565–569.
Villermet, G. M., and Weale, R. A. (1969). *J. Physiol.* **201,** 425–435.
Wachtel, H., and Kandel, E. R. (1967). *Science* **158,** 1206–1208.
Wagner, H. G., MacNichol, E. F., Jr., and Wolbarsht, M. L. (1960a). *Science* **131,** 1314.
Wagner, H. G., MacNichol, E. F., Jr., and Wolbarsht, M. L. (1960b). *J. Gen. Physiol.* **43** (Suppl.), 45–62.
Wagner, H. G., MacNichol, E. F., Jr., and Wolbarsht, M. L. (1963). *J. Opt. Soc. Am.* **53,** 66–70.
Wald, G. (1937). *Nature* **140,** 545–546.
Wald, G., and Brown, P. K. (1965). *Cold Spring Harbor Symp. Quant. Biol.* **30,** 345–359.
Wald, G., Brown, P. K., and Smith, P. H. (1953). *Science* **118,** 505–508.
Wald, G., Brown, P. K., and Gibbons, I. R. (1963). *J. Opt. Soc. Am.* **53,** 20–35.
Walls, G. L., and Judd, H. D. (1933). *Brit. J. Ophthal.* **17,** 641–675.
Walraven, P. L., and Bouman, M. A. (1960). *J. Opt. Soc. Am.* **50,** 780–784.
Watanabe, K., Tosaka, T., and Yokota, T. (1960). *Jap. J. Physiol.* **10,** 132–141.
Waterman, T. H. (1954). *Proc. Natl. Acad. Sci. U. S.* **40,** 258–262.
Weale, R. A. (1955a). *J. Physiol.* **127,** 572–586.
Weale, R. A. (1955b). *J. Physiol.* **127,** 587–591.
Weale, R. A. (1959). *Opt. Acta* **6,** 158–174.
Weale, R. A. (1962). *Vision Res.* **2,** 25–33.
Weale, R. A. (1964). *Science* **145,** 1205–1206.
Weale, R. A. (1965). *Photochem. Photobiol.* **4,** 67–87.

Weale, R. A. (1968a). *In* "Photophysiology" (A. C. Giese, ed.), **4**, 1–45, Academic Press Inc., New York.

Weale, R. A. (1968a). *Photophysiology* **4**, 1–45.

Weale, R. A. (1968b). *Nature* **220**, 583.

Weale, R. A. (1968c). *Nature* **218**, 238–240.

Weale, R. A. (1969a). *J. Physiol* **204**, 123–124.

Weale, R. A. (1969b). *In* "An Introduction to Photobiology" (C. P. Swanson, ed.). Prentice-Hall, New York.

Werblin, F., and Dowling, J. (1969). *J. Neurophysiol.* **32**, 339–355.

Wiesel, T. N., and Hubel, D. H. (1966). *J. Neurophysiol.* **29**, 1115–1156.

Witkovsky, P. (1967). *J. Neurophysiol.* **30**, 546–561.

Witkovsky, P. (1968). *Vision Res.* **8**, 823–837.

Yamada, E., and Ishikawa, T. (1965). *Cold Springs Harbor Symp. Quant. Biol.* **30**, 383–391.

Chapter 6

THE ACTION OF ULTRAVIOLET LIGHT ON MAMMALIAN CELLS*

Robert B. Painter

Laboratory of Radiobiology, University of California Medical Center, San Francisco, California

1. Introduction

As a result of the relatively rapid advances that have been made in studying the effects of ultraviolet light (UV) on bacteria, there has been a renewed interest in the effects of UV on mammalian cells. In 1960, the discovery of the thymine dimer by Beukers and Berends was followed by work showing that this and other pyrimidine dimers in UV-irradiated bacterial DNA were photoproducts which were obviated by either dark (excision) or light (photoreactivation) repair processes (Setlow and

* Work performed under the auspices of the U. S. Atomic Energy Commission.

Carrier, 1964; Boyce and Howard-Flanders, 1964; Goodgal *et al.*, 1957). The possibility that similar processes occur in mammalian cells has stimulated radiobiologists interested in the effects of ionizing radiation on mammalian cells to look for similar phenomena in mammalian cells.

Much of the remarkable progress in understanding the action of UV on bacterial systems has been a result of the use of mutant cell lines. The relatively advanced state of bacterial genetics permits great flexibility in choosing (or producing) strains that aid in making those correlations between repair and enhanced survival upon which most of our present understanding of repair processes depends. The present state of mammalian somatic cell genetics, however, is so primitive that such an approach is impossible. Recent advances in methods for inducing and detecting mutants in cultured mammalian cells (Puck and Kao, 1967; Chu and Malling, 1968) may be the prototypes of a kind of mammalian cell genetics that soon will permit experimental approaches of the kind now possible with bacteria.

2. UV-Induced Lesions in Mammalian Cells

The work with bacteria preceded the examination of mammalian DNA after UV-irradiation. It is noteworthy that little effort has been expended to search for photoproducts other than pyrimidine dimers. Evidently, it has been assumed that the same photoproducts occur in mammalian DNA as in bacteria. There is no *a priori* reason why this should be so, and the intimate association of protein with eucaryotic DNA suggests that photoproducts containing derivatives of both macromolecules, if they occur, would be easily uncovered in this kind of cell.

Trosko *et al.* (1965) irradiated Chinese hamster (CH461) cells with 2650 Å UV. They found that the dimer/thymine ratio in DNA was approximately 0.05% per 100 ergs/mm^2. The dimer content of DNA was not significantly affected by attempted photoreactivation with 3300–4000 Å or by incubation of the cells for 24 hours after the irradiation. Later, Trosko and Kasschau (1967) showed that no excision occurred in the form of oligonucleotides after 48 hours incubation. Klimek (1965, 1966) found dimers in mouse L cell DNA after UV irradiation, and neither the attempted photoreactivation nor the prolonged incubation of the cells reduced the dimer/thymine ratio. Pfefferkorn and Coady (1968), in conjunction with studies on photoreactivation of pseudorabies virus, found no loss of UV-induced dimers in rabbit cells exposed to photoreactivating light for 4.5 hours after irradiation.

Steward and Humphrey (1966) measured dimers produced by UV in Chinese hamster B14 cells at various stages of the cell cycle in synchronous cultures. Their data showed that cells irradiated with 500–100,000 ergs/mm^2 in G_1 contained about 20% fewer thymine dimers than did cells irradiated at any other time in the cycle. They preferred the interpretation that a greater attenuation of UV occurred in G_1 than in the other phases. Considering the fact that the G_1 cells were irradiated just 1½ hours after attachment of the mitotic cells, it is possible that attenuation resulted from the cells being partially rounded. The last sampling time in these experiments was 11 hours after irradiation, corresponding to late S or early G_2, and so the possibility of a recurring phase at which thymine dimers were less abundantly formed was not investigated.

Regan *et al.* (1968) found that the yield of dimers in three strains of human cells was almost identical to that observed in Chinese hamster cells, but that up to 50% of the dimers were removed in the first 24 hours after irradiation. This was the first report of dimer excision in mammalian cells. Neither caffeine (9 mM) nor hydroxyurea (2.5 mM) affected this excision.

Horikawa *et al.* (1968) showed that thymine dimers were formed with the same efficiency in three different kinds of mammalian cells: mouse L, mouse Ehrlich ascites, and porcine PS. The dimers were not removed from the DNA in either the L or PS cells, but about 30% of the dimers disappeared from the ascites cell DNA and appeared in the acid soluble fraction in the first 24 hours after irradiation with 400 ergs/mm^2; no more excision occurred over the next 96 hours. These data suggest that excision ability is not species specific, and leave in doubt what phylogenetic rules govern the ability of mammalian cells to excise dimers.

3. UV-Induced Chromosome Aberrations

Humphrey *et al.* (1963) examined the frequency of chromosome aberrations as a function of the position of Chinese hamster cells in the cell cycle, by pulse-labeling with ^{3}HTdR, irradiating with 2537 Å, and scoring the aberrations in autoradiographic preparations. By far the greatest frequency of aberrations occurred in labeled cells appearing at long intervals (12–16 hours) after irradiation. Presumably these were cells in early S at the time of the irradiation, but this is uncertain since differential delays could confuse the results. G_1 and G_2 cells showed about 10% the damage of these "early S" cells while labeled cells appearing in the first few hours showed the least damage, about 4% the damage of early S cells.

Chu (1965) examined the effects of 100 ergs/mm^2 of UV at seven wavelengths between 2250 Å and 3650 Å on chromatid aberrations in

Chinese hamster cells. From these experiments he plotted an action spectrum which showed a broad maximum between 2400 and 2800 Å. He also showed a provocative difference between the kind of chromosome aberrations observed after irradiation with 2650 Å and those seen after irradiation with 2800 Å. In the latter case, a high frequency of "chromosome demolition" occurred in the metaphases after relatively moderate doses; the phenomenon occurred only after high exposure to 2650 Å. Puromycin at concentrations which, in the absence of irradiation, produced no chromosome damage enhanced the effect. These results suggest that effects on proteins may play an important role in the induction of UV-induced chromosome aberrations.

Trosko and Brewen (1967) reported that BUdR greatly sensitized Chinese hamster cells to UV-induced aberrations, an observation that had also been made by Chu (1965). They found that cysteamine had no effect on the incidence of aberrations in cells that were grown normally, but reduced the frequency of aberrations in cells grown in BUdR.

4. The Action of UV on Reproductive Integrity

4.1 Asynchronous Populations

A summary of the survival parameters, D_0 and N (Elkind and Whitmore, 1967), of several kinds of cultured mammalian cells is shown in Table I. Several aspects of these data are worthy of discussion.

As first pointed out by Lee and Puck (1960), the value for the D_0 of mammalian cells seems surpisingly high, when compared with that of microorganisms and viruses, especially in contrast to the difference in sensitivity to X rays between these kinds of organisms. For instance, mammalian cells are more resistant to UV than viruses and yet almost infinitely more sensitive to X rays. Lee and Puck (1960) proposed that the polyploidy of HeLa cells might be responsible for this relative radioresistance. Because UV induces relatively few chromosomal aberrations compared to X rays, little loss of genetic material would occur from this process, and therefore killing by UV would depend mostly on direct inactivation of a necessary gene. Two or more of each gene exist in every mammalian cell, in contrast to haploid viruses or microorganisms, so both copies would have to be inactivated before death would occur. No experimental proof for this hypothesis exists, but it is an attractive idea.

With the exception of the data reported by the Japanese workers, the range of values for D_0 among mammalian cells thus far studied is only about four-fold and the range of values for N only about two-

TABLE I

Cell line	Species	D_0 ergs/mm²	N	Reference
HeLa	Human	108	2	Lee and Puck (1960)
D98/AG	Human	70 (280)[a]	—	Djordjevic and Szybalski (1960) Erikson and Szybalski (1961, 1963)
L	Mouse	97	1.8	Han *et al.* (1964)
L-60T	Mouse	70–80	4–5	Rauth and Whitmore (1966) Rauth (1967)
V79	Chinese hamster	50	4–5	Sinclair and Morton (1965)
F	Chinese hamster	50	4.5	Arlett (1967)
X		50	4.5	
M3-1	Chinese hamster	26–32[b]	3–5	Todd *et al.* (1968)
L	Mouse	400	1	Horikawa *et al.* (1968)
PS	Pig	210	1	Horikawa *et al.* (1968)
Ehrlich Ascites	Mouse	700–800	1	Horikawa *et al.* (1968)
T	Human	22–28	2–3	Bootsma and Humphrey (1968, and personal communication 1969)
DON	Chinese hamster	25	2–3	Bootsma and Humphrey (1968, and personal communication 1969)
V79	Chinese hamster	43	1.6	Cleaver (personal communication 1969)
V7979	Chinese hamster	30	1.3	Cleaver (personal communication 1969)
DFAF	Chinese hamster	60	3.0	Cleaver (personal communication 1969)

[a] 280 ergs/mm² when the cells were irradiated in suspension, but about 70 ergs/mm² when they were allowed to attach to glass and flatten before irradiation.

[b] Range of results with wavelengths from 2587 to 2803 Å.

fold. Thus, there has not been found an "exquisitely" UV-sensitive mammalian cell which would be of such obvious value for the study of repair processes.

The data of Horikawa *et al.* (1968) require special comment. Their data indicate that a four-fold difference exists between the PS (porcine, $D_0 = 210$ ergs/mm²) and Ehrlich cells (mouse, $D_0 = 800$ ergs/mm²). Moreover, their values for D_0 far exceed any other thus far reported, except for the 280 ergs/mm² reported by Erikson and Szybalski (1961) for D98AG (human) cells irradiated in suspension. Djordjevic and Szybalski (1960) found, however, that the D_0 for the latter cells was, in reality, less than 100 ergs/mm² when the cells were irradiated after attachment and flattening. The Japanese workers irradiated their cells in suspension, so that the high values they reported may be the result of protection (e.g., by cytoplasmic shielding) while in the rounded state in suspension. On the other hand, Rauth (1967) obtained a D_0 of 70–80 ergs/mm^2 for L cells in suspension. Thus, the possibility that rounding protects cells from UV is still not established.

4.2 Synchronous Populations

Erikson and Szybalski (1963) attempted to determine if the sensitivity of D98AG (human) cells varied with position in the cell cycle by treating with fluorouracil deoxyriboside (FUdR), a specific inhibitor of DNA synthesis, for 20 hours, and then measuring survival at various times after its removal. They reported somewhat complicated survival curves (see Fig. 1), but it was obvious that a considerable variation in survival existed at different times after the induction of "synchrony." The most resistant time was 8 hours after release from the FUdR block, after which the cells seemed to become progressively more radiosensitive. With these cells, 8 hours corresponds to the time just prior to the onset of cell division (late G_2). The most radiosensitive time, corresponding to late G_1 or early S, seemed to be immediately after removal of the blocking agent.

Although these results indicate high sensitivity in early S and high resistance in G_2, there are complicating factors which cloud the interpretation. FUdR, the agent used for induction of synchrony, is known

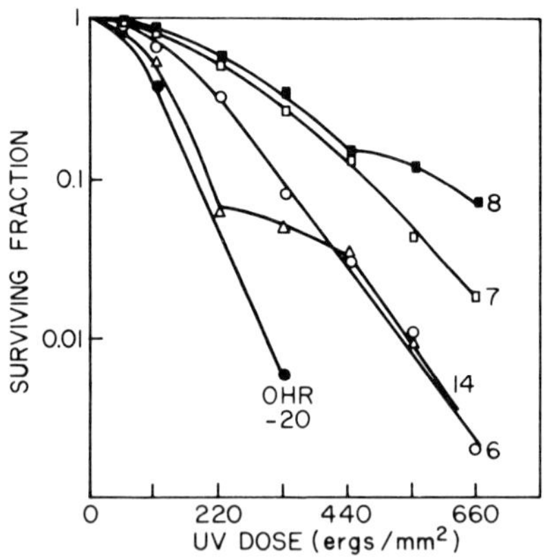

FIG. 1. Ultraviolet survival of colony-forming capacity of random (—20) and of synchronized populations of D98/AG cells, either 3 hours after attachment to the plastic surface (—20), or after 20 hours of exposure to 0.02 μg of FUdR per milliliter (0 hour) and at 6, 8, 14, and 20 hours after TdR (10^{-5} M) reversal. All irradiations were performed on surface-attached cells in a series of parallel dishes seeded with the same inoculum; the individual points correspond to the averages calculated from two duplicate dishes.

to cause some deleterious effects to mammalian cells (Taylor *et al.*, 1962). Moreover, the cells were blocked only once so that they were distributed throughout S and subsequent synchrony was imperfect. This may explain the two component survival curves they observed at 8 and 14 hours after FUdR removal.

Sinclair and Morton (1965) studied the phase sensitivity of synchronized Chinese hamster V79 cells to UV. In agreement with the results of Erikson and Szybalski (1963) they found the most radioresistant period to be G_2, which had a D_0 almost twice that of G_1 or early S cells. At the latter two times, the slopes of the survival curves were almost identical, but the early S cells had an extrapolation number of 10, compared to 3 for G_1 cells.

Rauth and Whitmore (1966) examined the stage sensitivity of mouse L cells to UV, using the "^{3}HTdR window" technique. Cells were synchronized by killing all cells except a few in late G_1 by incubation for a few hours with lethal concentrations of tritiated thymidine (^{3}HTdR). They obtained the same results by using FUdR to induce synchrony. These investigators found that the most radioresistant phase was also either late S or G_2, thus agreeing with the work on Chinese hamster and human (D98AG) cells.

Djordjevic and Tolmach (1967) measured the survival of HeLa S3 cells at various times after synchronization. Again, the most radioresistant stage was at 21 hours after mitotic cell collection, which corresponds to late G_2. In the case of the HeLa S3 cells, however, the most radiosensitive period was the middle of S whereas L cells were most sensitive in late G_1 or early S.

A number of theories have been advanced for the changing radiosensitivity to UV of mammalian cells through the cell cycle, including change in number of "targets" (DNA copies?), change in efficiency of photoproduct formation, and a change in the cells' capacity to repair damage. Each of these is contradicted by some evidence in one or another of the systems that have been studied, however. The basis of variability in response may be due only to a changing amount of UV-absorbing material between the source of UV and the radiosensitive nucleus. Djordjevic and Szybalski (1960) showed that flattened cells are much more radiosensitive than rounded cells in suspension. Mammalian cells begin to round when they approach mitosis and are rounded during mitosis. Since in all systems thus far studied the times just before or during division are those when mammalian cells are most resistant to UV, this resistance may be due simply to an increased amount of overlying cytoplasm, components of which partially shield the nucleus from the UV.

4.3 Action of Agents that Enhance UV-Induced Lethality

4.3.1 Base Analogues

Certain chemicals resemble naturally occurring compounds so closely that the enzymatic machinery of cells will use them in place of the natural material. In some cases, these analogues will be incorporated into macromolecules and subsequently affect their properties. An especially valuable group of analogues are the halogenated uridine deoxyribosides, which either are incorporated into DNA (bromouracil deoxyriboside, BUdR; chlorouracil deoxyriboside, CUdR; iodouracil deoxyriboside, IUdR) or attach to and block the enzymes responsible for methylation of uracil deoxyriboside derivates to form thymidine derivatives (fluorouracil deoxyriboside, FUdR). It has been known for some time that incorporation of BUdR or IUdR into DNA sensitizes bacteria to the action of ultraviolet light (Zamenhof, 1959).

Djordjevic and Szybalski (1960) first measured the effect of BUdR and IUdR on the sensitivity of mammalian (D98AG) cells to UV. They found that the extent of sensitization was proportional to the amount of substitution of BUdR for thymidine; however, the increase in radiosensitivity was relatively inconsequential until both strands of the DNA were substituted with the analogue. Erikson and Szybalski (1963) compared the ability of BUdR and IUdR, as well as bromocytidine deoxyriboside (BCdR) and iodocytidine deoxyriboside (ICdR), to sensitize D98AG cells to UV. The bromine-containing analogues were slightly more effective than the iodine-containing ones, but the pyrimidine upon which the halogen was substituted did not affect the extent of radiosensitization. At high substitutions by BUdR or BCdR, the shoulders of the survival curves were almost abolished and the extent of sensitization (D_0 control/D_0 analogue) reached about eight-fold.

4.3.2 Other Agents Enhancing Lethality

Caffeine enhances UV-induced killing of bacteria (Metzger, 1964) and its action is believed to result from an effect on DNA repair systems (Setlow and Carrier, 1967). Rauth (1967) showed that the inclusion of caffeine in media after UV irradiation markedly increased the sensitivity of L cells to UV, and suggested that this might result from an effect on a repair system. Domon and Rauth (1969) later showed that this synergism of caffeine and UV only occurred in cells in the S phase. For example, when G_2 cells were irradiated in the presence of caffeine, no enhanced killing occurred, if the medium was changed to one without caffeine before the cells reached the next S phase. Synergism did occur if the cells entered the S phase in the presence of caffeine. Although it is fashionable, at the time of this writing, to ascribe such synergism of radiation with chemicals to an effect of the latter on a repair system, it must be kept in mind that other mechanisms may be involved. Thus, although the bacterial experience suggests the caffeine effect to be on DNA repair, this may not be so, especially since L cells apparently do not excise dimers (Klimek, 1966) and caffeine does not affect mammalian repair replication (Cleaver, 1969a). That the problem in mammalian cells may be more complex is made obvious by the work of Arlett (1967) who showed that caffeine has an effect on the survival of UV-irradiated Chinese hamster cells only at concentrations where the drug itself is toxic.

4.4 The Effect of "Split Doses"

When X irradiation is delivered to mammalian cells in two doses, separated by one or two hours, the killing effect is less than when the same total dose is administered all at once. This effect, first reported by Elkind and Sutton (1960), shows that the cells recover from some of the damage inflicted by the first (conditioning) dose, and has been used as evidence for repair of sublethal damage in mammalian cells. Similar experiments have been performed with UV, but no evidence for repair of sublethal damage has been found. Han *et al.* (1964) irradiated L cells with 150 ergs/mm^2 at various times up to 12 hours after a first exposure to 250 ergs/mm^2. The effect of the total exposure of 400 ergs/mm^2 was no less when delivered in two parts than when delivered at one time; indeed, there appeared to be a lower survival when the second exposure was given five hours or more after the first. Very similar results were obtained by Todd *et al.* (1968) with Chinese hamster cells. These authors correctly cautioned, however, that it is inappropriate to assume that no recovery occurred between exposures, since asynchronous cultures were used, and the results could be due to a combination of recovery and the progression of cells to a more sensitive stage of the cell cycle. Nevertheless, at the present time there is no evidence to suggest that recovery from any kind of UV-induced sublethal damage occurs. It is important, however, to note that repair of potentially lethal damage could still occur.

5. Action of UV on RNA and Protein Synthesis

Little work has been done in mammalian cells on the effects of UV on RNA and protein synthesis. This is surprising because there is now a wealth of information on the sites of synthesis of the various RNA species and their roles in protein synthesis. This lack of attention to effects of UV on RNA synthesis reflects a tendency to work out the molecular events in procaryotic forms before application to higher cells. As will be discussed, however, the use of UV on mammalian cells may eventually be a powerful tool to help elucidate the fine details of RNA and protein synthesis and transport.

Perry and Errera (1960) and Perry *et al.* (1961) irradiated HeLa nucleoli with a high intensity UV microbeam which inhibited the forma-

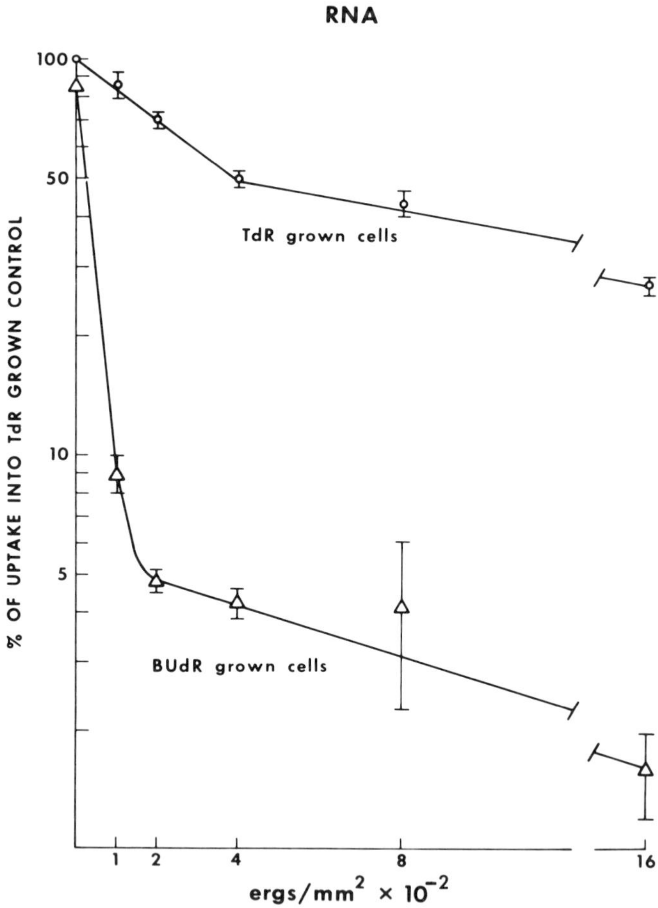

FIG. 2. Effect of UV on uptake of ^{3}H-uridine into HeLa RNA. ○——○, relative specific activity of RNA from cells grown for 48 hrs prior to irradiation in the presence of 1.6×10^{-5} M thymidine; △———△, relative specific activity of RNA from cells grown for 48 hrs prior to irradiation in the presence of 1.6×10^{-5} M BUdR. All data plotted as percent of specific activity of thymidine-grown control.

tion of approximately ⅔ of cytoplasmic RNA. Irradiation of areas equal to nucleoli in the chromatin portion of the nucleolus or in the cytoplasm showed much less effect. These experiments were the first to suggest the predominant role of the nucleolus in the formation of cytoplasmic RNA. Later, Perry (1964) and Penman *et al.* (1966) established the predominant role of the nucleolus in the formation of ribsomal RNA.

Recently, Painter (1969) has shown that the presence of BUdR in HeLa DNA greatly enhances the inhibitory effects of UV on RNA synthesis. When cells grown for two days in the presence of 1.6×10^{-5} M BUdR were irradiated with UV, there occurred a dose dependent inhibition of RNA synthesis which was an order of magnitude greater than in normally grown cells (Fig. 2). At exposures of 200 ergs/mm^2 or more, the incorporation of ^{3}H-uridine into RNA was less than 5% of that occurring in controls. After pulse labeling cells for 20 minutes prior to UV, there was a definite effect on nucleolar RNA synthesis, since nucleolar labeling was not increased in UV-treated cells at two hours after the pulse as it was in controls. Moreover, cytoplasmic labeling, which normally becomes obvious at two hours and intense at six hours after labeling, did not occur at all within the first six hours in BUdR-substituted cells. Thus, the synergistic effect of BUdR and UV appears to be on both synthesis and transport of RNA. Neither treatment with BUdR alone nor with 200 ergs/mm^2 UV alone gave the same results. Exploitation of these observations may lead to experiments that will improve our understanding of mammalian cell RNA transport, which is at present very poorly understood.

6. The Action of UV on DNA Synthesis

The first attempt to measure the effect of UV on mammalian DNA synthesis was that of Whitfield *et al.* (1961). They measured the total DNA of mouse L cells at rather long intervals (several hours) after the irradiation and concluded that DNA synthesis was "relatively insensitive to UV light; the principal effect . . . being the inhibition of cell divisions." In retrospect, it is obvious that the experimental design led to this interpretation, which is the antithesis of current interpretations.

Powell (1962a) also examined the DNA content per cell of L cells after UV. He found that low exposures (about 35 ergs/mm^2) inhibited DNA synthesis and confirmed this observation by autoradiography with tritiated thymidine. He also showed that this effect, which was dose dependent, did not result from an effect on the intracellular pools of DNA thymine precursors or on the phosphorylation of thymidine (Powell, 1962b).

Rasmussen and Painter (1964) showed an exposure-dependent inhibition of DNA synthesis in HeLa cells, as measured by incorporation of several precursors into DNA after UV. The plot of their results yielded a two (or more) component curve, the steeper portion of which was a simple exponential. These authors also examined the effect of prior incorporation of BUdR into the DNA. They found that this treatment enhanced the inhibition of DNA synthesis at exposures of 100 ergs/mm^2 or less, when measured by incorporation of either thymidine or guanine into DNA. At exposures above 100 ergs/mm^2, however, the two precursors yielded different results, even when used in double-labeling experiments. While the action of higher exposures of UV was to further depress the incorporation of guanine, exposures of 200 ergs/mm^2 or more stimulated thymidine incorporation (see Fig. 3). A similar result has been observed also in the diploid human cell WI-38 (Rasmussen 1968).

Cleaver (1965, 1967) has done the most extensive work on effects of UV on rate of mammalian DNA synthesis. The first of these experi-

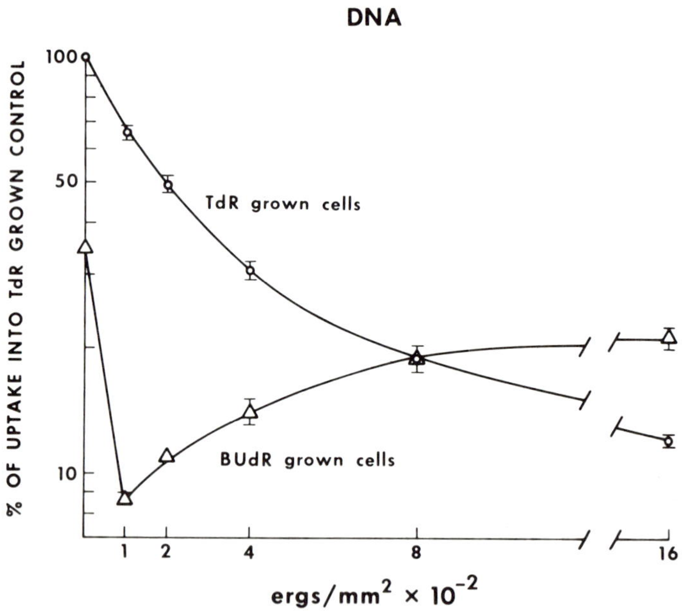

FIG. 3. Effect of UV on uptake of ^{14}C-thymidine into HeLa DNA. ○——○, relative specific activity of DNA from cells grown for 48 hrs prior to irradiation in the presence of 1.6×10^{-5} M thymidine; △——△, relative specific activity of DNA from cells grown for 48 hrs prior to irradiation in the presence of 1.6×10^{-5} M BUdR. All data plotted as percent of specific activity of thymidine-grown control.

ments was done by Dendy and Cleaver (1964) in which the effects of rather high exposures (240 and 960 ergs/mm^2) were measured as a function of the position of the mouse L cell in the S phase. The authors concluded that the sensitivity to UV increased as the DNA value of the cell increased; in other words, the inhibitory effect of UV on the rate of DNA synthesis increased as the cell progressed through the S phase.

Cleaver (1967) showed that the extent of inhibition of rate of DNA synthesis did not depend upon the temperature during irradiation, but was strongly dependent on the post-irradiation temperature of incubation —the higher the temperature, the greater the inhibition. At all tested incubation temperatures (25°, 31°, 37°C) a semilog plot of relative incorporation (as log of grain count) against UV exposure yielded a two (or more) component curve with a final slope of almost 0, for both mouse L cells and human amnion FL cells. These results, along with those of Rasmussen and Painter (1964), suggest that the steep component is a result of blocking of normal semiconservative DNA synthesis and that the resistant component may derive from a type of synthesis that does not normally precede mitosis. Although in some cell systems (especially of human origin) much of this UV-resistant DNA synthesis may be of the nonconservative kind now called "repair replication" (see Section 7.2), the latter occurs to such a low extent in rodent cells that it cannot be the explanation in these cells.

Cleaver (1967) also showed that DNA synthesis continued up to about one hour after irradiation with exposures up to 500 ergs/mm^2. From this observation, he argued that thymine dimers are not effective blocks to mammalian DNA synthesis because DNA growing points would encounter a dimer within a very few minutes after irradiation. A probability model for bypassing one or two dimers has been proposed by Klimek and Vanicek (1968), but the basis for UV inhibition of mammalian DNA synthesis still is in doubt.

Djordjevic and Tolmach (1967) also observed an exposure-dependent depression of DNA synthesis in synchronized HeLa S3 cells. Not only was the UV effective on cells in the S phase, but also cells irradiated in G_1 were affected once they entered the S phase. Domon and Rauth (1968) reported very similar results with L cells. Irradiation of asynchronous cultures resulted in an exposure-dependent depression of H^3TdR uptake. Cells accumulated in S phase due to the effect on DNA synthesis rate, but there was essentially no change in the rate of movement of cells from G_1 into S. At 20–30 hours after irradiation with 40 or 100 ergs/mm^2 recovery was almost complete, but after 200 ergs/mm^2 no recovery of DNA synthesis was evident.

7. Repair of Mammalian DNA Damaged by UV

7.1 Lack of Photoreactivation in Placental Animals

The occurrence of photoreactivation of UV-induced damage to DNA is now known to be widespread throughout the animal kingdom. Strangely, however, it seems to be absent in cell tissues of placental mammals. Although there had been reports of photoreactivation processes (Roe and Stevens 1965, Logan *et al.* 1959, Rieck and Carlson, 1955), these probably do not derive from a dimer reversing process. Cleaver (1966) reviewed the subject briefly, but the most definitive experiments have come from the Oak Ridge group and J. Cook has reviewed this subject in this volume (see Chapter 7).

7.2 Repair Replication and Unscheduled DNA Synthesis

Rasmussen and Painter (1964) first reported a phenomenon that has since been called "unscheduled DNA synthesis." This is the autoradiographic demonstration of uptake of labeled thymidine into DNA by cells not in the normal phase of DNA synthesis (S phase). Rasmussen and Painter (1966) reported that two mouse lines and one Chinese hamster line out of a total of 13 different cell lines did not show this phenomenon, but recently Painter and Cleaver (1969) have shown that all cell lines, including several species, exhibit this phenomenon but to greatly different extents. In general, human cells show this unscheduled synthesis to a much greater extent than do cells derived from rodents. There is one important exception—skin cells from humans with the De Sanctis-Cacchione syndrome of the disease xeroderma pigmentosum show absolutely no unscheduled DNA synthesis after UV (Cleaver, 1968a). With the exception of this unique pathological situation, unscheduled synthesis occurs in all mammalian cells thus far examined. Djordjevic and Tolmach (1967) confirmed the presence of unscheduled DNA synthesis in HeLa cells and showed that it occurred to roughly the same extent in G_1 and G_2 cells and Evans and Norman (1968a) reported unscheduled synthesis after both UV and ionizing radiation in human lymphocytes.

Rasmussen and Painter (1966) and Cleaver and Painter (1968) showed that much of the DNA synthesis occurring in HeLa cells after high exposures of UV was not of the normal semiconservative type. This synthesis was similar to that which occurs in repairing strains of *Escherichia coli* and that which has been termed "repair replication" by Pettijohn and Hanawalt (1964; also see Volume 4 of this series). Repair replication occurs in all those cells that show unscheduled synthesis and the correlation between the two phenomena suggests they both reflect the same fundamental process (Painter and Cleaver, 1969). As with unscheduled DNA synthesis, cell lines from human sources show repair replication to a much greater extent than do cells from rodent sources and the only cells thus far examined that show no repair replication after UV irradiation are those from patients with the De Sanctis-Cacchione syndrome of xeroderma pigmentosum.

A number of features of repair replication and unscheduled DNA synthesis are puzzling and even paradoxical. The presence of bromouracil deoxyriboside (BUdR) in the DNA at the time of UV irradiation increases the amount of unscheduled DNA synthesis (Rasmussen and Painter, 1964, 1966; Cleaver, 1968b). If this synthesis were only reflecting a successful repair process, one would expect less repair with BUdR in the DNA since it is known that BUdR greatly enhances the killing effect of UV (see Section 4.3). Apparently, the increased amount of unscheduled synthesis reflects the greater damage that occurs to BUdR-containing DNA (Smith, 1962), but most of the synthesis is unsuccessful in actually effecting repair.

Unscheduled synthesis is unaffected by most agents that inhibit normal DNA synthesis (Cleaver, 1969a; Djordjevic and Tolmach, 1967; Evans and Norman, 1968b). Hydroxyurea, fluorouracil deoxyriboside, cytosine arabinoside and amethopterin do not affect repair synthesis at concentrations which depress normal synthesis to 10–20% of controls (Cleaver, 1969a). Agents that bind to DNA, such as acriflavine and actinomycin D, inhibit repair synthesis to about the same extent as they do normal synthesis. Caffeine, which apparently inhibits excision repair in bacteria (Setlow and Carrier, 1967), is without effect up to a concentration of 10^{-2} M on repair synthesis in HeLa cells. Iodoacetate and cycloheximide inhibit repair synthesis, but not to the same extent that they affect normal synthesis. It must be kept in mind that the total amount of repair synthesis is less than one percent of the semiconservative synthesis that normally occurs in the same time period; thus, small amounts of precursors that completely satisfy the requirements for the former are totally inadequate for the latter.

There are other unusual features of this kind of repair synthesis. It occurs in cells that are in various stages of mitosis (Rasmussen and Painter, 1964, 1966). Condensed chromatin has always been considered metabolically inert and it is well known that RNA synthesis does not occur during metaphase and anaphase (Feinendegen *et al.*, 1960; Goldstein and Micou, 1959). Apparently, the enzymes involved in repair synthesis are bound to the DNA throughout mitosis and are capable of functioning immediately after the induction of damage by UV. Extensive degradation of DNA does not accompany repair replication in mammalian cells (Rasmussen and Painter, 1964; Painter, 1968; Cleaver, 1968b; Cleaver and Trosko, 1969) as it does in bacteria (Boyce, 1966). This lack of degradation suggests that the repair system in mammalian cells is much more specific and under better control than the bacterial system.

Information thus far accumulated permits one to draw a model for mammalian repair replication. After damage that is recognizable to the system, an enzyme, presumably an endonuclease, makes a "nick" in the strand containing the damage. Cleaver (1969b) has shown that cells from xeroderma pigmentosum, although incapable of undergoing repair synthesis after UV under normal circumstances, can repair if the cells are grown with BUdR prior to UV, or if they are irradiated with X rays. Since both of these treatments probably produce single strand breaks, it appears that these cells are deficient in the enzyme catalyzing the initial step (the incision) of repair replication. If the incision is accomplished by the action of the damaging agent itself, enzymes of the subsequent steps are present to carry out the rest of the process. The damaged portions are excised, probably at the level of the nucleotide, but excision does not continue very far past the damaged area. This lack of continued excision suggests that the enzyme is not a classical exonuclease. Nucleotides are then inserted into the strand where the damaged ones were removed and a ligase (Gellert, 1967; Lindahl and Edelman, 1968) directs the final sealing.

7.3 Significance of Repair Synthesis

At the time of this writing, it is impossible to definitely assert that repair synthesis is a mechanism that enhances the survival of mammalian cells. Thus far, no extremely UV-sensitive mammalian lines exist, although Todd and Hellewell (1969) have found that the slow-growing Chinese hamster subline V7979, isolated by Sinclair, is about 30% more sensitive to UV than to the parent V79 line. Cleaver (1969c) has found that V79 cells show slightly more unscheduled DNA synthesis than do V7979 after low exposures of UV. The difference is so small, however, that one hesitates to infer that this is the basis for the difference in radioresistance. Of course, the results with xeroderma pigmentosum strongly suggest that repair synthesis is somehow necessary for normal cell function, but the exact relationship between repair synthesis and viability remains obscure.

7.4 Other Kinds of DNA Repair

The search for an excision repair system has been so intense that little attention has been paid to other possible kinds of repair of UV-induced damage to DNA. This is unfortunate, since the work of Trosko, *et al.* (1965) and Klimek (1966), who could find no evidence for dimer removal from rodent DNA, strongly suggested that excision repair was lacking or operating at a very low level in these cells. Rupp and Howard-Flanders (1967) showed that another kind of DNA repair exists in excision-negative mutants of *E. coli.* Their data suggested that DNA synthesis occurs up to a dimer, then somehow "jumps" around the dimer and continues synthesis. This results in gaps in newly made single strands of DNA which are detectable as low molecular weight fragments on alkaline density gradients. If, after the first thirty minutes when the short pieces are detectable, the cells are allowed to continue growth for a few hours, the DNA becomes of normal high molecular weight. Therefore, the low molecular weight fragments formed immediately after irradiation are somehow joined subsequently. Cleaver and Thomas (1969) have found that Chinese hamster DNA first made after UV irradiation is also in the form of small pieces and that these pieces are later incorporated into DNA of high molecular weight. These preliminary observations suggest that synthesis around dimers also occurs in mammalian cells. This may be a very important repair system, since it is obvious that some kinds of mammalian cells can survive with dimers remaining in their DNA.

There may exist other kinds of DNA repair of which we are unaware; for instance, exchange of undamaged areas of redundant DNA or delayed reversal or elimination of damaged sites may occur. As yet, however, there is no experimental evidence for any kind of repair other than those reviewed here.

8. The Action of UV on Progression through the Cell Cycle

The evidence thus far accumulated indicates that the primary effect of UV on progression of cells through the cell cycle is secondary only to the effect that inhibits the rate of DNA synthesis. There is good general agreement that at doses below 200 ergs/mm^2 this is the only important effect on progression. Djordjevic and Tolmach (1967) found that 70 ergs/mm^2 to HeLa cells delayed movement through the S phase, as a result of the inhibition of DNA synthesis, but did not delay movement through G_1, G_2, or mitosis. Domon and Rauth (1968) reported similar findings with mouse L cells. At exposures up to 200 ergs/mm^2, the only significant effect on progression was again secondary to the inhibition of DNA synthesis, so that cells irradiated in G_1 or S were delayed in reaching division, but cells in late S and G_2 showed little or no delay. At 100 and 200 ergs/mm^2, however, the total number of cells reaching division was reduced, suggesting an all or none effect; either no delay occurred (other than the S effect) or the cell died soon after the irradiation. Bootsma and Humphrey (1968) examined the effect of UV on both a human kidney cell line (T cells) and a diploid Chinese hamster line (DON cells). They found little or no mitotic delay in either line in cells irradiated in late G_2 and only a small delay in cells irradiated in early or mid-G_2. The delay in arrival at mitosis by cells irradiated in S phase was exposure-dependent and the delay was also longer for cells irradiated in early S than for cells irradiated in late S. Thus, the main effect was again due to the inhibition of rate of DNA snythesis. These authors also observed a phenomenon not reported by others for UV irradiation; they found that exposure of the human T line during G_1 caused cells to enter S earlier than expected. This precocious entry into S is similar to reports by Das and Alfert (1961) and Feinendegen, *et al.* (1968) for cells irradiated with X rays, but its significance is unknown.

The first data on progression of cells after UV were actually reported by Cleaver (1965), but he used only relatively high exposures to UV (240 to 960 ergs/mm^2). In general, his results were similar to others except for the highest dose, where an immediate mitotic delay was observed and a G_1 to S block also seemed to occur.

9. The Effects of UV Microbeams

The work of Perry and Errera (1960) and Perry *et al.* (1961) which used UV microbeam irradiation of HeLa nucleoli to study RNA synthesis and transport has been covered in Section 5. Dendy and Smith (1964) irradiated the nuclei, cytoplasms, and areas outside but near embryonic mouse cells and mouse L cells and recorded the effect on DNA synthesis as measured autoradiographically after incubation with ^{3}HTdR. They found that irradiation of any part of the cell, or even close to the cell, inhibited DNA synthesis, but a few hours were required for the effect to develop fully. Later, Dendy *et al.* (1967) found that the effect of UV irradiation outside the cell could be reversed by catalase and inferred that this phenomenon was based on production of a peroxide which diffused into the cell and exerted its effect. Smith and Dendy (1968) also followed the development of colonies after irradiation with the UV microbeam and found that, for equivalent absorbed doses, the nucleus was more sensitive than the cytoplasm. At relatively high doses, irradiation of the nucleolus caused more cell loss than did irradiation of nuclear sap. The inhibition of colony formation resulted both from cells that failed to divide at all and from cells that divided at a slower rate than controls. These authors also reported the induction by UV of giant cells and polynucleate cells, which have not been reported by others for UV irradiation, although both phenomena are common after X irradiation.

10. Summary

The exact means by which UV kills mammalian cells is still in doubt. Although the evidence that DNA is the primary target is good, the point cannot be considered established. Effects on RNA may play an important role, and models by which an effect on RNA could result in a lethal process are easily constructed (e.g., changes in a RNA base resulting in coding for an invalid enzyme or repressor). Direct interaction with proteins is also a possibility, as might be inferred from the peculiar effect of 280 nm irradiation in inducing a specific kind of chromosomal damage.

Circumstantial evidence for repair of UV damage is relatively abundant, but the lack of a usable UV-sensitive mutant obviates the possibility of making those correlations necessary to establish repair-like phenomena with enhanced reproductive integrity. The importance of these phenomena, however, is manifested by their absence in cancerous cells from humans with xeroderma pigmentosum. Even when repair is better understood and a rational basis for UV killing is determined, UV will probably continue to be a valuable tool to delve into the problems of macromolecular synthesis and transport.

ACKNOWLEDGMENTS

I am indebted to Dr. J. E. Cleaver for reviewing this manuscript, and I am most grateful to Mrs. Sarah Pusey and other members of the administrative staff for their patience in assembling chaotic piles of scrap paper into this final product.

REFERENCES

Arlett, C. F. (1967). *Intern. J. Radiation Biol.* **13,** 369.
Beukers, R., and Berends, W. (1960). *Biochim. Biophys. Acta* **41,** 550.
Bootsma, D., and Humphrey, R. M. (1968). *Mutation Res.* **5,** 289.
Boyce, R. P. (1966). *Nature* **209,** 688.
Boyce, R. P., and Howard-Flanders, P. (1964). *Proc. Natl. Acad. Sci. U. S.* **51,** 293.
Chu, E. H. Y. (1965). *Mutation Res.* **2,** 75.
Chu, E. H. Y., and Malling, H. V. (1968). *Proc. Natl. Acad. Sci. U. S.* **61,** 1306.
Cleaver, J. E. (1965). *Biochim. Biophys. Acta* **108,** 42.
Cleaver, J. E. (1966). *Biochem. Biophys. Res. Comm.* **24,** 569.
Cleaver, J. E. (1967). *Radiation Res.* **30,** 795.
Cleaver, J. E. (1968a). *Nature* **218,** 652.
Cleaver, J. E. (1968b). *Biophys. J.* **8,** 775.
Cleaver, J. E. (1969a). *Radiation Res.* **37,** 334.

Cleaver, J. E. (1969b). *Proc. Natl. Acad. Sci. U. S.* **63,** 428.

Cleaver, J. E. (1969c). *Int. J. Radiat. Biol.* **16,** 277.

Cleaver, J. E., and Painter, R. B. (1968). *Biochim. Biophys. Acta* **161,** 552.

Cleaver, J. E., and Thomas, G. H. (1969). *Biochem. Biophys. Res. Comm.* **36,** 203.

Cleaver, J. E., and Trosko, J. E. (1969). *Intern. J. Radiation Biol.* **15,** 411.

Das, N. R., and Alfert, M. (1961). *Proc. Natl. Acad. Sci. U. S.* **47,** 1.

Dendy, P. P., and Cleaver, J. E. (1964). *Intern. J. Radiation Biol.* **8,** 301.

Dendy, P. P., and Smith, C. L. (1964). *Proc. Roy. Soc. B* **160,** 328.

Dendy, P. P., Smith, C. L., and Aebi, H. E. (1967). *Photochem. Photobiol.* **6,** 461.

Djordjevic, B., and Szybalski, W. J. (1960). *J. Exptl. Med.* **112,** 509.

Djordjevic, B., and Tolmach, L. J. (1967). *Radiation Res.,* **32,** 327.

Domon, M., and Rauth, A. M. (1968). *Radiation Res.* **35,** 350.

Domon, M., and Rauth, A. M. (1969). *Radiation Res.* **39,** 207.

Elkind, M. M., and Sutton, H. (1960). *Radiation Res.* **13,** 556.

Elkind, M. M., and Whitmore, G. F. (1967). *In* "The Radiobiology of Cultured Mammalian Cells." Gordon and Breach, New York.

Erikson, R. L., and Szybalski, W. (1961). *Biochem. Biophys. Res. Comm.* **4,** 258.

Erikson, R. L., and Szybalski, W. (1963). *Radiation Res.* **18,** 200.

Evans, R. G., and Norman, A. (1968a). *Nature* **217,** 455.

Evans, R. G., and Norman, A. (1968b). *Radiation Res.* **36,** 287.

Feinendegen, L. E., Bond, V. P., Shreeve, W. W., and Painter, R. B. (1960). *Exptl. Cell Res.* **19,** 443.

Feinendegen, L. E., Drew, R. M., Bond, V. P., and Heiniger, H. J. (1968). *In* "Effects of Radiation on Cellular Proliferation and Differentiation," p. 39. IAEA, Vienna.

Gellert, M. (1967). *Proc. Natl. Acad. Sci. U. S.* **57,** 148.

Goldstein, L., and Micou, L. (1959). *J. Biophys. Biochem. Cytol.* **6,** 1.

Goodgal, S. H., Rupert, C. S., and Herriott, R. M. (1957). *In* "The Chemical Basis of Heredity" (W. D. McElroy and B. Glass, eds.). Johns Hopkins Press, Baltimore, Maryland.

Han, A., Miletic, B., and Petrovic, D. (1964). *Intern. J. Radiation Biol.* **8,** 187.

Horikawa, M., Nikaido, O., and Sugahara, T. (1968). *Nature* **218,** 489.

Humphrey, R. M., Dewey, W. C., and Cork, A. (1963). *Radiation Res.* **19,** 247.

Klimek, M. (1965). *Neoplasma* **12,** 559.

Klimek, M. (1966). *Photochem. Photobiol.* **5,** 603.

Klimek, M., and Vanicek, J. (1968). *Studia Biophysica* **12,** 83.

Lee, H., and Puck, T. T. (1960). *Radiation Res.* **12,** 340.

Lindahl, T., and Edelman, G. M. (1968). *Proc. Natl. Acad. Sci. U. S.* **61,** 680.

Logan, R., Errera, M., and Ficq, A. (1959). *Biochim. Biophys. Acta* **32,** 147.

Metzger, K. (1964). *Biochem. Biophys. Res. Comm.* **15,** 101.

Painter, R. B. (1968). *In* "Effects of Radiation on Cellular-Proliferation and Differentiation," p. 91. IAEA, Vienna.

Painter, R. B. (1969). *Proc. Radiation Res. Soc.,* 17th Ann. Meeting.

Painter, R. B., and Cleaver, J. E. (1969). *Radiation Res.* **37,** 451.

Penman, S., Smith, I., and Holtzman, E. (1966). *Science* **154,** 786.

Perry, R. P. (1964). *Natl. Cancer Inst. Monograph* **14,** 73.

Perry, R. P., and Errera, M. (1960). *In* "The Cell Nucleus" (J. S. Mitchell, ed.), p. 24. Academic Press, New York.

Perry, R. P., Hell, A., and Errera, M. (1961). *Biochim. Biophys. Acta* **49,** 47.

Pettijohn, D. E., and Hanawalt, P. C. (1964). *J. Mol. Biol.* **9,** 395.

Pfefferkorn, E. R., and Coady, H. M. (1968). *J. Virology* **2,** 474.
Powell, W. F. (1962a). *Biochim. Biophys. Acta* **55,** 969.
Powell, W. F. (1962b). *Biochim. Biophys. Acta* **55,** 979.
Puck, T. T., and Kao, F-t. (1967). *Proc. Natl. Acad. Sci. U. S.* **58,** 1227.
Rasmussen, R. E. (1968). Ph.D. Dissertation, University of California.
Rasmussen, R. E., and Painter, R. B. (1964). *Nature* **203,** 1360.
Rasmussen, R. E., and Painter, R. B. (1966). *J. Cell Biol.* **29,** 11.
Rauth, A. M. (1967). *Radiation Res.* **31,** 121.
Rauth, A. M., and Whitmore, G. F. (1966). *Radiation Res.* **28,** 84.
Regan, J. D., Trosko, J. E., and Carrier, W. L. (1968). *Biophys. J.* **8,** 319.
Rieck, A. F., and Carlson, S. D. (1955). *J. Cell. Comp. Physiol.* **46,** 301.
Roe, E. M. F., and Stevens, G. H. (1965). *Photochem. Photobiol.* **4,** 759.
Rupp, W. D., and Howard-Flanders, P. (1967). *J. Mol. Biol.* **31,** 291.
Setlow, R. B., and Carrier, W. L. (1964). *Proc. Natl. Acad. Sci. U. S.* **51,** 226.
Setlow, R. B., and Carrier, W. L. (1967). *In* "Replication and Recombination of Genetic Material" (W. J. Peacock and R. D. Brock, eds.), p. 134. Australian Acad. Sci., Canberra.
Sinclair, W. K., and Morton, R. A. (1965). *Biophys. J.* **5,** 1.
Smith, C. L., and Dendy, P. P. (1968). *Cell and Tissue Kinet.* **1,** 225.
Smith, K. C. (1962). *Biochem. Biophys. Res. Comm.* **8,** 157.
Steward, D. L., and Humphrey, R. M. (1966). *Nature* **212,** 298.
Taylor, J. H., Haut, W. F., and Tung, J. (1962). *Proc. Natl. Acad. Sci. U. S.* **48,** 190.
Todd, P. W., and Hellewell, A. B. (1969). *Mutation Res.* **7,** 129.
Todd, P. W., Coohill, T. P., and Mahoney, J. (1968). *Radiation Res.* **35,** 390.
Trosko, J. E., and Brewen, J. G. (1967). *Radiation Res.* **32,** 200.
Trosko, J. E., and Kasschau, M. R. (1967). *Photochem. Photobiol.* **6,** 215.
Trosko, J. E., Chu, E. H. Y., and Carrier, W. L. (1965). *Radiation Res.* **24,** 667.
Whitfield, J. F., Rixon, R. H., and Youdale, T. (1961). *Exptl. Cell Res.* **22,** 450.
Zamenhof, S. (1959). *Ann. N. Y. Acad. Sci.* **81,** 784.

Chapter 7

PHOTOREACTIVATION IN ANIMAL CELLS*

John S. Cook

University of Tennessee-Oak Ridge Graduate School of Biomedical Sciences and Biology Division, Oak Ridge National Laboratory, Oak Ridge, Tennessee

1. Introduction

Photoreactivation is the mitigation of the effects of far ultraviolet (UV) radiation (usually wavelengths between 220–310 nm) on a biological system by subsequent illumination of the system with light of longer wavelengths (usually in the range 310–500 nm). The significance of the phenomenon was first clearly recognized and described by Kelner (1949) and independently by Dulbecco (1949). Following their papers,

* Research supported by the U. S. Atomic Energy Commission under contract with the Union Carbide Corporation.

the first descriptions of photoreactivation in animal cells (sea urchin gametes) appeared in that same year (Blum *et al.*, 1949, Marshak, 1949) and by the end of 1952 (see Dulbecco, 1955 and Jagger, 1958) it was clear that the phenomenon had a very broad phylogenetic distribution.

The enzymatic basis (Rupert, 1960) of many photoreactivation phenomena is now well established; the enzyme acts in the light to split the dimers which are formed between adjacent pyrimidines in UV-irradiated DNA. Unrepaired, such dimers can result in important biological effects (Setlow, 1964). The enzyme has no other known function, and pyrimidine dimers in polydeoxynucleotides are its only known substrates. The biochemistry of photoreactivation has been reviewed extensively in recent years (Rupert, 1964; J. Setlow, 1966a, 1967) and in this chapter only the most salient facts are outlined (Section 2). One of the most important of these facts is that *biological* photoreactivation, as defined above, can come about by very indirect means. The occurrence of biological photoreactivation does not necessarily require the functioning or even the existence of a photoreactivating enzyme. (See Section 2.)

Our knowledge of how photoreactivating enzyme acts is derived almost entirely from experience with microbiological systems. There are a number of reasons for this, apart from the simple fact that the study of any enzyme of importance to molecular biology is most profitably carried out with those cells whose molecular biology is most thoroughly understood. (*a*) The systems of greatest interest among higher organisms are mammalian, and it is now reasonably certain that photoreactivating enzyme does not occur in placental mammals (Section 4.1.3). (*b*) The detection and quantitation of photochemical lesions is easiest after high UV doses. At doses which most animal cells can survive, the known DNA lesions are difficult to quantitate by the methods currently available. The most profitable single system for studying both photochemistry and photobiology simultaneously has been transforming DNA in bacteria, in which biological effects at high doses can be measured (see Section 3). (*c*) Finally, insofar as we know, the molecular mechanism of photoenzymatic repair is in any case the same in animal cells as in microorganisms. Rupert (1962b) showed that photoreactivating enzyme from yeast could repair UV-irradiated transforming DNA from *Haemophilus influenzae* and that this repair was competitively inhibited by UV-irradiated calf-thymus DNA. In a similar experiment, we (Cook and Setlow, 1966) were able to show that the photorepair of UV-irradiated *Haemophilus* DNA by photoreactivating enzyme from sea urchin eggs was not further enhanced by subsequent treatment with the yeast enzyme. The fact that the *sources* of either DNA or enzyme were immaterial in these experiments supports the idea that the basic mechanism for photoenzymatic repair is the same wherever it is found.

The successes in identifying UV-lesions and repair mechanisms in microbiological systems inevitably color, and perhaps restrict, our approach to the study of similar problems in eucaryotic cells. In fact, all of the assays (Section 3) used in identifying the photorepair of UV lesions in higher organisms are based on assays developed with microorganisms, and will not detect other types of direct enzymatic photoreactivation, if such exist. One of the underlying themes of this chapter is that our ultimate understanding of UV-lesions and their repair in nucleated cells will require an appreciation of the molecular organization of the chromosome and nucleus, and their changes in the course of the cell cycle. In recent years it has become clear that in the very large amounts of DNA (compared to the amounts in bacteria) found in the cells of higher organisms there exist redundant or reiterated sequences (Britten and Kohne, 1968; Walker, 1968). These may represent multiple copies of single genes, such as has been shown for ribosomal DNA in several species (Birnstiel *et al.*, 1968; Gall, 1968). For purposes of this discussion, we may assume that DNA is organized in the chromosome according to the "master/slave" concept of Callan (1967), whereby there exists a "master" DNA template responsible for genetic continuity and from which "slave" copies are made, whereas the "slaves" are copied in DNA-RNA transcription. In a structure of this kind, UV-lesions in "slave" DNA may be of no significance to cell survival. Similarly, in differentiated systems, photochemical lesions in DNA which contains information not utilized by the irradiated cells (e.g., genes for hemoglobin in skin epithelial cells) may never be expressed. From the biological point of view then, some DNA molecules may be more sensitive than others. Moreover, the chromosome may be so organized that only the "sensitive" DNA is actually repaired after irradiation or other deleterious treatment. This proposal is a corollary to Callan's hypothesis that "the master genes are thus supposedly sequestered from the remainder of the genetic material, and less exposed to degradative processes than otherwise they would be." Such a scheme of restricted repair would be economical. It could also explain such observations as those of Trosko and Kasschau (1967) who found that UV-irradiated Chinese hamster cells could survive, without apparently repairing, tens of thousands of pyrimidine dimers per nucleus whereas on the order of 1–5 such unrepaired dimers are lethal to *E. coli* B_{s-1} (Swenson and Setlow, 1966). Other observations which may be accounted for by preferential repair, especially with respect to photoreactivation, will be pointed out in the final section of the chapter.

A second theme underlying the chapter concerns the recurring question of the significance of photoreactivation in nature. Most animals are protected from the sun either by habitat and/or a virtually opaque external covering of biologically inert material such as chitin, scales, hair, or feathers. For organisms not so protected it would seem reasonable that photoreactivation would be adaptively advantageous, since whenever an organism is exposed to the shorter deleterious wavelengths of the solar spectrum, the longer wavelengths utilized by the photoreactivating enzyme are also present. In a model experiment, Rupert and Harm (1966) have shown that *E. coli* B phr$^-$, which differs from its parent organism only in its lack of photoreactivating enzyme, is many times more sensitive to sunlight than the parent. In other words, they have demonstrated that photoreactivation can work effectively under natural conditions. On the other hand, the enzyme appears to be absent in places where it might be useful (human skin, some plant leaves) and present where it probably cannot fuction as photoreactivating enzyme (turtle heart, opossum liver). It is possible that the enzyme may have a second function to which its distribution may give a clue. Therefore Section 4 contains a review of the distribution of biological photoreactivation and of the enzyme activity.

The final section (Section 5) of the chapter is concerned with a number of biological lesions in animal cells which are subject to photoreactivation. Under favorable circumstances, the experimentalist can introduce lesions (with UV) and remove them at will (by photoreactivation). In Section 5 the emphasis is on the use of photoreactivation as a tool in understanding the photophysiology of a number of cell processes.

2. Photoreactivable Lesions (Chemical)

Since the occurrence of photoreactivation is identified on both the biological and biochemical levels as the mitigation in light of a UV-induced lesion, it is appropriate first to consider the identification of UV lesions. On the molecular level, these may be alterations produced in any molecule by the absorption of one or more quanta of UV radiation. Whether such alterations bear any direct relation to physiological lesions is frequently difficult to ascertain (Setlow, 1968a). In the most commonly explored range of deleterious wavelengths (220–310 nm), both proteins and nucleic acids absorb strongly. The ultraviolet photochemistry of these compounds has been reviewed by McLaren and Shugar (1964). From the point of view of cell survival following whole-cell irradiation, the inactivation of a few enzymes, a few RNA messages, or a few ribosomes is probably of trivial consequence; such molecules may be re-

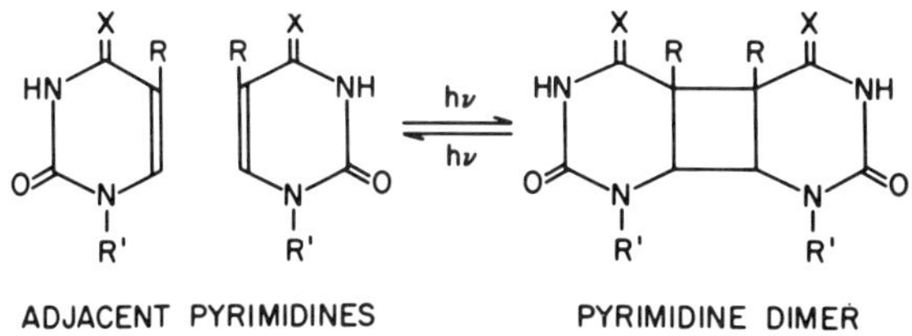

FIG. 1. Photochemically reversible dimerization of adjacent pyrimidines with the formation of the cyclobutane ring and saturation of the 5,6-double bonds. X is either O (thymidine, uridine) or NH_2 (cytidine); R is either H (cytidine, uridine) or CH_3 (thymidine); R′ is either ribose (RNA) or deoxyribose (DNA).

placed.[1] The intactness of DNA is of much greater importance to the cell's reproductive capacity. Almost all of the recognized repair systems, including photoreactivation, are concerned with the maintenance of intact DNA.

UV radiation causes a number of lesions in DNA. One of these, now well known, is the cyclobutyl dimer formed between adjacent pyrimidines (Fig. 1) which was first described by Beukers and Berends (1960). The work of R. B. Setlow and J. K. Setlow and their co-workers (Setlow and Setlow, 1962; J. K. Setlow, 1966a, 1967; R. B. Setlow, 1968a,b) established that such dimers are of great importance in the inactivation of biological activity of DNA. Pyrimidine dimers are also the only known substrate for photoreactivating enzyme (see below).

An important lesion to photoreactivation studies is that formed in cells which have incorporated bromouracil (BU) in place of thymine in their DNA. (More commonly, in animal cells, bromodeoxyuridine is used as an analog for thymidine, because the nucleosides penetrate into the cells more readily than the bases.) Cells containing BU-DNA have an enhanced UV-sensitivity but a substantially reduced or lost photoreactivability (Greer, 1960). Rupert (1961) has found that UV-irradiated BU-DNA is not photoreactivated by photoreactivating enzyme *in vitro*. A number of UV-lesions other than pyrimidine dimers, including cross-links between DNA and protein, have also been identified (reviewed in Smith, 1967) and the biological effectiveness of some of them seems

[1] A possible exception to this generalization is the structural integrity of the cell membrane. A quite small number of UV lesions here can lead to leaks of small molecules into and out of cells, and such leaks can lead to cell lysis (Cook, 1956, 1965). There is evidence that ionizing-radiation damage to cell membranes may be repaired to some extent (Myers and Bide, 1966; R. M. Sutherland *et al.*, 1967). The problem of membrane-turnover and repair in UV-irradiated animal cells deserves more attention than it has received. The UV dose relations are such that this is an important consideration for survival in animal cells, although it is much less critical in bacteria.

well established (Donnellan and Stafford, 1968; Rahn *et al.*, 1969), but for this discussion the nondimer lesions may be grouped together on the grounds that they appear to share the common property of non-photoreactivability.

UV-lesions in DNA may be repaired by several mechanisms. In the past decade, this subject, and its genetic implications, have attracted very wide interest and have been reviewed many times, recently by Hanawalt (1968) in this series. (See also Haynes *et al.*, 1968; Howard-Flanders, 1968; Read, 1968; Setlow, 1968b, and Strauss, 1968.) Repair, in the dark, is thought to involve a number of steps: (1) First is recognition of the UV lesion (perhaps only as a region of local denaturation) by an endonuclease which cuts the damaged DNA strand near the lesion. (2*a*) This is followed by an excision step, the operational result of which is that (if pyrimidine dimers are the lesion being followed) the dimers appear in an acid-soluble form. This step seems to involve, at some point in the overall process, the activity of a nuclease which cuts on the other side of the lesion being excised. (2*b*) A polymerase which uses the undamaged DNA-strand as template synthesizes a new strand-segment which replaces the excised bases, thereby restoring the original (pre-UV) sequence. In *E. coli*, DNA-polymerase is capable of carrying out both steps 2*a* and 2*b* *in vitro* (Deutscher and Kornberg, 1969; Englund *et al.*, 1969; Kelly *et al.*, 1969), although whether it actually does so in dark repair *in vivo*, and in what order, is not certain. (3) Finally, the newly synthesized segment is attached to the old strand by the action of a ligase.

All the necessary enzymes required by this mechanism have been isolated in various laboratories (see Setlow, 1968b; Strauss, 1968 for recent references). Consideration of this sort of mechanism leads to a number of predictions, of which two have assumed great importance in the photobiology of higher organisms. (*a*) *Repair replication.* If cells are allowed to incorporate density markers such as bromodeoxyuridine in place of thymidine, or ^{15}N- and/or ^{2}H-containing DNA precursors, then new, semiconservatively synthesized DNA will be heavier than the parental DNA and can be separated from it by buoyant density centrifugation. If the heavy precursors are labeled, the label appears only in the heavy strand. In the dark-repair scheme outlined above, however, a few bases are inserted into the repaired regions of damaged parental strands. These may be too few to affect significantly the density of the parent (light) strand, but enough to introduce detectable label into the light strand. Label so found in the parental DNA from UV-irradiated cells is said to represent "repair replication." Since direct photoreactivation repairs pyrimidine dimers *in situ* without excision (see following), photoreactivation should reduce the extent of observable repair replication. Experimental results conforming to these predictions were first obtained in *E. coli* by Pettijohn and Hanawalt (1964) and more recently, insofar as the interrelationship with photoreactivation is concerned, in the eucaryotic *Tetrahymena* (Brunk and Hanawalt, 1967). The technique has been used extensively to study repair in human cells (see Chapter 4 by Fork and Amesz in this volume). (*b*) *Unscheduled synthesis.* If repair occurs in the absence of net DNA synthesis during the usual S period of the cell cycle, it should be detectable radioautographically by the uptake of tritiated thymidine in G_1 or G_2 cells; hence the term "unscheduled synthesis." McGrath (1960, 1963) suggested that the incorporation of ^{3}H-thymidine into DNAase sensitive molecules in X-irradiated intestinal-crypt cells of mice or in grasshopper neuroblasts represented repair *in situ.* Similar, but more extensive, observations were made by Rasmussen and Painter (1966) with UV-irradiated mammalian cells in culture. They demonstrated that unscheduled synthesis was not semiconservative, was dose-dependent, and was enhanced by the presence of BUdR in the DNA.

Although repair replication and unscheduled synthesis can be predicted by the dark-repair scheme, it is important to remember that the converse may not be true. The demonstration of repair replication and/or unscheduled synthesis, as operationally defined, in any system does not necessarily prove the existence of dark repair in that system. It is easy to construct plausible hypotheses of aberrant synthesis, which do not lead to any true repair, in irradiated cells. The demonstration of genuine repair is difficult to make and requires, at least, correlative evidence that what was once a lesion is, after repair, no longer a lesion.

Apart from being able to carry out dark repair, many cells are capable of repair by photoreactivation. Direct photoreactivation is mediated by a single enzyme; the only known substrates are pyrimidine dimers in DNA (J. Setlow, 1966a, 1966b, 1967; Setlow *et al.*, 1965; Rupert, 1964). In brief, the reaction involves the reversible binding of enzyme to UV-irradiated DNA, a step which can take place in the dark. In the light, the dimers are split (monomerized) *in situ,* regenerating native DNA, and the enzyme is released. The reaction directly reverses the UV-lesion; no bases are excised or inserted as in "dark repair" (Wulff and Rupert, 1962; Rupert, 1964; Cook, 1967). In most experimental situations the number of dimers greatly exceeds the number of available enzyme molecules, and at high light intensities the reaction may become saturated, i.e., the rate may become limited by the light-independent steps. Under these conditions the reaction is also strongly temperature-dependent. If the light intensity is reduced to the point where light itself is rate limiting, the temperature dependence is very much less. The practical consequence of these relations is that, in most cases, photoreactivation is slow and may require minutes to hours in order to achieve maximum photoreversal of UV lesions which may have been inflicted in seconds. In a dynamic living system, maximum photoreactivation may not be achievable before the lesion is expressed.

Action spectra for photoreactivation have been worked out for a number of systems, both *in vivo* and *in vitro* (see Section 3, below, for measurements of photoreactivation *in vitro*). There seem to be real differences between the spectral responses of photoreactivating enzyme from various sources (see, for example, Terry *et al.*, 1967), but the nature of the chromophore has not been identified in any of these cases. Nevertheless, the action spectra overlap substantially; in general, they all lie in the range between 320–500 nm, i.e., in the near-ultraviolet-to-blue.

The study of biological photoreactivation with intact organisms is complicated by the possibility that not just one, but any one or more of four very distinct mechanisms may underlie the phenomenon. (*a*) Direct photoenzymatic repair of UV-induced lesions in DNA, as outlined immediately above, may be responsible. (*b*) Indirect photoreactivation has been defined by Jagger and Stafford (1965) as the case "in which the molecular site of action of the far ultraviolet radiation is not immediately and directly altered by the energy of the photoreactivating photon." They describe a case in *E. coli* B phr$^-$, a mutant with no photoreactivating enzyme. The effects of 254 nm irradiation on these cells could be substantially mitigated by illuminating them with 334 nm radiation either before ("photoprotection") or after ("indirect photoreactivation") the far ultraviolet dose was given. Jagger and Stafford showed that 334 nm irradiation induced a very substantial division delay, and suggested that during this delay period dark-repair mechanisms would have had more time to carry out effective repair than would have been the case if the cells started growth immediately after the far ultraviolet irradiation. In more complex eucaryotic cells and multicellular organisms there are many possibilities for indirect photoreactivation; ordinarily such indirect mechanisms can be detected by treatment of the system with the photoreactivating light *before* the UV. Such pretreatment has no effect on direct photoenzymatic repair but, in indirect photoreactivation, pretreatment is likely to be as effective as posttreatment. (*c*) Direct (nonenzymatic) photoreversal of UV lesions may in some cases be accomplished by wavelengths other than the one used for the initial irradiation. Thus pyrimidine dimers in DNA are both formed and broken by UV photons. Since the dimers have different absorption spectra from the monomers, the equilibrium state is wavelength dependent (Johns *et al.*, 1962). Thus it is possible, at wavelengths where dimer absorption is small (say 280 nm), to form a higher proportion of dimers than exist in the photosteady state at wavelengths where the dimers absorb strongly (say 240 nm), resulting in a substantial rate for the back (monomerization) reaction. The Setlows have exploited these relationships in the inactivation at 280 nm followed by reactivation at 240 nm of *Haemophilus* transforming DNA, but the doses required in this "short wavelength reversal" are very high (Setlow and Setlow, 1965), probably two orders of magnitude greater than those which any animal cells can survive. Direct photoreversal of other UV lesions (see for example, Takebe and Jagger, 1968) is possible, but no case of significance to animal cells has been described. (*d*) It is possible that biological photoreactivation may be observed if the "reactivating" light induces a response antagonistic to the response induced by the first irradiation, thereby reducing the initial response without repairing the initial lesion. A mechanism of this kind was proposed by Blum and Terus (1946) who observed a diminution of erythema in humans when 254 nm irradiation was followed by 290–320 nm irradiation, provided the second UV was given before the erythema from the first had appeared. They suggested that the longer wavelengths had a direct effect on the blood vessels of the skin which interfered with their response to the vasodilators released in the skin after irradiation with the shorter wavelengths. Although this mechanism is by no means established, it does point up the possibility of biological photoreactivation without repair of the initial lesion.

3. Assays for Photoreactivation in Animal Cells

The simplest qualitative assay for biological photoreactivation is perfectly straightforward: a population of cells (or organisms) is UV-irradiated and divided into two aliquots; one is incubated in the dark and the other in photoreactivating light. After an appropriate interval, the two populations are examined for whatever effect has been selected as the criterion for UV action. In this, as in any assay, it is important to show that the photoreactivating light alone does not bring about the effect selected as the end point. Many cells of higher animals are sensitive to black lights (320–460 nm), which are the most commonly used fluorescent sources for photoreactivation studies. The sensitivity is very likely, although not certainly, due to photodynamic effects in which endogenous heme-containing compounds are the light absorbers. We have found, for example (Regan, Cook and Takeda, unpublished), that the growth of reptilian cells in culture, including iguana heart, turtle heart, and gekko lung, is in each case inhibited by either small doses of UV at 265 nm or large doses of blacklight. In the proper dose-range, blacklight will substantially photoreactivate the growth-inhibition of 265 nm radiation (Regan *et al.*, 1968b), but at higher blacklight doses the photoreactivating effect disappears and the inhibition of growth may even become exacerbated.

Frequently experiments with higher organisms are sufficiently complex so that, together with the proper controls, it is technically feasible to use only a single UV-dose schedule and a single photoreactivating-dose (vs. dark) schedule. Such experiments can yield only a "yes-or-no" answer to the question of photoreactivability and the "no" answers are, as in any negative experiment, difficult to assess (see below). Much more information can be obtained by the use of multiple doses of both UV and PRL. From the latter, one can find conditions for maximum photoreactivation. From the former one can compute dose reduction factors (DRF). The computation is simply the ratio of UV doses x/y, where x is a UV dose producing a given effect in the dark and y is the dose producing the same effect in the light. The population given dose y, and then photoreactivated, behaves as though it had been given the smaller dose x without photoreactivation. Photoreactivation thus appears to have reduced the dose by the factor x/y. The so-called "photoreactivable sector" for a given UV effect is $1 - x/y$. If there is no photoreactivation, x and y are equal and $1 - x/y$ is zero. On the other hand, if photoreactivation is extensive, x is small (i.e., the photoreactivated population behaves as though it had been treated with a low UV dose) and the photoreactivable sector approaches 1.0. Since photoreactivation is a repair phenomenon, the concept of the photoreactivable section is useful in that this sector is one indication of the extent of repair.

Any biological effect of UV (killing, mutagenesis, carcinogenesis, mitotic delay, etc., etc.) may be assayed for photoreactivability, but a positive result does not necessarily imply that the mechanism is the direct photoenzymatic repair of the UV lesion. With our increased understanding of UV lesions in DNA and of the mechanism of direct photoreactivation, four additional assays have become available.

1. *Light-dependent, heat-sensitive repair by cell-free extracts of UV-irradiated transforming DNA.* This assay was first introduced by Goodgal *et al.* (1957), using the transformation of *Haemophilus influenzae* to streptomycin resistance (see also Muhammed, 1966). Briefly the assay employs as substrate *Haemophilus* DNA which has been UV-irradiated to about 1% of its transforming activity. Such DNA, when irradiated at 254 nm, has a large photoreactivable sector (J. Setlow, 1966b). The DNA is mixed with a cell-free extract of the tissue to be tested, and the mixture incubated in the light. If the tissue contains photoreactivating-enzyme activity, the transforming ability of the UV-irradiated DNA

may be substantially restored, and such restoration will be light-dependent. Under standard conditions, the rate of restoration of transforming ability is proportional to the quantity of enzyme in the reaction mixture (see Muhammed, 1966 for details of the assay, and for units). In the many tissues which have now been tested, photoreactivating enzyme is quite heat labile, and holding the cell-free extract at 65°C for 10 minutes prior to assay may be expected to abolish all photoreactivating activity.

The *Haemophilus* system requires only about 25 nanograms of irradiated DNA, and the repair of a small number of UV lesions yields a readily measurable increase in number of transformants. At the present time this system is the most sensitive assay available for photoreactivating-enzyme activity, and the high sensitivity is its chief asset. Its chief liability (apart from the difficulties of growing and maintaining competent *Haemophilus* cells) is that the DNA is sensitive to nucleases which occur in most crude tissue extracts. Frequently the nuclease problem may be overcome by addition to the reaction mix of an excess of unirradiated calf-thymus DNA, with which PR enzyme does not react (Rupert, 1962), but occasionally, as with our experiments with Drosophila (unpublished) the nuclease activity was so great that we were not able to assay for PR enzyme (see Muhammed *et al.*, 1967).

2. *Light-dependent, heat-sensitive ability of cell-free extracts to monomerize pyrimidine dimers in UV-irradiated E. coli DNA.* The substrate in this assay is *E. coli* DNA, labeled with ^{3}H-thymine, which has been UV-irradiated until about 4–6% of the total thymine is in the form of cyclobutyl pyrimidine dimers. In principle, any DNA which can be labeled at high specific activity and which can be easily isolated would make, when irradiated, a satisfactory substrate. The DNA is incubated under photoreactivating light with the extract to be tested and at intervals samples are taken and analyzed for pyrimidine dimers by chromatographic methods (Setlow and Carrier, 1966). This assay requires somewhat more DNA than the *Haemophilus* assay (depending on the specific activity of the labeled DNA) and, since more lesions must be repaired to yield a definitively positive result, is correspondingly less sensitive. On the other hand, since photoreactivating enzyme from yeast acts with approximately equal efficiency on dimers in any polydeoxynucleotide greater than 18 bases long (Setlow and Bollum, 1968), this assay should be less affected by the action of endonucleases in the preparation. Again the demonstration of light-dependence and heat-lability are the minimum necessary controls.

3. *Light-dependent monomerization of pyrimidine dimers in vivo.* If the cells in question can be conveniently labeled with ^{3}H-thymidine, it

is frequently possible by chromatographic methods to show that UV-irradiation induces cyclobutyl dimers in the DNA and that photoreactivating illumination leads to their disappearance, i.e., to their monomerization. Ideally, this assay, performed on intact cells, can be correlated with the biological effects of the radiation and photoreactivation and thus give insights into the mechanisms of both the UV and the repair processes. In practice, the assay is very difficult to carry out and in no case has the "ideal" been approached with eucaryotic cells. This point is of great importance, and will be reiterated below; the lack of reliable biochemical data of this kind is a great impediment to our understanding of photoreactivation in animal cells.

The difficulty with the assay is easily described. A dose of UV at 254 or 265 nm of 150 ergs/mm^2 is lethal to most cells in tissue culture. Although large cells, such as paramecia or Tetrahymena, are not killed until the dose reaches some 10 times this value, their nuclei are sufficiently shielded by cytoplasm that only about one-tenth of the incident dose reaches the nuclear DNA (see Sutherland *et al.*, 1968). These doses convert less than 0.1% of the total thymine into pyrimidine dimers, and as photoreactivation progresses the number becomes still smaller. A sample containing 10,000 cpm in thymine will have less than 10 cpm in the dimer region. The problem is compounded by the fact that most ^{3}H-thymidine labeled cells of higher organisms contain a radioactive, TCA-insoluble contaminant which co-chromatographs with pyrimidine dimers, but which is not a dimer (by the criterion of its nonreversibility to thymine on extensive irradiation in aqueous solution). The contaminant is commonly present in unirradiated cells in variable amounts, although usually not exceeding 0.2%. That is, however, precisely the range in which the dimer measurements must be made, and the resulting uncertainty in background makes very difficult a precise correlation of the fate of dimers with biological effects.

The problem may be and has been side-stepped by irradiating with very high (in fact, sterilizing) doses which induce relatively large numbers of photoproducts and hoping that the cells live long enough to photoreactivate a distinctly measurable quantity of dimers. Experiments of this kind may be successful in demonstrating photoreversibility of dimers *in vivo* (see Fig. 2), but they are not useful in illuminating the biology of photoreactivation.

4. *Competition of DNA from UV-irradiated cells with heterologous DNA in an in-vitro system.* If the pyrimidines in the DNA of the system under study cannot be readily labeled to quite high specific activity, as is a drawback with, say, *Neurospora*, or unfertilized frog eggs, there remains nevertheless an assay for the photoreactivability of such DNA *in vivo*. The basis of the assay is a standard reaction mix-

ture of UV-irradiated *Haemophilus* transforming DNA and yeast photoreactivating enzyme which, under given constant conditions of illumination, temperature, pH, etc., will give a certain rate of restoration of transforming ability as outlined in assay 1 above. Ultraviolet-irradiated DNA from any source, when added to this standard reaction mixture, competes for the enzymic activity and inhibits the rate of reactivation of *Haemophilus* transformation (Rupert, 1962a). The inhibitory capacity of such DNA is reduced or abolished if the competing DNA is itself photoreactivated before being added to the reaction mixture. It is therefore possible to irradiate and photoreactivate a population of cells, extract their DNA, and test for both the UV and PR effects by this competition technique (see Rupert and Harm, 1966, p. 23, and Setlow and Bollum, 1968 for successful applications of the method).

Competition techniques are also effective when the reaction measured is the rate of monomerization of pyrimidine dimers (assay 2, above). The statements already made with regard to relative sensitivity and interference by nuclease in the two assays apply equally well to the cases of competitive inhibition.

It would appear that assays based on competitive inhibition would be extraordinarily useful in studies on inactivation and photoreactivation in animal cells, and it is regrettable that they have not been exploited. For example, only 25 nanograms of *Haemophilus* DNA are required for the transformation assay. The effect, and repair, of UV doses as low as 1000 ergs (254 nm) are readily measurable in this system. For purposes of an order of magnitude calculation, we may assume that the addition to the reaction of an equal number of UV lesions in heterologous DNA will halve the rate of repair of the assayed DNA. Thus a dose of 100 ergs to 0.25 μgms, or even only 10 ergs to 2.5 μgms of DNA should yield easily measurable competition. Two and one-half micrograms is the amount of DNA in only 150,000 HeLa S3 cells (Lee and Puck, 1960). The use of such a potentially sensitive assay warrants more attention and development than it has received. Although such assays are not simple to perform, they offer a real opportunity to study repair of DNA in animal cells at the usual "biological" dose levels.

In assays 3 and 4 it is important to use appropriate dark controls in order to distinguish between the light-dependent monomerization of dimers and excision of dimers from DNA characteristic of dark-repair (Setlow and Carrier, 1964; Regan *et al.*, 1968c). In the former case, the dimers are converted back to pyrimidine monomers, and therefore disappear from both TCA-soluble and -insoluble cell fractions. In the latter case, the dimers may be lost not only from the DNA by excision but also may escape from the cell into a large volume of external medium where the problems of finding them become that much greater. In cells where both dark repair and photoreactivation occur, the excision-repair appears to be much slower than photoreactivation if sufficient light is provided. (See for example, Sutherland *et al.*, 1968.)

To recapitulate, the assays described in this section measure three aspects of photoreactivation: amelioration of a biological effect, occurrence in cells of a photoreactivating enzyme, and repair of DNA by photoreactivation. Although there are no known counter examples, the presence of the enzyme (assays 1 and 2) does not guarantee that the enzyme functions to repair DNA (assays 3 and 4). Although a start has been made with some systems, particularly protozoa (see following) we do not yet have for any animal cell a reasonable description for the biological effects of UV-radiation in terms of UV lesions and their repair.

4. Distribution of Photoreactivating Activity

4.1 Phylogenetic Distribution

Photoreactivation has been found in every phylum where it has been sought (see Jagger, 1958 and Blum, 1959, p. 77, for earlier references) and in a very broad range of cell types. Two major groups of organisms are exceptional in that they lack direct photoreactivation; they are transformable bacteria (*Bacillus subtilis*, *Haemophilus influenzae*, *Diplococcus pneumoniae*, Kelner, 1964; to which may now be added *Micrococcus luteus* [*lysodeikticus*], Elder and Beers, 1965; Mahler and Grossman, 1968, and *Micrococcus radiodurans*, Moseley and Setlow, 1968)[2] and placental mammals. Up until the past few years, nearly all of the descriptions of photoreactivation in eucaryotes have been in terms of biological effects; they will be considered in a later section. Only recently have attempts been made to identify chemically the lesions in DNA, to correlate their occurrence and/or repair with the observed biological effects, and to search for photoreactivating-enzyme activity in higher organisms. Three groups of such studies are described in this section.

4.1.1 Studies in Protozoa

Photoreactivation of UV-induced mutagenesis in *Paramecium aurelia* was shown by Kimball and Gaither in 1951. Recently Berger and Kimball (1964) have developed an ingenious method for the labeling of paramecium DNA by feeding the organisms with *E. coli* 15T$^-$ which had been grown in a medium containing ^{3}H-thymidine. Using this technique,

[2] This question would be worth re-examination in view of the recent demonstration of transformation in certain strains of *E. coli* by Avadhani *et al.* (1969).

B. M. Sutherland *et al.* (1967, 1968) showed by CsCl-buoyant density determinations that the label was indeed incorporated into paramecium DNA, and that after UV-irradiation of such labeled paramecia with high UV doses, but ones which permitted 100% survival of the organisms for at least 48 hours, a readily measurable number of pyrimidine dimers were formed in the DNA. In the dark, after a lag of about 6 hours, the dimers were excised during the subsequent 18–20 hours. At any time during the lag or excision phase, virtually all of the remaining dimers could be monomerized if the organisms were given sufficient doses of photoreactivating light.

Similar observations have been made with *Tetrahymena*, an organism in which biological photoreactivation had been demonstrated by Christensen and Giese in 1956 (for a review of photoreactivation in Protozoa, see Giese, 1967). As with *Paramecium*, UV radiation produced pyrimidine dimers in the TCA-insoluble fraction of the cells, and 80 percent or more of the dimers could be monomerized in two hours with a reasonably intense photoreactivating source at 23°C (Francis and Whitson, 1969). In the dark, about the same percent of the dimers were excised from DNA and lost to the medium in two days (Whitson *et al.*, 1968). The action spectrum and temperature dependence of photoreactivation in this organism point to a direct photoenzymatic repair mechanism, although nucleases in *Tetrahymena* extracts have so far precluded an *in vitro* assay for the enzyme with the *Haemophilus* transformation system (Cook, unpublished).

Brunk and Hanawalt (1967, 1969) have also, indirectly, shown photoreactivation in *Tetrahymena* DNA. Following treatment of the cells with UV, they allowed a complete round of DNA replication (indicating that they were working with surviving cells) and showed that the *parental* DNA strands contained large sections of newly-synthesized material, presumably a consequence of excision of UV lesions and repair replication. No such nonconservative replication was observed in the controls, and the effect could be almost entirely abolished if the irradiated cells were photoreactivated immediately after UV. Presumably the photoreactivation had abolished the lesions which would otherwise have been substrates for the dark-repair system.

In protozoa then, the mechanisms of photoreactivation appear similar to those already described for other microorganisms.

4.1.2 Studies on the Phylogenetic Distribution of Photoreactivating Enzyme

First observed in bacteria (Goodgal *et al.*, 1957) and then in yeast (Rupert, 1960), photoreactivating enzyme activity has now been found in tissue extracts from algae (Werbin and Rupert, 1968; Diamond *et al.*,

1969), *Neurospora* (Terry *et al.*, 1967) and, as representative of higher plants, the seedling of the pinto bean, *Phaseolus vulgaris* (Saito and Werbin, 1969). The latter authors could find no trace of activity in the closely related mung bean, *Phaseolus aureus Roxbg.*

Among the metazoa, we have found activity in echinoderms (Cook and Setlow, 1966), arthropods and vertebrates (Cook and McGrath, 1967). The most extensive distribution studies have been among the classes of vertebrates, where we have found activity in fish (Regan and Cook, 1967), amphibia (Cook and McGrath, 1967; Regan *et al.*, 1968a), reptilia (Regan *et al.*, 1968b), birds (Cook and McGrath, 1967), and a lower order of mammals, viz., marsupials (Cook and Regan, 1969). The special case of placental mammals is considered in Section 4.1.3.

In all of these studies, the enzyme was detected by the *Haemophilus* transformation assay. In many of them, the light-dependent monomerization of pyrimidine dimers *in vivo* was demonstrated as well. Dimer-splitting has also been shown in cultured plant cells from *Nicotiana* and ginkgo (but not in cultured *Haplopappus* cells) by Trosko and Mansour (1968, 1969) and in chick embryo fibroblasts (but not rabbit-kidney cells) by Pfefferkorn and Coady (1968).

Other assays performed in these various studies included photoreactivation of cell-survival, survival of infecting viruses, cell growth, rate of DNA synthesis, and/or the ability of tissue extracts to monomerize pyrimidine dimers in UV-irradiated *E. coli* DNA *in vitro*. Not all of these assays were performed in every case, but in no case was there any discrepancy among the assays which were performed. For any given cell, the assays were either all positive or all negative.

4.1.3 Studies on Mammals and Mammalian Cells

It now appears reasonably certain that direct photoenzymatic photoreactivation does not occur in placental mammals. Nevertheless, negative conclusions are notoriously difficult to establish, and this one has not yet been unanimously accepted. Because of the special interest in mammals, the arguments for, and the apparent limitations on, this negative conclusion are reviewed here. The photobiology of mammalian cells in general is discussed by Painter in Chapter 6 of this volume.

After the definitive descriptions of photoreactivation by Kelner (1949) and Dulbecco (1949), the phenomenon was found in so many plant and animal as well as microbial systems that Rieck and Carlson's (1955) description of the photoreactivation of lethality in mice seemed to be a clear-cut example of the same phenomenon. Subsequently several more papers describing photoreactivation in mammals appeared (reviewed in Cleaver, 1966), but none gave unequivocal results. During the same period more and more papers appeared in which photoreactivation was *not* observed in mammalian systems until in 1966 Cleaver described it as a "radiation repair mechanism absent from mammalian cells." His conclusion was based on a review of the literature plus his own observations on DNA synthesis, as determined radioautographically, and cell killing in a number of UV-irradiated cell lines or strains in culture.

In addition to these negative results with biological assays, Trosko *et al.* (1965) showed that although thymine-containing dimers were formed by UV in DNA of Chinese hamster cells (see also Klimek, 1966) there was no indication of monomerization of such dimers after a photoreactivating dose of blacklight which would have given a substantial effect in bacteria irradiated with the same UV dose. Working with normal human skin cells in culture, Regan and I increased this photoreactivating dose by a factor of 18 and still failed to observe dimer monomerization (Fig. 2).

Finally, J. McGrath and I (Cook and McGrath, 1967) used the very sensitive *Haemophilus*-transformation assay to look for photoreactivating-enzyme activity in a number of mammalian tissues, both adult and fetal, skin as well as internal organs; in some 30 attempts we were unable to detect any activity. It appeared then that, although UV induced the usual photoreactivable substrate (dimers) in mammalian

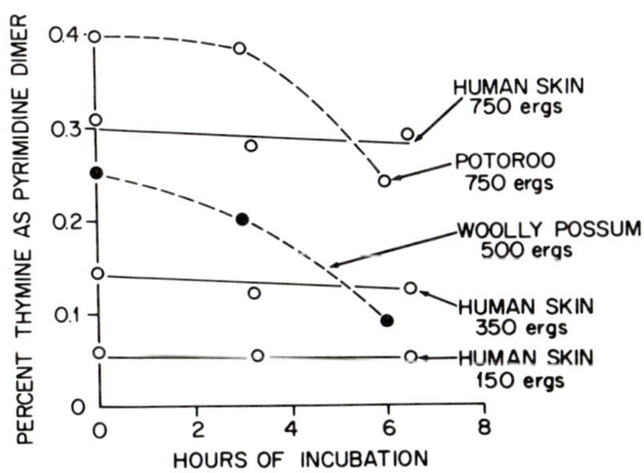

FIG. 2. Photoreactivation *in vivo* of thymine-containing pyrimidine dimers in cultured marsupial cells (potoroo, woolly possum) as opposed to the apparent lack of such photoreactivation in cultured cells from normal human skin. Cells irradiated at 254 nm with the doses shown (in ergs/nm^2) and subsequently illuminated at 37°C under blacklight with Corning filter #7380 (cutoff at 350 nm). Intensity of blacklight was about 6000 ergs/mm^2-min. For reasons of clarity, the dark controls have been omitted from the figure; there was no detectable disappearance of dimers in the dark in the 6-hour period (Cook and Regan, unpublished).

cells, the failure to observe biological photoreactivation was probably due to the fact that these cells did not contain the repair enzyme.

An important exception to this mammalian pattern was found in the marsupials. Regan and I (Cook and Regan, 1969) examined tissues from three sources: cultured cells from the kidneys of South American woolly 'possum and the Tasmanian rat kangaroo (potoroo) as well as primary tissue from the North American opossum. In the first two we looked for biological photoreactivation (of growth rate) and *in vivo* monomerization of dimers (Fig. 2). In extracts from all three sources we looked for photoreactivation of *Haemophilus* transforming activity and *E. coli* dimers. All assays were positive. Of particular relevance to this discussion was the fact that although the assays with extracts yielded about the same specific activity of the enzyme as we had found in lower vertebrates or invertebrate tissue, the *in vivo* assays required very long exposures to photoreactivating light. The rate constant for monomerization of dimers by the woolly 'possum cells in Fig. 2, for example, is less than one-sixth that of fish cells under similar conditions (Regan and Cook, 1967) even though the latter was determined at a temperature 12°C lower. The reasons for the slow rate *in vivo* are not known. The interior of a cell nucleus is not a simple solution where enzyme and substrate encounter each other by random diffusion. Although there are a large number of substrate lesions (for the curves in Fig. 2, assuming roughly 2×10^{-11} gms DNA/cell [Lee and Puck, 1960], there are on the order of 10^7 pyrimidine dimers per cell) they are not necessarily all equally available to the photoreactivating enzyme at all stages of the cell cycle.

In any event, these observations raise the question of the adequacy of the controls in many of the previously reported negative experiments. Cleaver (1966) and Trosko *et al.* (1965) used photoreactivating conditions very similar to those shown in Fig. 2, yet Cleaver illuminated his cells for only 10 minutes in all but one two-hour experiment, and Trosko *et al.* illuminated for only 20 minutes. With these short doses, the very definite photoreactivability of marsupial cells might not have been observed.

Of special interest to the problem of proper controls for the negative experiment are the data of Pfefferkorn and Coady (1968) who infected primary chick embryo fibroblasts and primary rabbit kidney cultures with pseudorabies virus. The titers of the unirradiated virus were virtually identical on the two cell types, and were reduced to approximately 1% of initial activity after 100 ergs/mm^2 irradiation with a germicidal lamp. The irradiated virus could be photoreactivated in chick cells with a dose-reduction factor of about 0.6, i.e., about 40% of the inactivating lesions were repaired. Under identical conditions there was no detectable photoreactivation of the virus in rabbit cells, although the assay appears sensitive enough to have clearly revealed as little as 5% repair (Pfefferkorn *et al.*, 1966). Pfefferkorn and co-workers were also unable to show photoreversal of UV-induced pyrimidine dimers in rabbit cells whereas such photoreversal could be shown in chick cells; the assay in this case however was much less sensitive than that utilizing the virus-survival technique.

In our experience, the demonstration of photoreactivating enzyme activity in cell extracts by the *Haemophilus*-transformation assay has been far more sensitive and less equivocal than any of our biological assays with intact cells. With this assay we could easily have recognized an activity only one-twentieth that observed in homogenates of the marsupial cells, and yet we never saw even this small amount of activity in preparations from any of the placental mammals. We therefore agree with the generally-held conclusion that direct photoenzymatic photoreactivation does not occur in the Placentalia.

This conclusion leaves unresolved the few positive reports in the literature. In the case of survival of UV-irradiated mice (Rieck and Carlson, 1955), it is very likely that the cause of death was irreversible shock, a condition which is markedly affected by endocrine balance; the latter in turn is affected by the constant-light conditions. Photoreactivation in this case could easily be indirect.

Johnson *et al.* (1968, p. 180) cite several other cases, including photoreactivation of sunburn and of the inhibition of growth of hair follicles in UV-irradiated skin. There is no evidence that DNA is the responsible light absorber in sunburn. The latter case is more suggestive, but indirect mechanisms have not been ruled out.

4.2 Tissue Distribution of Photoreactivating Activity

Excepting some of the early work with gametes, the tissue distribution of photoreactivating activity has been of particular concern in only two sets of studies: one in our laboratory with animal cells and the other in Werbin's laboratory with plant tissues.

In the 5-day-old sprout of the pinto bean, a variety of *Phaseolus vulgaris*, Saito and Werbin (1969) found the highest specific activities of photoreactivating enzyme in the plumule and hypocotyl, structures which in the natural habitat are exposed to sunlight; on the other hand, they detected no activity in the radicle which ordinarily remains below ground. This rather straightforward-appearing distribution becomes harder to understand as the plant develops; some activity was demonstrable in the leaves of 2-week-old plants, but the activity was negligible in 3- to 6-week-old leaves. The distribution appears to have less to do with the tissue's being exposed to sunlight than with its stage of development.

In our studies with animal tissues we also found little relationship between the activity in an organ and its potential exposure to solar irradiation. Four examples make this clear. (*a*) Although the shed sperm of the sea urchin, *Arbacia punctualata*, are not photoreactivable (Blum *et al.*, 1951), the testis contains a high level of photoreactivating enzyme activity (Cook and McGrath, 1967). The shed sperm of course are free in the open sea (although it is not at all certain how much they are threatened by radiation at the depths at which fertilization occurs), whereas the testes within the adult's body cavity are shielded by several layers of tissues which make a virtually opaque covering. (*b*) The adult land crab, *Gecarcinus lateralis*, is a nocturnal animal, remaining in deep burrows during the day, and is covered with a heavily pigmented exoskeleton. Nevertheless, epithelial tissues from the gill chambers, somatic muscle, and heart muscle as well as testis and ovary all show photoreactivating-enzyme activity whereas the midgut gland (hepatopancreas), which is physically adjacent to the heart, does not. (*c*) Five internal organs of the frog (Cook and McGrath, 1967) and six internal organs of the opossum (Cook and Regan, 1969) were all active. In the vertebrate tissues tested, the only inactive ones (excepting those from placental mammals) were blood serum and red blood cells of the toad, *Bufo marinus*. The white blood cells from this species were very active. (*d*) Unless special methods are adopted (Sato and Yasumura, 1966), cells in tissue culture tend to lose the biochemical and physiological properties of the parent tissue while retaining those enzymatic properties which are "shared by all cells of an organism regardless of differentiation" (Davidson, 1964). In this regard, it is of interest that all of the cultured vertebrate cells which we examined (again excepting placental mammals but including fish, amphibia, reptiles, birds, and marsupials) were active in all of the various photoreactivation assays which we employed.

In general, then, where photoreactivating enzyme occurs at all, it is widely distributed throughout the entire organism and not restricted to one or a few tissues or organs.

4.3 Distribution with Respect to Age of the Organism

Excepting a few studies with larval insects, very little has been done on the changes in UV-sensitivity and photoreactivability of organisms or tissues with age or embryological development.

The most thoroughly studied case is that of the parasitic wasp

Bracon hebetor Say (=*Habrobracon juglandis* Ashmead). Von Borstel and Wolff (1955) differentiated between failure of these eggs to hatch because of UV-damage to the cytoplasm and nonhatching because of UV-effects on the nucleus, and showed that the latter effect was photoreactivable, whereas the former was not. Shortly after the eggs are deposited, the nucleus migrates from the ventral surface to the interior where it is protected by the cytoplasm, and undergoes a number of mitoses. During this time, UV-irradiation of any surface leads to nonphotoreactivable "cytoplasmic" death. After three hours the axial nuclei migrate back to the periphery; UV sensitivity increases 30-fold, the type of death becomes "nuclear" again, and photoreactivability returns (von Borstel and Rekemeyer, 1958). Using the criterion of "percent-hatchability," Amy (1964, 1965) has followed the sensitivity changes until the time of hatching at 29 hours. He finds that despite complex UV-sensitivity changes during gastrulation and organogenesis, after the first three hours the dose reduction factor for maximum photoreactivation remains quite constant at about 0.35 until it rises slightly in the last few hours. This late rise might be due to there being simply not enough time, even at saturating intensities, for photoreactivation to repair the DNA lesion before the lesion is expressed. It would be interesting to know whether photoreactivation is as effective after eclosion as it is in these larval stages.

Sensitivity patterns roughly similar to that of *Bracon* have also been observed in *Drosophila*, although the data are not quite as complete (Goldman and Setlow, 1956; Ghelelovitch, 1966). Of particular interest in this genus is the presence of a nonreplicating cytoplasmic UV-sensitivity factor which is transmitted with the second chromosome (Ghelelovitch, 1968) but may be controlled by numerous genes. The interaction of the sensitivity factor and photoreactivability is not established.

In our survey of metazoan tissues, we found that photoreactivating-enzyme activity was very feeble if not absent in most adult chicken tissues whereas embryonic tissues (including primary chick embryo fibroblasts in culture) were quite active (Cook and McGrath, 1967). As mentioned above, a similar observation was made in the plant kingdom by Saito and Werbin (1969) who found that the specific activity of photoreactivating-enzyme was considerably greater in the plumules of pinto bean seedlings than in the adult leaves. Such observations of higher activity in developing tissues led us to look for the enzyme in calf fetus, but, as before, we could find no activity in this mammalian system.

Experiments on changes in UV-sensitivity and photoreactivability of developing systems are difficult to control because of the ever-changing optical properties of the system, changes which are not the same at all wavelengths (Amy, 1964; Goldman and Setlow, 1956); i.e., they are not the same for inactivating and reactivating wavelengths. Moreover, photoreactivation is a relatively slow process, and rapidly developing systems can be expected to change substantially in the time required to achieve maximum photoreversal. The DNA-lesion may be expressed before it can be repaired. Hence the time-resolution of such experiments may be poor. Nevertheless, more quantitative studies of both the biological and biochemical changes in photoreactivability during embryogenesis would be welcome since they may provide clues to the physiology of photoreactivation and the function(s) of the photoreactivating enzyme.

4.4 Subcelluar Distribution of Photoreactivating Activity

Direct photoreactivation requires of course the physical juxtaposition of the repairable lesion and the repairing enzyme. It is now well known that DNA may be found in a number of cell organelles in addition to the nucleus, and in most but not all cases it appears available to photoreactivating activity. In this section are reviewed some experiments relating to intracellular distribution of UV-sensitive structures and their photoreactivability.

4.4.1 Nucleus

Since in most cells (some eggs are exceptional) the majority of the DNA is localized in the nucleus, the nucleus is the prime target in UV killing and mutagenesis. For this reason several studies have been directed to the question of whether photoreactivation is a function concentrated in nuclei. The first such studies gave a negative or near-negative answer. Sperm, which were regarded as haploid nuclei nearly devoid of cytoplasm, from a number of marine invertebrates were shown to be very UV-sensitive when assayed by the criterion of mitotic delay (Wells and Giese, 1950; Blum *et al.*, 1951; Iverson and Giese, 1954; Cook and Rieck, 1962) but in only one case (the California sea urchin, *Strongylocentrotus purpuratus*, according to Wells and Giese) and then only feebly, could sperm be photoreactivated prior to insemination. Sperm, however, are a highly specialized cell with very limited viability once they are shed. Using a more "complete" cell, i.e., one with an indefinite viability (the alga, *Netrium digitus*) Marcenko (1968) studied photoreactivation with a microbeam of visible light following whole-cell UV-irradiation. Survival could be photoreactivated only when the nucleus was included in or very close to the beam. These results indicate that the critical lesion was in the nucleus and that it was repaired *in situ*. Jagger et al. (1969) have also used microbeam irradiation with amebae. They flattened the cells with a quartz coverslip during irradiation and illumination. This preparation is optically more satisfactory than most single-cell preparations and permits irradiation of all of the nucleus or all of the cytoplasm with little overlap. They found that both division delay and killing could be induced with nuclear irradiation, but that the photoreactivable sector was small (less than 0.3) for both effects.

We have looked for photoreactivating-enzyme activity in subcellular fractions of frog liver (Cook and McGrath, 1967). In our experience, more than 80% of the cell's total activity was found in the nuclear fraction. The activity was easily dissociated from the nuclei and was then recovered in the soluble fraction.

4.4.2 MITOCHONDRIA

The biogenesis of mitochondria has recently been reviewed by Roodyn and Wilkie (1968). That these organelles contain DNA has been well established in many laboratories. The DNA appears to carry genetic information for at least insoluble mitochondrial-protein.

The most thoroughly studied case of the ultraviolet photobiology involving mitochondria is that of induction of "petite" mutations in yeast. These mutants are respiration-deficient, and their mitochondria are lacking cytochromes *a* and *b* and some dehydrogenases. The UV-induction of such mutations is photoreactivable (Pittman *et al.*, 1959). Photoreactivable induction of petite mutants is also possible with anaerobically grown cells which contain no recognizable mitochondria (Wilkie, 1963; Maroudas and Wilkie, 1968) so that these authors postulate a master template system which "ensures genetic continuity during mitochondrial repression." This master template, itself a cytoplasmic determinant of the yeast mitochondrion, is a photoreactivable target in the UV studies.

Not in yeast and certainly not in any animal cells have there been any direct measurements on the photobiology and repair of mitochondrial DNA. Although we (Cook and McGrath, 1967) were not able to detect any photoreactivating activity in a mitochondrial fraction from frog liver, this negative result means only that less than 5% of the total cellular activity was to be found in this fraction. Kelner (personal communication) has suggested that the most meaningful expression of photoreactivating activity would be as enzyme units per milligram DNA. On this basis there might be as much activity in the mitochondrial fraction of frog liver as in the nuclear fraction.

In view of the expanding interest in mitochondrial genetics and development, this is a profitable area for direct photobiological and photochemical study in animal cells, particularly in tissue-culture systems.

4.4.3 Chloroplasts

Although chloroplasts do not occur in animal cells, some of the photobiological studies in *Euglena gracilis* are germane to the question of subcellular distribution of photoreactivating activity. The continuity of chloroplasts from generation to generation, and chloroplast inheritance in this organism have been reviewed recently by Schiff and Epstein (1968) and by Gibor and Herron (1968). At sublethal doses, UV radiation prevents green-colony forming ability in this organism apparently by either preventing replication of the plastid (see Schiff and Epstein, 1968) or by causing a permanent heritable block (mutation) impairing some early step in the ability of the proplastid to form a mature chloroplast (Gibor and Herron, 1968). Whatever the mechanism, the action spectrum for bleaching is consistent with the light absorber being nucleic acid, presumably DNA (Lyman *et al.*, 1961), and the effect shows a photoreactivable sector approaching 1.0. The action spectrum for photoreactivation is very similar to that for photoreactivation in *E. coli* (Schiff *et al.*, 1961). In a microbeam study, Gibor and Granick (1962) showed that irradiation of the nucleus alone could kill the cell but did not cause bleaching, whereas irradiation of the entire cell with the nucleus shielded did cause bleaching.

These studies, taken together, represent one of the most clear-cut cases of a non-nuclear UV effect being photoreactivated *in situ*. Photoreactivating-enzyme activity has been found in extracts of both green and bleached *Euglena* (Diamond *et al.*, 1969) and the simplest interpretation of all the data is that the enzyme has access not only to nuclear but also to chloroplast DNA.

4.4.4 "Cytoplasm"

Photoreactivation in cytoplasm, without any clear association of the activity with a particular organelle, has been observed in a number of systems. Blum *et al.* (1951) separated *Arbacia* eggs into halves by a centrifugation technique and found that although the anucleate half was not sensitive to UV radiation (mitotic delay being the criterion for sensitivity), it was capable of photoreactivating irradiated sperm. We (Cook and Rieck, 1962) obtained a similar result with *Echinarachinus* gametes by showing that irradiated sperm could be substantially photoreactivated during the first ten minutes after fertilization, even though the sperm pronucleus does not fuse with the egg pronucleus until 30 minutes after fertilization (Simmel and Karnofsky, 1961). Presumably the egg cytoplasm contains photoreactivating enzyme in these species.

In a comparable but much more difficult series of experiments Rustad and his co-workers (Rustad, 1964; Rustad *et al.*, 1965, 1966) have divided in half sea urchin eggs by cutting them individually. Although more laborious than the centrifugation technique, this method minimizes the stratification and compartmentalization of cell organelles other than the nucleus. In contrast to Blum's studies, they found that UV (and gamma) irradiation of the non-nucleated half causes a mitotic delay which, after UV, is photoreactivable. There is some evidence that mitochondria or mitochondria-associated structures might be the radiation-sensitive target in the enucleate half-cells.

Also working with cut cells, Skreb and Errera (1957) and later Skreb and Bevilacqua (1962) showed that the nonnucleated halves of *Amoeba proteus* survived longer after UV-irradiation if photoreactivated than if kept in the dark. The effect was also observed with the nucleated half.

Using microbeam UV-irradiation followed by either dark incubation or whole-cell exposure to photoreactivating light, Jagger *et al.* (1969) have investigated cytoplasmic sensitivity in their flattened-ameba preparation. They find that, with division delay as an index, the cells are more sensitive to cytoplasmic irradiation than nuclear. If killing is the criterion, cytoplasm and nucleus are about equally UV-sensitive. For both effects, the photoreactivable sector is much larger after cytoplasmic irradiation. The authors point out that there is a substantial amount of nucleic acid, both DNA and RNA, in ameba cytoplasm.

From all of these experiments it is clear that the cytoplasm of many cells has a substantial photoreactivating capacity. What is not certain is whether photoreactivating enzyme exists as a soluble enzyme in the cytoplasm of any cell, or whether it will always be found localized in DNA-containing organelles.

4.4.5 Surface Membranes

Before it became clear that direct photoreactivation was a DNA-repair mechanism, a number of attempts were made to photoreactivate UV effects on cell membranes. UV-activation of unfertilized sea urchin eggs, presumably a membrane effect which leads to several irregular and abortive cleavages, was not photoreactivable (Blum *et al.*, 1954). Attempts to prevent UV-induced hemolysis, a clear-cut membrane effect, in human and rabbit erythrocytes, led to an acceleration of hemolysis due to a photodynamic effect of the "photoreactivating" wavelengths (Cook, 1955). Booth and von Muralt (1951) were also unsuccessful in photoreactivating UV-induced conduction block in frog neurons, although these neurons contain photoreactivating enzyme (Cook and McGrath, 1967). An exception to these negative results was that of Pierce and

Giese (1957) who succeeded in photoreversing the decline in the magnitude of the action potential in UV-irradiated crab and frog nerves. The photoreversal was observed only under special conditions and with low UV doses; the mechanism is unknown, but it now seems very unlikely to be direct photoenzymatic repair.

4.5 Conclusions

Since photoreactivating-enzyme activity is readily demonstrable in the internal organs of metazoa, many of which are nocturnal, the enzyme clearly does not function in nature as it does in our laboratory assays. Since as far as we know pyrimidine dimers are formed in DNA only by ultraviolet irradiation or photosensitized reactions (Lamola and Yamane, 1967; Menningmann and Wacker, 1968) there is no apparent substrate for photoreactivating enzyme in many tissues where it is found. Although energy supplied by light is used for the enzymatic monomerization of pyrimidine dimers, both Jane Setlow (personal communication) and I (unpublished) have attempted to demonstrate repair of *Haemophilus* transforming DNA in the dark using cell extracts known to contain photoreactivating enzyme activity. Instead of light, potential energy sources for the reaction in these experiments could have come from either the complement of cellular enzymes present in crude extracts, or from ATP-regenerating systems which we added to the reaction mixture. Nevertheless, in no case was repair observed in the dark.

Blum and Matthews (1952) suggested that the persistence of photoreactivating activity in amphibia might represent an evolutionary vestigial function. On the other hand, Kelner (1961) suggested that photoreactivation might serve a function other than repair of UV-damage, a function as yet unidentified. The negative experiments described in the preceding paragraph do not support either alternative. Assuming conservatively, however, that any enzyme so widely found is probably of adaptive importance to the parent organisms, we may look to its distribution for clues as to its function. Here again there are no obvious correlations. J. D. Regan (unpublished) has pointed out that among the metazoa the placental mammals form the only *class* of organisms (there are many *specific* exceptions) which spend their entire embryonic development within the mother, and this is the only class of organisms in which photoreactivation is lacking. His suggestion implies that the principal function of photoreactivation is a protective one during embryogenesis, and that the persistence of the activity into adult life is of less significance and in some cases, such as the chicken (and pinto bean, among plants) may be more or less dispensed with in adult stages.

A second correlation, not entirely unrelated to Regan's proposition, is concerned with regenerative competence. Since Mizell's (1968) demonstration that newborn opossums are capable of regenerating limbs, provided that the regenerating stump is well supplied with nervous tissue, it now appears that placental mammals are the only class of organisms in which complete regeneration of complex tissues is not possible (this does not include, of course, such processes as the turnover of skin tissue, intestinal epithelia etc., nor the regeneration of liver, which is not possible if the entire liver is extirpated). The participation of nervous tissue in regeneration (Singer, 1959) is of special interest because in our experiments brain always appears to be the tissue with the highest photoreactivating activity in any given vertebrate, including the opossum. At the other extreme, circulating erythrocytes are the example *par excellence* of terminally differentiated cells with low developmental potential, and among amphibian tissues, for example, the erythrocytes, even though nucleated and DNA-containing, are the only cells in which we found no PR activity.

In conclusion, if there is a second function for the photoreactivating enzyme, we do not know what it is, nor even what its natural substrate may be. That such a second function exists at all is suggested by the wide distribution of the activity and its relatively poor correlation with a potentially protective role in exposure to sunlight. Although alternative correlations are not much more satisfactory, the more interesting of these seem to be related to the developmental potential of photoreactivable tissues. At this time, the correlations remain only correlations.

5. Photoreactivable Lesions (Biological)

Any of the cellular reactions in which DNA participates is vulnerable to UV and, if the enzyme is present, is potentiallly photoreactivable. Several classes of such UV-sensitive processes which have been studied in animal cells are discussed in this section. The arrangement is somewhat arbitrary since the categories overlap; in some cells, for example, the *inhibition of DNA synthesis* may be more-or-less directly responsible for a subsequent *delay in cell division* (Bootsma and Humphrey, 1968), although these subjects are treated separately below. Moreover, it is not feasible to attempt here a complete review of each biological phenomenon in which UV lesions may be important. Rather, the emphasis is on the types of information which UV-inactivation/photoreactivation studies have contributed, or may be expected to contribute, in studies of DNA-dependent phenomena.

5.1 Cell Killing

In a discussion of DNA repair, R. B. Setlow (1967) has listed a number of causes of death in irradiated bacterial cells (in the sense that they do not form colonies). His list, somewhat modified, may also be applied to eucaryotic cells: (*a*) cells may die because DNA synthesis is permanently inhibited, or their DNA is broken down in abortive repair. Such responses to DNA lesions may occur in either chromosomal or non-chromosomal DNA. In chromosomal DNA, abortive repair *may* yield chromosomal aberrations. (*b*) UV lesions in DNA may lead to synthesis of incorrect DNA (mutagenesis). (*c*) UV lesions in DNA may inhibit synthesis of RNA or lead to synthesis of incorrect RNA and protein, upsetting the physiological balance of the cells. (*d*) Irradiated cells may fail to divide, although not necessarily because of failure in DNA synthesis. Even a successful cytokinesis can yield nonviable daughter cells if the chromosomes fail to separate at anaphase. (*e*) Cells may die for reasons not directly traceable to a DNA-related malfunction, but rather for reasons associated with the loss of a vital and irreplaceable activity. For example the maintenance of cell volume may not be controllable if the cell membrane is damaged (Cook, 1965; see also footnote number 1, Section 2).

The broad categories may be divided into a very large number of specific subdivisions, particularly when one considers, for example, the number of different mutations which may be induced by UV. Although one or another of these effects may predominate at a given dose, they are probably all operative to some degree in any population of irradiated cells. To survive a potentially lethal lesion, the cell must either repair or replace the damaged molecules before these molecules are required to carry out their normal cellular functions. Until the moment they participate in and/or interfere with the physiological chemistry of the cell, UV-altered molecules cannot be recognized as biological lesions. Given our present understanding of the mechanisms of photoreactivation, we are now able in a number of instances to pinpoint the time-of-action of certain DNA functions by determining when in the life cycle of the cell these functions are photoreactivable and when they are not. Specific cases, related to possible lethal effects as listed above, are described in the subsequent sections.

5.2 Inhibition of DNA Synthesis

In an early study of the mechanism of photoreactivation in bacteria (*E. coli* B/r) Kelner (1953) showed that an immediate effect of UV was the cessation of DNA synthesis, and that synthesis was resumed after exposure of the cells to photoreactivating light. This observation was greatly extended in a series of papers by R. B. Setlow and his co-workers, whose results, in brief, were: (*a*) UV-irradiation inhibits, but does not totally prevent, the template activity of DNA when tested with calf-thymus-DNA-polymerase. The most satisfactory model for the data was that UV-lesions, identifiable as pyrimidine dimers by their short-wavelength reversibility, acted as blocks around which polymerization (incorporating random bases) could occur very slowly (Bollum and Setlow, 1963). (*b*) In *E. coli* B_{s-1}, which is incapable of dark repair (Setlow and Carrier, 1964), one dimer every 200 microns (20% of the *E. coli*-chromosome length) blocks DNA synthesis. The rate of DNA synthesis in this organism at very low UV doses can be photoreactivated with a dose-reduction factor of about 0.5 which, under the photoreactivating conditions used, corresponded to the dose reduction for both colony formation and dimer splitting (Setlow *et al.*, 1963). If the photoreactivation was delayed until DNA-polymerization had proceeded up to a block, the photoreactivable sector was greatly reduced (Swenson and Setlow, 1966). The authors postulated that the blocked site might be inaccessible to the photoreactivating enzyme, and photoreactivation elsewhere on the chromosome would be of no significance in alleviating the inhibition. These relationships were, of course, greatly modified in cells capable of dark repair.

Equivalent studies have not been carried out in any eucaryotic cells. In mammalian cells it is clear that UV inhibits DNA synthesis, but even in cells which appear to have no capacity for dimer excision the block to DNA synthesis is not absolute (Trosko *et al.*, 1965; Cleaver, 1965, 1966; Klimek and Vlasinova, 1966; Trosko and Kasschau, 1967; see also Chapter 6). Since human cells *are* capable of dimer excision (Regan *et al.*, 1968c), which might in fact occur in preferential (i.e., nonrandom) parts of the genome, the nature of the blocks to DNA synthesis are less clear in this case. The nonphotoreactivability of mammalian cells deprives workers in this field of a useful technique in the study of this problem.

Regan, Lee and I (Regan and Cook, 1967; Regan *et al.*, 1968a) found that in cultured fish and amphibian cells, UV doses which the cells could just survive (75 ergs/mm^2; 265 nm) led to a marked depression of DNA synthesis, and the effect was highly photoreactivable. This observation argues strongly that pyrimidine dimers are inhibitory to DNA synthesis in these cells, and it seems unlikely that the mechanisms of DNA synthesis in mammalian cells are so different from those of other vertebrates that the inhibition observed in them is unrelated to the pyrimidine dimers.

A curiously different observation has been made in echinoderm zygotes which, either just before or just after fertilization, have been sufficiently X-irradiated or UV-irradiated to induce a substantial mitotic delay (Rao and Hinegardner, 1965; Zeitz *et al.*, 1968). In both cases the first DNA-synthetic period is not detectably inhibited; the second synthetic period occurs at the normal rate at the normal time with respect to the first division (although, since the first division is delayed, the second DNA-synthetic period is necessarily delayed to the same extent; Section 5.3). The UV doses (50 ergs/mm^2; 254 nm) to naked sperm were of the same order as those used in the fish and amphibian experiments above. Since mitotic delays are highly photoreactivable in these species, there is no doubt that dimers have been introduced into the DNA but they have no apparent inhibitory effect on the priming activity. It would be interesting to explore whether this difference is related to the fact that DNA polymerase from echinoderms differs from bacterial and mammalian polymerases in that it has a marked preference for double-stranded rather than single-standed DNA as a template (Loeb *et al.*, 1967).

The effects of UV on DNA synthesis in Tetrahymena are complex, and appear to reflect a dissociation of the control of synthesis and the control of cell division. When the cells are irradiated in G_2, DNA synthesis begins again without division, which may in fact be delayed as much as 15 hours under conditions in which the control cells are dividing every 3 hours. During the delay period the cells may synthesize up to 3½ times the normal amount of DNA; the excess is lost in extrusion bodies in subsequent generations (Shepard, 1965). A prolonged, slow S period is also found after UV irradiation in G_1 (Shepard and Chu, 1967). Photoreactivation was specifically avoided in these experiments.

5.3 Mitotic Delay

A photoreactivable division delay, which may persist for more than one generation, has been observed in many experiments with UV-irradiated cells. It is probably the reason for the decreased slopes in growth curves of cell populations irradiated with low doses (see, e.g., Regan and Cook, 1967; Regan *et al.*, 1968a), although it is not possible to draw firm conclusions about delay solely from observations on log-phase populations. The analysis of mitotic inhibition has been attempted in only a few photoreactivable systems.

One of the most favorable, and one of the earliest studied, is that of echinoderm zygotes, in which either parental nucleus can be irradiated independently, in which parts of cells can be studied, in which a high degree of synchrony can be achieved by the simultaneous fertilization of large numbers of eggs, and for which there is a large supporting literature on the physiology and biochemistry of cell division. The literature through 1964 on UV-induced mitotic delay has been summarized by Rustad (1964).

The principal observations are that any division (usually only the first two, or at most the first four, are studied) may be delayed by a low ultraviolet dose which all of the cells will survive. The radiation is effective up to about the end of the S period or very early prophase (there is virtually no G_2 in these cells). Once mitosis begins, the radiation will not stop it except at very high lethal doses. The delayed phase of the cell cycle is prophase, which is prolonged to an extent equal to the eventual delay in cytokinesis (Keller *et al.*, 1970). No other phase of mitosis is affected and, as pointed out above, DNA synthesis itself is not prolonged. The delay is photoreactivable, but photoreactivability is also lost at the end of S (Cook and Rieck, 1962). During the prolonged prophase after UV, the zygotes are *not* sensitive to a second UV dose (Cook, 1968), although the period of UV sensitivity may be prolonged by puromycin treatment (Rustad and Burchill, 1966).

The light absorber in the phenomenon seems clearly to be DNA since the inhibition is photorepairable on the one hand and hypersensitized following BUdR incorporation on the other.[3] If the cells are irradiated during S while incorporating BUdR, the corresponding mitosis is delayed but not hypersensitive, although later divisions are both hypersensitized and nonphotoreactivable (Cook, 1968). This and related observations are most easily accounted for by a model in which only lesions in *un*replicated DNA affect the onset of the subsequent mitosis.

It is not at all necessary to presume that the sensitive DNA is chromosomal. Rustad (1964) has argued that the primary target may be the centrioles (which, for example, may fail to separate at the onset of mitosis). This is an attractive idea, in line with the observations of Mazia *et al.* (1960) on the "mitotic centers" in sea urchin zygotes, as well as compatible with a number of other observations summarized by Rustad; there is no compelling evidence against it.

Another useful preparation in the study of mitosis, particularly from the cytological point of view is the grasshopper neuroblast, in which 8 or more subdivisions of prophase alone can be distinguished in the living cell (Carlson, 1969). Gaulden and Perry (1958) showed that microbeam irradiation of either one or both of the two prominent nucleoli

[3] Among echinoderms this and other BUdR effects seem to be restricted to sand dollars, (L. Zeitz, personal communication; see also Rao, 1963; Mazia and Gontcharoff, 1964; and Gontcharoff and Mazia, 1967).

led to a substantial delay in or blockage of mitosis, provided that the irradiation was carried out prior to middle prophase; in late prophase or after there was no effect. The effect is greater at 265 nm than 280 nm (Carlson *et al.*, 1961) which implies a nucleic acid absorption, but the extra parameter of photoreactivation was expressly avoided throughout. Irradiation of non-nucleolar regions of the nucleus also resulted in division delay. Despite many superficial similarities, it is not yet possible to compare directly the data from the neuroblasts with those from marine eggs because the types of information so far obtained in the two systems overlap at very few points.

The UV-lesions underlaying photoreactivable mitotic delay are not always repaired in a single cell cycle. A number of divisions may be required before the interdivision interval returns to normal, and, in sand dollars at least (Cook and Rieck, 1962) the delaying effects may be photorepaired in cell cycles subsequent to the one in which the UV treatment was given.

Division-retarding effects of UV over several cell cycles are most remarkable in ciliated protozoa. Commonly in paramecia (Giese and Reed, 1940; Kimball and Gaither, 1951) in Blepharisma (Giese and Lusignan, 1961) and in Tetrahymena (Christensen and Giese, 1956), following UV-irradiation there are one or two retarded divisions which may then be followed by no divisions at all in a "cessation" period which can last for weeks. Cells which survive this cessation period will, after one or two slow divisions, return to the normal growth rate. The pattern of response to UV is dependent on the nutritional state of the organism (Giese and Reed, 1940), but the effects in any event are highly photoreactivable. It is not known whether the "cessation" period is due to lesions persisting in DNA after one or more rounds of replication, or whether cessation is a delayed cellular response which has been determined a generation or more before it occurs. The question, which could be easily explored by attempting photoreactivation one or two generations post-UV, arises because of Brunk and Hanawalt's (1969) observation that most dark repair-replication in Tetrahymena is completed before the first normal round of replication occurs. Delayed photoreactivation experiments might give a clue to the physiology of the cessation phenomenon.

The photoreactivable sector for division delay is large and may exceed 0.9 in some ciliates (Giese *et al.*, 1952). Francis and Whitson (1969) have made preliminary correlations of the photoreactivation of division delay and pyrimidine-dimer monomerization in Tetrahymena; they state that a 90 percent dose-reduction of the biological effect is achieved by illumination which photorepairs only 5 percent of the dimers. This may be another example of preferential repair, and deserves further study.

5.4 Mutagenesis

The induction of mutations by UV, with particular emphasis on the wavelengths responsible, has recently been reviewed by Giese (1968). Since metazoa are protected from their environment by an external covering which is usually opaque, or nearly opaque, to far ultraviolet, and since their germ cells are usually internal, such organisms make very unfavorable material for studies on ultraviolet mutagenesis. The problem is complicated by the fact that, at any given time, the gametes are commonly in a wide range of developmental stages and UV-sensitivity (Kimball, 1966). Once gametes are shed as exposed cells they are frequently much more vulnerable to UV, and irradiation of the early embryological stages may lead to developmental anomalies. Although these may by some criteria be considered mutations, they are treated here in a separate section (5.7; see also Perlitsh and Kelner, 1953 on the photoreactivation of UV-induced phenocopies in Drosophila).

Nevertheless, the first demonstration of a UV-induced mutation was in a metazoan. Altenburg (1930, 1933) irradiated Drosophila larvae at the "polar cap" stage, when there are near the surface of the embryo about 20 pole cells which will become the germ line. He looked for, and found, x-chromosome lethals which were transmitted to 5–10 percent of the daughters of irradiated males. Shortly after the discovery of photoreactivation, the Altenburgs (1952) and independently H. V. Meyer (1951) were able to photoreactivate chromosome-II lethals in Drosophila. The effect is very large; from the data of Browning and Altenburg (1962) the photoreactivable sector appears to be on the order of .85. The same paper describes very graphically the difficulty of performing such experiments in a well controlled manner.

A eucaryotic cell which is more amenable to such studies is the ciliate *Paramecium aurelia*, in which R. F. Kimball and his associates have studied the induction of recessive lethals and "slow growth" mutations. The fundamental problem with which they have been concerned is the nature of the reparable premutational lesion, the nature of its repair, and its final fixation as a permanently heritable trait. The simplest hypothesis for most of the data is that premutational UV-lesions may be repaired very effectively, given sufficient time before DNA synthesis, but that the lesion becomes fixed during S. Presumably the fixation occurs by mis-replication at the site of the lesion, and the subsequent faithful copying of the mistake by the usual semiconservative mode of synthesis. (See Kimball, 1966 and 1969 for discussion of hard-to-reconcile data and comparison with other mutagenic systems.) Thus the mutation yield is very low if the paramecia are irradiated in G_2; it rises sharply with irradiation in G_1, and reaches a maximum at the end of G_1 or early S when there is no time for repair before replication. These data are of particular interest because they may be compared with the observations of B. Sutherland *et al.* (1967, 1968) on pyrimidine dimers in the same organism. It is important to remember that the premutational lesions are in the micronuclei, and that the mutation is not expressed until after autogamy when both new micronuclei and the new macronucleus develop from information in one of the original micronuclei; on the other hand, most of the pyrimidine dimers as determined chemically are in the macronucleus. In the dark, after 400 ergs/mm^2 (254 nm) there is a lag of some 6 hours before detectable dimer excision begins. Kimball (1969) has shown that after ¼ that dose, DNA synthesis is delayed at least 4 hours, and at the higher dose is undoubtedly delayed very much longer. There is thus no particular difficulty in reconciling the lag in excision and the ability of the cells to carry out dark repair of potential mutations. In the light, the photoreactivation of dimers is very efficient (B. Sutherland *et al.*, 1967); all of the detectable dimers appear capable of being monomerized. The photoreactivable sector for mutation yield, however, is only about 0.5. It is not certain whether the nonphotoreactivable lesions are nondimer lesions, or are dimers in the micronucleus which for some reason are not available to photoreactivating enzyme.

Working with log phase cells, Sutherland *et al.* (1968) showed that at any time up to 10 hours after UV, the nondark-repaired dimers could be photoreactivated. This result is to be compared with one of Kimball's in which he showed that if the cells were irradiated a half-hour after division, 5 hours later (still G_1) photoreversibility of mutagenesis was lost in irradiated cells (but not in unirradiated). The mechanism whereby these premutational lesions are isolated from photorepair is unknown.

5.5 Virus Inactivation

Although there is a substantial literature on the UV-inactivation of animal viruses, the great majority of it is concerned with infections in mammalian cells. In all such studies, photoreactivation has either been avoided or looked for and not found (Carp and Gilden, 1965; Pfefferkorn and Coady, 1968).

Pfefferkorn and his co-workers (1965, 1966, 1968; see also Section 4.1.3) utilizing either pseudorabies or herpes simplex in chick embryo fibroblasts, have given the most complete description of a photoreactivable system with animal viruses. Following UV at 254 nm, the maximum photoreactivable sector was approximately 0.5. They found that, at 37°C, the photoreactivability of irradiated virus was negligible for the first 20 minutes after absorption to the chick cells, then rose to a maximum at about 2 hours and decayed again thereafter. The development of the "photoreactivable state" for the virus was sensitive to cycloheximide, i.e., required protein synthesis, although cycloheximide did not alter the ability of the host cells to photoreactivate pyrimidine dimers in their own DNA. The authors conclude that the inhibition is not affecting the synthesis of the repair enzyme, but is affecting some step (uncoating of the viral DNA? transport of the virus to the cell nucleus?) required to make the UV-lesions susceptible to the host's photorepair system. Vaccinia virus, which is not uncoated, is not photoreactivated in chick cells.

An unusual type of photoreactivation has been described by David and Magnus (1967) in an insect virus, the granulosis virus which infects the European cabbageworm. In this case the free (and highly purified) virus particle could be photoreactivated while still free in suspension, i.e., before infection. The only other case in which free virus can be photoreactivated without host-cell enzyme is that of the clear-plaque mutation in the "Kappa" bacteriophage infecting *Serratia marcescens* (Kaplan *et al.*, 1960). The assay for the insect virus is complex (virus is applied to cabbage leaves, leaves are fed to larvae, larvae are scored for survival) and dose reduction factors cannot yet be ascertained from the data; hence the magnitude of the photoreactivation effect is not clear. Nothing is known of the mechanism.

5.6 Chromosome Aberrations

Strictly speaking, this subject should not be included here since the most thorough attempts to photoreactivate chromosome aberrations have been negative in both plant (*Tradescantia*, Kirby-Smith and Craig, 1957) and mammalian cells (Chinese hamster, Chu, 1965). There is however no independent evidence for photoreactivation of any function in *Tradescantia*, and there is now good reason not to expect it in Chinese hamster cells (Section 4.1.3). The question has not yet been put to a rigorous test.

The possibility of photoreactivating chromosome aberrations in a favorable cell-type is an intriguing one. That DNA is the light absorber in this type of damage is strongly suggested by the action spectra of Kirby-Smith and Craig (1957) and Chu (1965). Moreover, Chu also showed that, following the incorporation of BUdR, not only did the chromosomes exhibit greater UV-sensitivity, but the action spectrum was also shifted to longer wavelengths; both effects are characteristic of lesions in BU-DNA (Boyce and Setlow, 1963). It has been suggested by Evans (1966, 1967) working with X-rays, and independently by Kihlman and Hartley (1967) working with 2′-deoxyadenosine that chromosome aberrations do not arise directly from the initial lesions in DNA, but rather as a result of mistakes in the dark-repair of these lesions, leading to "cross polymerization" or "illegitimate rejoining" of opened DNA strands. A similar model has also been proposed by Grigg (1968). This view has been disputed by Wolff and Scott (1969), partly on the grounds that aberrations can be induced by UV in cells that are deficient, and possibly even incapable, of dark repair.

This is not the place to review this complex subject; I wish only to point out here that the possibility of introducing and later reducing DNA lesions by UV and subsequent photoreactivation offers an expanded range of experimental approaches to this problem. It would be well worth looking for in cells known to be capable of photoreactivation.

5.7 Developmental Effects

5.7.1 Echinoderms

A number of dose-dependent UV-effects may be differentiated in echinoderm zygotes (Blum *et al.*, 1954; Rieck, 1965). At low doses, corresponding to less than 1000 ergs/mm^2 at 254 nm, unfertilized eggs may be fertilized normally; they will undergo normal division, with the photoreactivable mitotic delay discussed above (Section 5.3), and virtually 100 percent of the population will develop into normal plutei. At doses in excess of 10^4 ergs/mm^2, the eggs undergo nonphotoreactivable cytolysis without any normal cell divisions. In the intermediate range of doses, the zygotes go through slow but fairly normal-appearing division until the blastula stage, and then a large fraction fail to gastrulate normally; such embryos subsequently cytolyze without further development. The effect is photoreactivable (Rieck, 1965). This developmental pattern bears a striking resemblance to that seen in echinoderm zygotes grown in the presence of Actinomycin D (Gross and Cousineau, 1963; Gross *et al.*, 1964). Gross and his co-workers have shown that these embryos develop in the absence of RNA synthesis up until gastrulation. Postfertilization protein and DNA synthesis take place as usual; the cells divide with normal cleavage planes and all cells contain nuclei. Apparently all the RNA required for early development is present in the unfertilized egg. Because of the morphogenetic similarities between actinomycin-D-treated and UV-irradiated zygotes, it is not unlikely that the UV effect is a photorepairable DNA-RNA transcription failure, which becomes manifest as a lethal event at gastrulation, rather than a DNA replication failure. Appropriate measurements on RNA synthesis have not been made in the irradiated cells.

5.7.2 Insects

Photoreactivable stages in insect development have been described above (Section 4.3). The criterion of UV-sensitivity in almost all of the published data is survival at eclosion. Attempts to relate more specific effects to the stages of organogenesis at the time of irradiation have been made, but have not yet yielded very fruitful results (cf. discussion in Amy, 1964). Microbeam studies on pole cells in Drosophila larvae (Hathaway and Selman, 1961) or on the genital disc at a later stage (third instar larvae; Ursprung, 1957) have demonstrated a mosaic pattern in development in that unirradiated areas adjacent to the beam were unable to replace the killed cells. Use of photoreactivation, as in mutation fixation studies, would be of interest in this type of experiment.

5.7.3 Amphibia

One hemisphere of the frog's egg is darkly pigmented. In what may be an adaptation to the exposed environment in which early development takes place, the fertilized egg floats with the pigmented half upward, facing the sun. Smith (1966) showed that irradiation of this the upper surface (animal pole) of the large eggs of *Rana pipiens* with doses in excess of 30,000 ergs/mm^2 (254 nm) had essentially no effect on development, despite the fact that the nucleus is near this surface. Smith also confirmed the earlier observation of Bounoure *et al.* (1954) that irradiation of the under surface (vegetal pole) of the egg resulted in animals with gonads either greatly reduced in size or absent. Although the action spectrum for the effect indicated nucleic acid absorption, the effect was not photoreactivable and the UV-sensitive target was certainly not the nucleus. Smith was able to replace, with fair success, the UV-inactivated "germinal plasm" by transfer of vegetal-pole cytoplasm from unirradiated eggs into the irradiated cells.

As amphibian larvae develop, only the surface structures appear to be affected by UV. Irradiation of limb buds can prevent development of the limb; irradiation of the developing limb can cause either regression or induction of supernumerary structures, and prior irradiation of a limb which is later amputated greatly inhibits regeneration (Rieck, 1954; Butler and Blum, 1955; Blum *et al.*, 1957). All of these effects are photoreactivable to some extent, although the experiments are too difficult and complex to establish definitive dose relations. From the cytological data given in these studies, it appears that the UV may have penetrated below the ectoderm into mesodermal structures (see also Kraft, 1968). Thus the cells which are the UV targets in these experiments have not been positively identified.

References

Altenburg, E. (1930). *Anat. Record* **47**, Abstr. p. 383.

Altenburg, E. (1933). *Science* **78**, 587.

Altenburg, L. S., and Altenburg, E. (1952). *Genetics* **37**, 545–553.

Amy, R. L. (1964). *J. Exp. Zool.* **155**, 43–56.

Amy, R. L. (1965). *Radiation Res.* **25**, 674–683.

Avadhani, N-G., Mehta, B. M., and Rege, D. V. (1969). *J. Mol. Biol.* **42**, 413–423.

Berger, J. D., and Kimball, R. F. (1964). *J. Protozool.* **11**, 534–537.

Beukers, R., and Berends, W. (1960). *Biochim. Biophys. Acta* **41**, 550–551.

Birnstiel, M. L., Speirs, J., Purdom, I., Jones, K., and Loening, U. E. (1968). *Nature* **219**, 454–463.

Blum, H. F. (1959). "Carcinogenesis by Ultraviolet Light." Princeton Univ. Press, New Jersey.

Blum, H. F., Cook, J. S., and Loos, G. M. (1954). *J. Gen. Physiol.* **37**, 313–324.

Blum, H. F., Loos, G. M., Price, J. P., and Robinson, J. C. (1949). *Nature* **164**, 1011.

Blum, H. F., and Matthews, M. R. (1952). *J. Cellular Comp. Physiol.* **39**, 57–71.

Blum, H. F., Butler, E. G., Chang, J. J., Mawe, R. C., and Schmidt, S. E. (1957). *J. Cellular Comp. Physiol.* **47**, 153–169.

Blum, H. F., Robinson, J. C., and Loos, G. M. (1951). *J. Gen. Physiol.* **35**, 323–342.

Blum, H. F., and Terus, W. S. (1946). *Am. Physiol.* **146**,97–106.

Bollum, F. J., and Setlow, R. B. (1963). *Biochim. Biophys. Acta* **68**, 599–607.

Booth, J. A., and von Muralt, A. (1951). Unpublished observations.

Bootsma, D., and Humphrey, R. M. (1968). *Mutation Res.* **5**, 289–298.

Bounoure, L., Aubry, R., and Huck, M. L. (1954). *J. Embryol. Exptl. Morphol.* **2**, 245–263.

Boyce, R., and Setlow, R. B. (1963). *Biochim. Biophys. Acta* **68**, 446–454.

Britten, R. J., and Kohne, D. E. (1968). *Science* **161**, 529–540.

Browning, L. S., and Altenburg, E. (1962). *Genetics* **47**, 361–366.

Brunk, C. F., and Hanawalt, P. C. (1967). *Science* **158**, 663–664.

Brunk, C. F., and Hanawalt, P. C. (1969). *Radiation Res.* **38**, 285–295.

Butler, E. G., and Blum, H. F. (1955). *J. Natl. Cancer Inst.* **15**, 877–889.

Callan, H. G. (1967). *J. Cell Sci.* **2**, 1–7.

Carp, R. I., and Gilden, R. V. (1965). *Virology* **27**, 639–641.

Carlson, J. G. (1969). *Radiation Res.* **37**, 1–14.

Carlson, J. G., Gaulden, M. E., and Jagger, J. (1961). *In* "Progress in Photobiology" (B. Chr. Christensen and B. Buchmann, eds.), pp. 251–253. Elsevier, Amsterdam.

Christensen, E., and Giese, A. C. (1956). *J. Gen. Physiol.* **39**, 513–526.

Chu, E. H. Y. (1965). *Mutation Res.* **2**, 75–94.

Cleaver, J. E. (1965). *Biochim. Biophys. Acta* **108**, 45–52.

Cleaver, J. E. (1966). *Biochem. Biophys. Res. Commun.* **24**, 569–576.

Cleaver, J. E. (1967). *Radiation Res.* **30**, 795–810.

Cook, J. S. (1955). Unpublished observations.

Cook, J. S. (1956). *J. Cellular Comp. Physiol.* **47**, 55–84.

Cook, J. S. (1965). *J. Gen. Physiol.* **48**, 719–734.

Cook, J. S. (1967). *Photochem. Photobiol.* **6**, 97–101.

Cook, J. S. (1968). *Exptl. Cell Res.* **50**, 627–638.
Cook, J. S., and McGrath, J. R. (1967). *Proc. Natl. Acad. Sci. U. S.* **58**, 1359–1365.
Cook, J. S., and Regan, J. D. (1969). *Nature* **223**, 1066–1067.
Cook, J. S., and Rieck, A. F. (1962). *J. Cellular Comp. Physiol.* **59**, 77–84.
Cook, J. S., and Setlow, J. K. (1966). *Biochem. Biophys. Res. Commun.* **24**, 285–289.
David, W. A. L., and Magnus, I. A. (1967). *J. Invert. Pathol.* **9**, 266–268.
Davidson, E. (1964). *Advan. Genet.* **12**, 144–280.
Deutscher, M. P., and Kornberg, A. (1969). *J. Biol. Chem.* **244**, 3029–3037.
Diamond, J., Schiff, J. A., and Kelner, A. (1969). Personal communication.
Donnellan, J. E., and Stafford, R. S. (1968). *Biophys. J.* **8**, 17–28.
Dulbecco, R. (1949). *Nature* **163**, 949–950.
Dulbecco, R. (1955). *In* "Radiation Biology" (A. Hollaender, ed.), Vol. 2, pp. 455–486. McGraw-Hill, New York.
Elder, R. L., and Beers, R. F., Jr. (1965). *J. Bacteriol.* **89**, 1225–1230.
Englund, P. T., Kelly, R. B., and Kornberg, A. (1969). *J. Biol. Chem.* **244**, 3045–3052.
Evans, H. J. (1966). *In* "Genetical Aspects of Radiosensitivity: Mechanisms of Repair" pp. 31–47. Intern. Atomic Energy Agency, Vienna.
Evans, H. J. (1967). *In* "Radiation Research 1966" (G. Silini, ed.), pp. 482–501. North-Holland, Amsterdam.
Francis, A. A., and Whitson, G. L. (1969). *Biochim. Biophys. Acta* **179**, 253–257.
Gall, J. G. (1968). *Proc. Natl. Acad. Sci. U. S.* **60**, 553–560.
Gaulden, M. E., and Perry, R. P. (1958). *Proc. Natl. Acad. Sci. U. S.* **44**, 553–559.
Ghelelovitch, S. (1966). *Intern. J. Radiation Biol.* **11**, 255–271.
Ghelelovitch, S. (1968). *Mutation Res.* **5**, 117–131.
Gibor, A., and Granick, S. (1962). *J. Cell Biol.* **15**, 590–603.
Gibor, A., and Herron, H. A. (1968). *In* "The Biology of *Euglena*" (D. E. Buetow, ed.), Vol. II, pp. 335–349. Academic Press, New York.
Giese, A. C. (1967). *In* "Research in Protozoology" (T. T. Chen, ed.), Vol. 2, pp. 267–356. Pergamon Press, Oxford.
Giese, A. C. (1968). *Photochem. Photobiol.* **8**, 527–546.
Giese, A. C., Brandt, C. L., Iverson, R., and Wells, P. H. (1952). *Biol. Bull.* **103**, 336–344.
Giese, A. C., and Lusignan, M. (1961). *J. Gen. Physiol.* **44**, 543–554.
Giese, A. C., and Reed, E. A. (1940). *J. Cellular Comp. Physiol.* **15**, 395–408.
Goldman, A. S., and Setlow, R. B. (1956). *Exptl. Cell Res.* **11**, 146–159.
Gontcharoff, M., and Mazia, D. (1967). *Exptl. Cell Res.* **46**, 315–327.
Goodgal, S. H., Rupert, C. S., and Herriott, R. M. (1957). *In* "The Chemical Basis of Heredity" (W. D. McElroy and B. Glass, eds.), pp. 341 343. Johns Hopkins Press, Baltimore, Maryland.
Greer, S. (1960). *J. Gen. Microbiol.* **22**, 618–634.
Grigg, G. W. (1968). *In* "Replication and Recombination of Genetic Material" (W. J. Peacock and R. D. Brock, eds.), pp. 120–133. Australian Acad. Sci., Canberra.
Gross, P. R., and Cousineau, G. H. (1963). *Biochem. Biophys. Res. Commun.* **10**, 321–326.
Gross, P. R., Malkin, L. I., and Moyer, W. A. (1964). *Proc. Natl. Acad. Sci. U. S.* **51**, 407–413.
Hanawalt, P. C. (1968). *In* "Photophysiology" (A. C. Giese, ed.), Vol. 4, pp. 204–251. Academic Press, New York.
Hathaway, D. S., and Selman, G. G. (1961). *J. Embryol. Exptl. Morphol.* **9**, 310–325.

Haynes, R. H., Baker, R. M., and Jones, G. E. (1968). *In* "Energetics and Mechanisms in Radiation Biology" (G. O. Phillips, ed.), pp. 425–465. Academic Press, New York.

Howard-Flanders, P. (1968). *Ann. Rev. Biochem.* **37,** 175–200.

Iverson, R. M., and Giese, A. C. (1954). *Science* **120,** 504.

Jagger, J. (1958). *Bacteriol. Rev.* **22,** 99–142.

Jagger, J., Prescott, D. M., and Gaulden, M. E. (1969). *Exptl. Cell Res.* **58,** 35–54.

Jagger, J., and Stafford, R. S. (1965). *Biophys. J.* **5,** 75–88.

Johns, H. E., Rapaport, S. A., and Delbrück, M. (1962). *J. Mol. Biol.* **4,** 104–114.

Johnson, B. E., Daniels, F., Jr., and Magnus, I. A. (1968). *In* "Photophysiology" (A. C. Giese, ed.), Vol. 4, pp. 139–202. Academic Press, New York.

Kaplan, R. W., Winkler, U., and Wolf-Ellmauer, H. (1960). *Nature* **186,** 330–331.

Keller, J., Rieck, A. F., and Schmid, C. (1970). *J. Cell. Physiol.* (in press).

Kelly, R. B., Atkinson, M. R., Huberman, J. A., and Kornberg, A. (1969). *Nature* **224,** 495–501.

Kelner, A. (1949). *Proc. Natl. Acad. Sci. U. S.* **35,** 73–79.

Kelner, A. (1953). *J. Bacteriol.* **65,** 252–262.

Kelner, A. (1961). *In* "Progress in Photobiology" (B. Chr. Christensen, and B. Buchmann, eds.), pp. 276–278. Elsevier, New York.

Kelner, A. (1964). *J. Bacteriol.* **87,** 1295–1303.

Kihlman, B. A., and Hartley, B. (1967). *Mutation Res.* **4,** 771–782.

Kimball, R. F. (1966). *In* "Advances in Radiation Biology" (L. G. Augenstein, R. Mason, and M. Zelle, eds.), Vol. 2, pp. 135–166. Academic Press, New York.

Kimball, R. F. (1969). *Mutation Res.* **8,** 79–89.

Kimball, R. F., and Gaither, N. T. (1951). *J. Cellular Comp. Physiol.* **37,** 211–231.

Kirby-Smith, J. S., and Craig, D. L. (1957). *Genetics* **42,** 176–187.

Klimek, M. (1966). *Photochem. Photobiol.* **5,** 603–607.

Klimek, M., and Vlasinova, M. (1966). *Intern. J. Radiation Biol.* **11,** 329–337.

Kraft, A. v. (1968). *Wilhelm Roux' Arch. Entwickl.-Mech. Org.* **160,** 259–297.

Lamola, A. A., and Yamane, T. (1967). *Proc. Natl. Acad. Sci. U. S.* **58,** 443–446.

Lee, H. H., and Puck, T. T. (1960). *Radiation Res.* **12,** 340–348.

Loeb, L. A., Mazia, D., and Ruby, A. D. (1967). *Proc. Natl. Acad. Sci. U. S.* **57,** 841–848.

Lyman, H., Epstein, H. T., and Schiff, J. A. (1961). *Biochim. Biophys. Acta* **50,** 301–309.

Mahler, I., and Grossman, L. (1968). *Biochem. Biophys. Res. Commun.* **32,** 776–781.

Marcenko, E. (1968). *Radiation Botany* **8,** 325–338.

Marshak, A. (1949). *Biol. Bull.* **97,** 315–322.

Maroudas, N. G., and Wilkie, D. (1968). *Biochim. Biophys. Acta* **166,** 681–688.

Mazia, D., and Gontcharoff, M. (1964). *Exptl. Cell Res.* **35,** 14–25.

Mazia, D., Harris, P. J., and Bibring, T. (1960). *J. Biophys. Biochem. Cytol.* **7,** 1–20.

McGrath, R. A. (1960). *Intern. J. Radiation Biol.* **2,** 177–185.

McGrath, R. A. (1963). *Radiation Res.* **19,** 526–537.

McLaren, A. D., and Shugar, D. (1964). "Photochemistry of Proteins and Nucleic Acids." Pergamon Press, Oxford.

Menningmann, H. D., and Wacker, A. (1968). Fifth International Congress Photobiology, p. 128. Hanover.

Meyer, H. U. (1951). *Genetics* **36,** 565.

Mizell, M. (1968). *Science* **161,** 283–286.

Moseley, B. E. B., and Setlow, J. K. (1968). *Proc. Natl. Acad. Sci. U. S.* **61,** 176–183.

Muhammed, A. (1966). *J. Biol. Chem.* **241,** 516–523.

Muhammed, A., Gonçalves, J. M., and Trosko, J. E. (1967). *Develop. Biol.* **15,** 23–32.

Myers, D. K., and Bide, R. W. (1966). *Radiation Res.* **27,** 250–263.

Perlitsh, M., and Kelner, A. (1953). *Science* **118,** 165–166.

Pettijohn, D., and Hanawalt, P. C. (1964). *J. Mol. Biol.* **9,** 395–410.

Pfefferkorn, E. R., Burge, B. W., and Coady, H. M. (1966). *J. Bacteriol.* **92,** 856–861.

Pfefferkorn, E. R., and Coady, H. M. (1968). *J. Virol.* **2,** 474–479.

Pfefferkorn, E. R., Rutstein, C., and Burge, B. W. (1965). *Virology* **27,** 457–459.

Pierce, S., and Giese, A. C. (1957). *J. Cellular Comp. Physiol.* **49,** 303–317.

Pittman, D., Ranganathan, D., and Wilson, F. (1959). *Exptl. Cell Res.* **17,** 368–377.

Rahn, R. O., Setlow, J. K., and Hosszu, J. L. (1969). *Biophys. J.* **9,** 510–517.

Rao, B. (1963). Univ. California Lawrence Radiation Lab. Publ. UCRL–10979.

Rao, B., and Hinegardner, R. T. (1965). *Radiation Res.* **26,** 534–537.

Rasmussen, R. E., and Painter, R. B. (1966). *J. Cell Biol.* **29,** 11–19.

Read, J. (1968). *In* "Actions Chimiques et Biologique des Radiations" (M. Haissinsky, ed.), Vol. 12, pp. 145–196. Masson, Paris.

Regan, J. D., and Cook, J. S. (1967). *Proc. Natl. Acad. Sci. U. S.* **58,** 2274–2279.

Regan, J. D., Cook, J. S., and Lee, W. H. (1968a). *J. Cell. Physiol.* **71,** 173–176.

Regan, J. D., Cook, J. S., and Takeda, S. (1968b). 19th Annual Meeting Tissue Culture Assoc. p. 58.

Regan, J. D., Trosko, J. E., and Carrier, W. L. (1968c). *Biophys. J.* **8,** 319–325.

Rieck, A. F. (1954). *J. Morphol.* **94,** 367–407.

Rieck, A. F. (1965). *Bulletin Mount Desert Island Biological Laboratory* **5,** 36.

Rieck, A. F., and Carlson, S. (1955). *J. Cellular Comp. Physiol.* **46,** 301–305.

Roodyn, D. B., and Wilkie, D. (1968). "The Biogenesis of Mitochondria." Methuen, London.

Rupert, C. S. (1960). *J. Gen. Physiol.* **43,** 573–595.

Rupert, C. S. (1961). *J. Cellular Comp. Physiol.* **58,** Suppl. 1, 57–68.

Rupert, C. S. (1962a). *J. Gen. Physiol.* **45,** 703–724.

Rupert, C. S. (1962b). *J. Gen. Physiol.* **45,** 725–741.

Rupert, C. S. (1964). *In* "Photophysiology" (A. C. Giese, ed.), Vol. 2, pp. 283–327. Academic Press, New York.

Rupert, C. S., and Harm, W. (1966). *In* "Advances in Radiation Biology" (L. G. Augenstein, R. Mason, and M. Zelle, eds.), Vol. 2, pp. 1–81. Academic Press, New York.

Rustad, R. C. (1964). *Photochem. Photobiol.* **3,** 529–538.

Rustad, R. C., and Burchill, B. R. (1966). *Radiation Res.* **29,** 203–210.

Rustad, R. C., McGurn, E., Yuyama, S., and Rustad, L. C. (1966). *Radiation Res.* **27,** 543.

Rustad, R. C., Yuyama, S., and Rustad, L. C. (1965). *Radiation Res.* **25,** 234.

Saito, N., and Werbin, H. (1969). *Photochem. Photobiol.* **9,** 389–393.

Sato, G. H., and Yasumura, Y. (1966). *Trans. N. Y. Acad. Sci.* Ser. II, **28,** 1063–1079.

Schiff, J. A., and Epstein, H. T. (1968). *In* "The Biology of *Euglena*" (D. E. Buetow, ed.), Vol. II, pp. 286–333. Academic Press, New York.

Schiff, J. A., Lyman, H., and Epstein, H. T. (1961). *Biochim. Biophys. Acta* **50,** 310–318.

Setlow, J. K. (1966a). *In* "Current Topics in Radiation Research" (M. Ebert and A. Howard, eds.), Vol. 2, pp. 197–248. North-Holland, Amsterdam.

Setlow, J. K. (1966b). *Radiation Res. Suppl.* **6,** 141–155.
Setlow, J. K. (1967). *In* "Comprehensive Biochemistry" (M. Florkin and E. H. Stotz, eds.), pp. 157–209. Elsevier, New York.
Setlow, J. K., Boling, M. E., and Bollum, F. J. (1965). *Proc. Natl. Acad. Sci. U. S.* **53,** 1430–1436.
Setlow, J. K., and Bollum, F. J. (1968). *Biochim. Biophys. Acta* **157,** 233–237.
Setlow, R. B. (1964). *In* "Mammalian Cytogenetics and Related Problems in Radiobiology" (C. Pavan, C. Chagas, O. Frota-Pessoa, and L. R. Caldas, eds.), pp. 291–307. Pergamon Press, New York.
Setlow, R. B. (1967). *In* "Radiation Research 1966" (G. Silini, ed.), pp. 525–537. North-Holland, Amsterdam.
Setlow, R. B. (1968a). *Photochem. Photobiol.* **7,** 643–649.
Setlow, R. B. (1968b). *In* "Progress in Nucleic Acid Research and Molecular Biology" (J. N. Davidson and W. E. Cohn, eds.), Vol. 8, pp. 257–295. Academic Press, New York.
Setlow, R. B., and Carrier, W. L. (1964). *Proc. Natl. Acad. Sci. U. S.* **51,** 226–231.
Setlow, R. B., and Carrier, W. L. (1966). *J. Mol. Biol.* **17,** 237–254.
Setlow, R. B., and Setlow, J. K. (1962). *Proc. Natl. Acad. Sci. U. S.* **48,** 1250–1257.
Setlow, R. B., and Setlow, J. K. (1965). *Photochem. Photobiol.* **4,** 939–940.
Setlow, R. B., Swenson, P. A., and Carrier, W. L. (1963). *Science* **142,** 1464–1466.
Shepard, D. C. (1965). *Exptl. Cell Res.* **38,** 570–579.
Shepard, D. C., and Chu, L. K. (1967). *J. Cell Biol.* **35,** 124 A.
Simmel, E. B., and Karnofsky, D. A. (1961). *J. Biophys. Biochem. Cytol.* **10,** 59–65.
Singer, M. (1959). *In* "Regeneration in Vertebrates" (C. S. Thornton, ed.), p. 59–81. Univ. of Chicago Press.
Skreb, Y., and Bevilacqua, L. (1962). *Exptl. Cell Res.* **28,** 430–432.
Skreb, Y., and Errera, M. (1957). *Exptl. Cell Res.* **12,** 649–656.
Smith, K. C. (1967). *In* "Radiation Research 1966" (G. Silini, ed.), pp. 756–770. North-Holland, Amsterdam.
Smith, L. D. (1966). *Develop. Biol.* **14,** 330–347.
Strauss, B. S. (1968). *In* "Current Topics in Microbiology and Immunology" Vol. 44, pp. 1–85. Springer-Verlag, New York.
Sutherland, B. M., Carrier, W. L., and Setlow, R. B. (1967). *Science* **158,** 1699–1700.
Sutherland, B. M., Carrier, W. L., and Setlow, R. B. (1968). *Biophys. J.* **8,** 490–499.
Sutherland, R. M., Stannard, J. N., and Weed, R. I. (1967). *Intern. J. Radiation Biol.* **12,** 551–564.
Swenson, P. A., and Setlow, R. B. (1966). *J. Mol. Biol.* **15,** 201–219.
Takebe, H., and Jagger, J. (1968). 5th Intern. Congr. Photobiol. Hanover, p. 46.
Terry, C. E., Kilbey, B. J., and Howe, H. B., Jr. (1967). *Radiation Res.* **30,** 739–747.
Trosko, J. E., Chu, E. H., and Carrier, W. L. (1965). *Radiation Res.* **24,** 667–672.
Trosko, J. E., and Kasschau, M. R. (1967). *Photochem. Photobiol.* **6,** 215–219.
Trosko, J. E., and Mansour, V. H. (1968). *Radiation Res.* **36,** 333–343.
Trosko, J. E., and Mansour, V. H. (1969). *Mutation Res.* **7,** 120–121.
Ursprung, H. (1957). *Rev. Suisse Zool.* **64,** 303–311.
Von Borstel, R. C., and Rekemeyer, M. L. P. (1958). *Radiation Res.* **9,** 199–200.
Von Borstel, R. C., and Wolff, S. (1955). *Proc. Natl. Acad. Sci. U. S.* **41,** 1004–1009.
Walker, P. M. B. (1968). *Nature* **219,** 228–232.
Wells, P. H., and Giese, A. C. (1950). *Biol. Bull.* **99,** 163–172.
Werbin, H., and Rupert, C. S. (1968). *Photochem. Photobiol.* **7,** 225–230.

Whitson, G. L., Francis, A. A., and Carrier, W. L. (1968). *Biochem. Biophys. Acta* **161,** 285–290.
Wilkie, D. (1963). *J. Mol. Biol.* **7,** 527–533.
Wolff, S., and Scott, D. (1969). *Exptl. Cell Res.* **55,** 9–16.
Wulff, D. L., and Rupert, C. S. (1962). *Biochem. Biophys. Res. Commun.* **7,** 237–240.
Zeitz, L., Ferguson, R., and Garfinkel, E. (1968). *Radiation Res.* **34,** 200–208.

Chapter 8

ULTRAVIOLET CARCINOGENESIS

John H. Epstein, M.D.

Department of Dermatology, University of California School of Medicine, San Francisco, California

1. Introduction

The etiological role of sunlight in human skin cancer formation was suggested by a number of astute clinical observations around the turn of the 20th century (Unna, 1894; Shield, 1899; Hyde, 1906; Dubreuilh,

1907). Since that time, a large body of circumstantial evidence has been accumulated supporting this concept (Blum, 1941, 1959). However studies confirming this association, as well as defining energy levels, action spectra, and the like, have been confined to animal experimentation not only because of the impropriety of such investigations in people, but also because of their impracticability. Such tumor induction would most likely require 10 to 20 years or more of repeated exposures. Therefore, much of the knowledge that we have concerning the carcinogenic effects of nonionizing radiation stems from animal studies.

The present review is designed to place in focus the experimental and epidemiological evidence for the carcinogenic properties of nonionizing radiation. Since no correlation between the various acute responses to this light energy and eventual cancer formation has been made, most of the discussion will deal with the chronic effects of these rays.

2. Carcinogenic Action Spectrum

Determination of the offending wavelengths or action spectrum is one of the primary objectives in the study of photobiological responses. Such data have been compiled for a number of normal and abnormal reactions of the skin with the use of monochromatic light sources. However, it is not available for ultraviolet light (UV) induced cancer formation as yet. The paucity of this information for one of the most extensively studied photobiological reactions is due to a number of factors including the large number of potential wavelengths, the considerable number of animals necessary and the length of time (a matter of many months) required for exposure to each wavelength as well as the difficulties in immobilization of the mice and the need for an especially good monochrometer with practically no stray light contamination. Though the complete curve of the carcinogenic spectrum is not known, certain aspects of this carcinogenic spectrum have been determined with less sophisticated methods. Roffo (1934), reported that windowglass filtration eliminated the carcinogenic effects of sunlight on white rats. Thus the offending rays of the sun would fall approximately between 290 nm and 320 nm. A number of investigators utilizing mercury arc and fluorescent sun lamps with filters confirmed that 320 nm represents the longer wavelength limit for cancer formation under the experimental conditions used (Griffin *et al.*, 1958; Blum, 1959; Winkelmann *et al.*, 1963). Further, carcinogenic responses have been produced by rays as short as 230.2 nm (Roffo, 1934). Thus under the conditions utilized, the action spectrum appears to include wavelengths between 230 and 320 nm. In addition, the rays between 280 and 320 nm have been shown to have significantly greater carcinogenic effect than the rays shorter than 260 nm (Rusch *et al.*, 1941; Blum and Lippincott, 1942; Blum, 1943a; Kelner and Taft, 1956). These cancer producing rays are also responsible for the normal phototoxic sunburn reaction. Longer UV and visible rays are neither erythrogenic nor carcinogenic under ordinary conditions; though acute phototoxic responses (Blum, 1941; Pathak, 1961) and eventual cancer formation (Büngeler, 1937; Griffin *et al.*, 1958; Urbach, 1959) can be produced with longer rays when a photosensitizer with an appropriate action spectrum is administered concomitantly.

2.1 Summary

Determination of a sophisticated action spectrum for UV-induced cancer formation utilizing monochrometer techniques is not feasible at the present time. However, studies using gross methods have established that the carcinogenic rays range between 230 nm and 320 nm. Further, wavelengths between 280 nm and 320 nm are apparently much more efficient in producing tumors than are the shorter UV rays. These are also the wavelengths that are responsible for the so-called acute sunburn response. No definite conclusion can be made from this data concerning the light absorbing molecule or molecules in the skin or the specific early UV induced events which lead to eventual cancer formation.

3. Experimental Model and Tumor Types

The understanding of human skin cancer is the ultimate concern of investigations into the carcinogenic effects of UV. As noted previously, the skin of man is not an appropriate experimental model, not only because of the obvious moral and sociological objections, but also because of the prolonged tumor induction time. UV-induced cancer formation in man is a chronic process requiring repeated exposures over many years. Therefore the experimental production of tumors has been confined to appropriate laboratory animals.

Fortunately, rats and mice were found to be susceptible to the production of tumors by UV within a relatively short period of time. Thus the housing and handling of the large number of animals needed was feasible. Three specific factors: pigment, hair, and thickness of the stratum corneum were found to alter susceptibility to tumor induction. Rusch and Baumann (1939) found that pigmented mice required significantly more radiation to produce tumors than albino animals. Hair offered even greater protection (Blum, 1959). Thus the hairless rat appeared to be a likely subject. The results of Heuper's (1941) extensive studies indicated that this animal was in fact, quite resistant due to the protection from UV penetration by its thick stratum corneum. Since pigment, hair, and the stratum corneum were limiting factors, the ears of albino mice and rats became the traditional test sites for experimental production of cancer by UV radiation. Utilizing this system a remarkable amount of quantitative data was accumulated. The usual tumor produced in this tissue was a sarcoma (Roffo, 1934; Grady *et al.*, 1941, 1943). Therefore, the albino ear model could not be used for evaluation of qualitative changes associated with epidermal carcinogenesis which is the primary process induced by UV in human skin.

Winkelmann and his co-workers (1960) reported the production of squamous cell carcinomas on the backs of hairless mice with UV energy. Further studies established that carcinomas could be induced in this animal almost to the exclusion of sarcomas (Epstein and Epstein, 1963; Epstein, 1965). In addition, UV-induced melanoma formation has been recently reported in pigmented hairless mice (Epstein *et al.*, 1967). Thus the hairless mouse has provided an excellent model for examination of qualitative as well as quantitative carcinogenic effects of UV.

3.1 Tumor Types

As noted, the type of tumor produced by UV varied with the experimental animal studied.

3.1.1 Sarcomas and Squamous Cell Carcinoma.

Roffo (1934) found that 40% of the tumors produced in his rats were sarcomas. A few years later, Grady and his co-workers (1941, 1943) reported that 90% of the tumors induced in albino mice ears were sarcomas and only about one fourth of the growths contained epidermal cancers. Recently Epstein and Epstein (1963) reported that 40% of the malignant tumors produced on the backs of albino Swiss strain hairy mice by the combined effects of the chemical carcinogen 7,12-dimethylbenz (a)anthracene and UV were sarcomas (Figs. 1 and 2). In addition, fibroplasia was a common response in these animals. Blum (1959) postulated that the discrepancy between UV-induced tumor types in man and in the experimental systems was due to the greater penetration of the carcinogenic rays through the thinner animal epidermis, and thus the connective tissues became the primary target of the irradiation. In support of this concept, Blum and Lippincott (1942) and Kelner and Taft (1956) reported that the less penetrating shorter UV wavelengths (253.7 nm) resulted in sarcomas primarily. Heuper (1941) noted that the few tumors which appeared in his study with the hairless rats that had very thick stratum corneum were epidermal in type. Adenomas have also been produced in fetal tissue with UV further em-

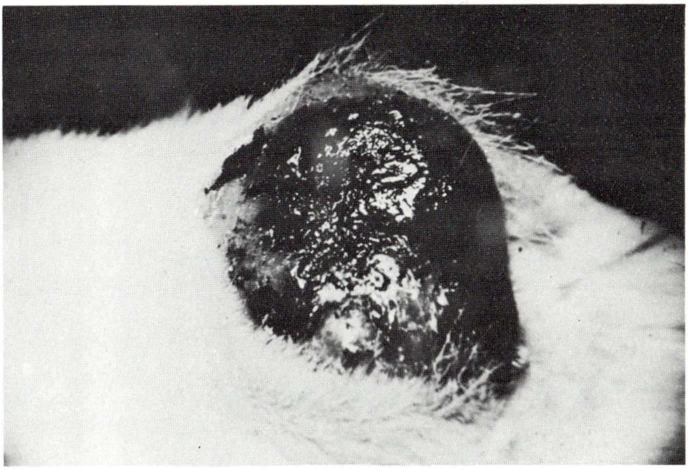

FIG. 1. Fibrosarcoma produced with UV in an albino hairy mouse (Epstein and Epstein, 1963).

phasizing the role of the tissue receiving the greatest carcinogenic exposure (Rogers, 1955) in determining the type of tumor produced.

Though wavelength penetration appears to be of obvious importance, other factors also influence the type of growth induced by UV energy. Grady *et al.* (1943), using the albino hairy mouse, found that the magni-

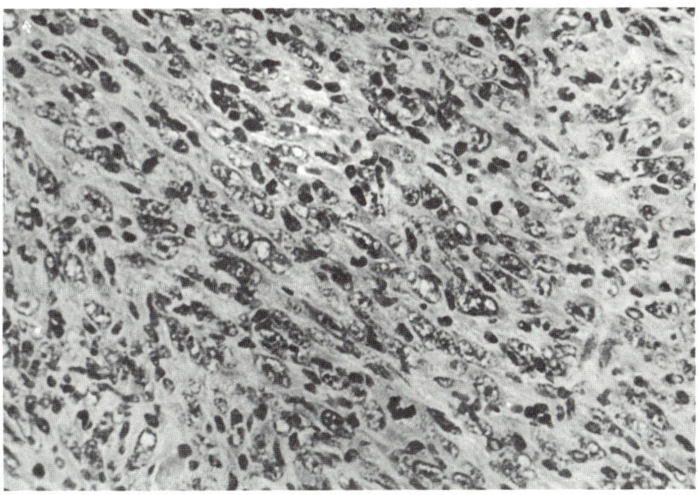

FIG. 2. Histology of the fibrosarcoma produced by UV (Epstein and Epstein, 1963).

tude of individual doses had no effect on the carcinoma/sarcoma ratio, but reducing the intervals between exposures increased the number of epidermal carcinomas. These findings suggest that various tissues respond differently to the carcinogenic rays.

Studies with the hairless mouse, who has an epidermis comparable to the hairy animal in thickness, demonstrate a variation in the species susceptibility as well, since epidermal squamous cell carcinomas have been produced almost to the exclusion of sarcoma formation in these animals (Epstein and Epstein, 1963; Epstein, 1965) (Figs. 3 and 4).

3.1.2 Melanoma

The role of sunlight in melanoma formation has not been established. The results of some surveys have been suggestive but a number of factors continue to cloud the picture (Section 7.3). Recent data indicate that light energy is capable of producing such tumors.

The experimental production of melanocytic tumors in animals has been examined most extensively in hamsters and pigmented mice (Della Porta *et al.*, 1956; Ghadially and Barker, 1960; Quevedo *et al.*, 1961; Nakai and Rappaport, 1963; Epstein and Epstein, 1963; Forbes, 1965; Klaus and Winkelmann, 1965). These growths were induced by the application of chemical carcinogens and consisted of dense collections of dermal melanocytes which were not invasive and simulated cellular blue nevi. Attempts to stimulate malignant transformation in these lesions by a number of traumatic mechanical procedures excluding UV

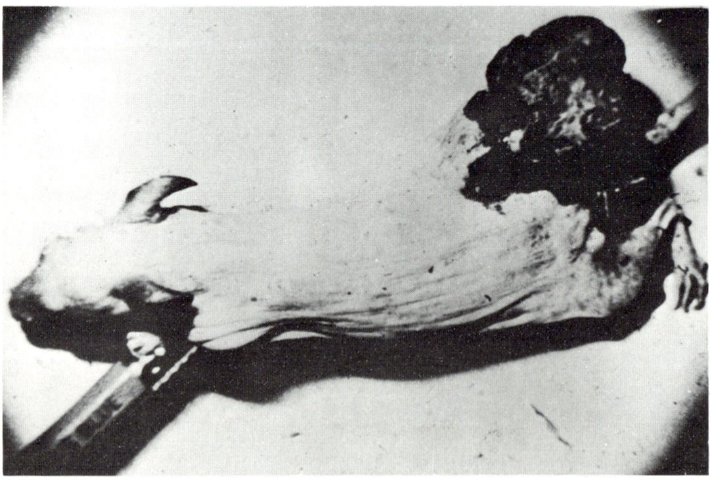

FIG. 3. Squamous cell carcinoma produced by UV in a hairless mouse (Epstein and Epstein, 1963).

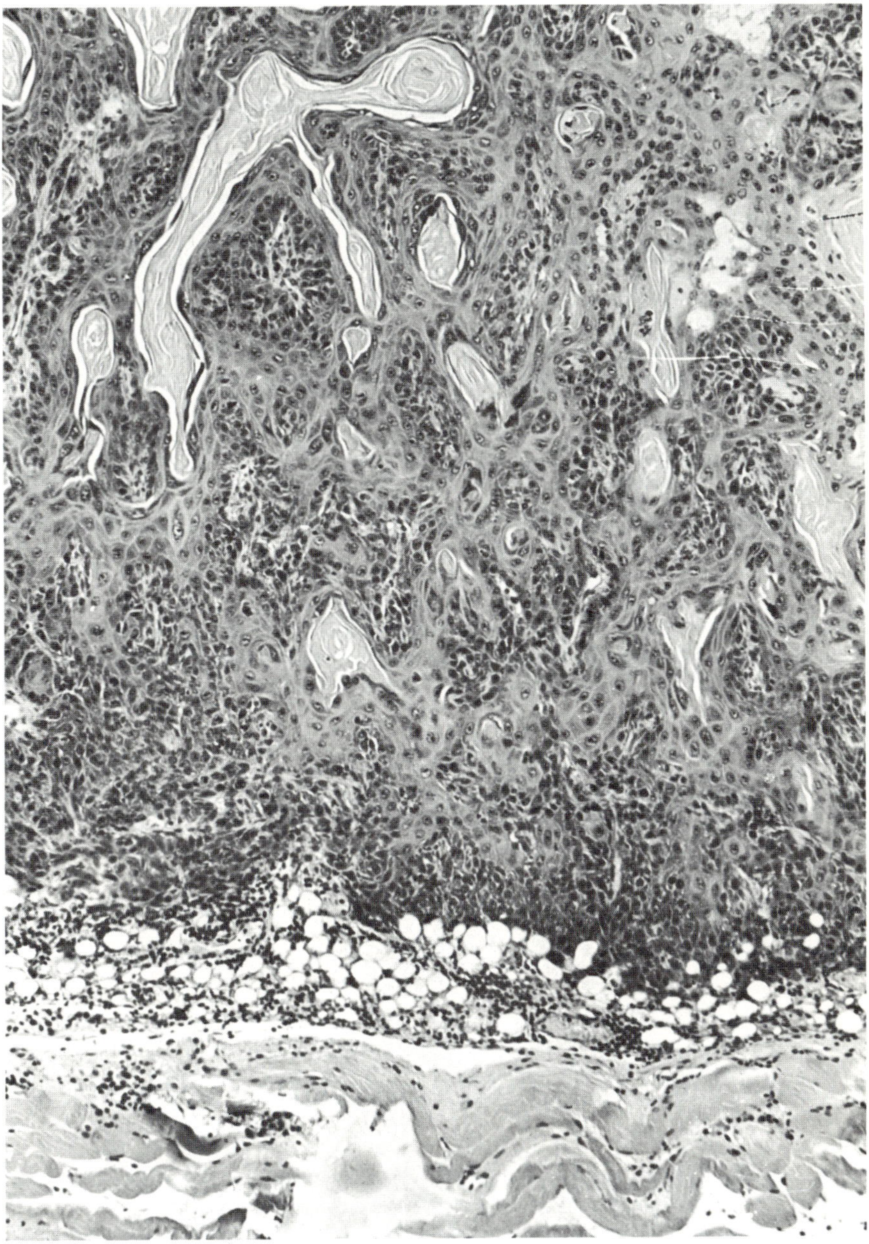

FIG. 4. Histology of the invasive well-differentiated squamous cell carcinoma produced by UV (Epstein and Epstein, 1963).

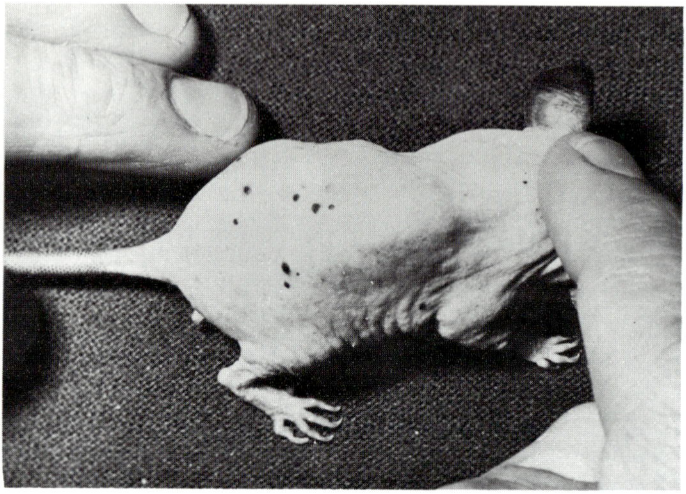

FIG. 5. "Blue nevi" on the back of a pigmented hairless mouse painted with a single application of 7,12-dimethylbenz (a) anthracene (Epstein *et al.*, 1967).

injury have been unsuccessful (Ghadially *et al.*, 1960). In contrast, Epstein and his co-workers (1967) demonstrated that UV would induce such changes. These authors produced small benign pigmented growths in a series of pigmented hairless mice with a single application of DMBA (200 μg) (Figs. 5 and 6). Chronic low dose UV exposures (1.2×10^6

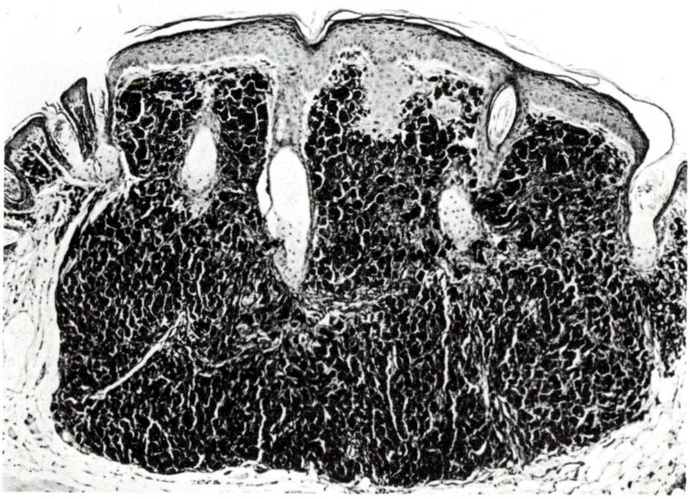

FIG. 6. Histology of the "blue nevus" showing the dense subepidermal accumulation of dermal melanocytes (Epstein *et al.*, 1967).

FIG. 7. Large melanoma resulting from UV stimulation of a benign "blue nevus" (Epstein *et al.*, 1967).

ergs/cm^2 of 280 nm to 320 nm energy) initiated several months afterwards resulted in the production of large invasive melanocytic tumors which frequently metastasized to regional lymph nodes (Figs. 7 and 8). The functioning neoplastic nature of the tumor melanocytes was estab-

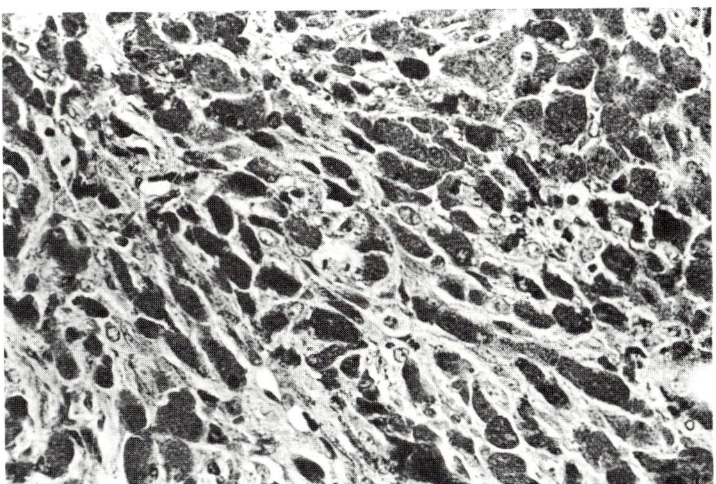

FIG. 8. Histology of the invasive melanoma. The nuclei of the melanoma cells are abnormal in size, shape and chromatin pattern (Epstein *et al.*, 1967).

lished by histologic, electron microscopic, and autoradiographic studies. Similar tumors did not occur in the nonirradiated control animals.

These results indicate that benign pigmented growths can be transformed into melanomas by UV energy. Whether similar responses occur in human skin remains to be determined.

3.2 Summary

For many years the ears of albino rats and mice were the primary sites for study of experimental UV-induced tumors because hair, pigment, and a thick stratum corneum inhibited the penetration of these rays. The basic tumor produced in these animals was a sarcoma which was quite adequate for the monumental quantitative investigations being carried out. Recently, the hairless mouse was found to provide an excellent model for evaluation of qualitative as well as quantitative development patterns in epidermal cancer formation which is the type that occurs in human skin. This animal also proved useful for studying UV stimulation of melanoma formation from benign pigmented lesions under controlled conditions.

4. Tumor Development

Study of the developmental patterns of tumors can be divided into two basic categories: (*a*) the quantitative evaluation of energy requirements; and (*b*) the examination of qualitative changes associated with cancer formation.

4.1 Quantitative Studies of Tumor Development

The primary role of UV rays shorter than 320 nm in the experimental production of skin cancer had been established by the work of several investigators between 1928 and 1941 and has been confirmed by several others more recently (Section 2). With this early information as a background, a number of remarkable studies concerning energy requirements and dose rate and time relationships were accomplished between 1940 and 1960 by Blum and his co-workers.

4.1.1 Dose-Time Relationships

Beginning in 1940, Blum and his co-workers (Blum, 1959) initiated and carried out a series of remarkable experiments concerning the energy requirement, dose rate, and time relationships that remain as monumental guide-posts in the field of experimental UV carcinogenesis. These authors utilized a set exposure procedure monitored by a stable photoelectric cell and an integrating meter (Kuper *et al.*, 1941) and confined their observations to tumor formation on the ears of an inbred strain of albino mice. Though the exposure of the test tissue was not exceptionally efficient, the reproducibility of their results attests to the significance of their findings. In addition, they used very large numbers of mice, further enhancing the validity of their results. Utilizing this system, these authors first examined the problem of dose-rate dependence. They found that reciprocity held within the accuracy of the experiments when doses greater than 0.35×10^4 to 0.4×10^4 ergs/cm^2/sec were employed (Blum, 1943b; Blum, 1959). Below this rate the UV became less efficient in producing tumors in their system.

In subsequent studies using energies that conformed to dose-rate dependence, they reported that increasing the dose and/or shortening the interval accelerated tumor formation but did not alter the shape of the incidence curve (Figs. 9 and 10). Discontinuing the exposures

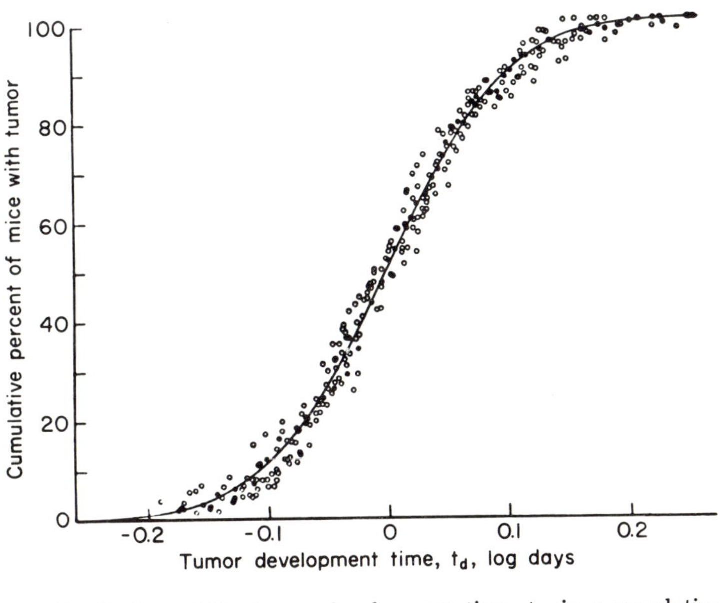

FIG. 9. Distribution of the tumor development time, t_d, in a population of genetically homogenous (Strain A) male mice, exposed to UV, the exposures being placed at regular intervals up to the time of tumor appearance. The data are from 12 experiments including a total of 676 mice. The dose and interval between doses was the same for a given experiment but varied between experiments. The mean values of t_d varied accordingly and, in the figure, all the points have been corrected to a common, mean value for each experiment, i.e., log $t_d = 0$. The drawn curve is the integral of a normal distribution with standard deviation 0.081 log days. The same distribution was found for a number of additional experiments in which smaller numbers of mice were used (Blum, 1950).

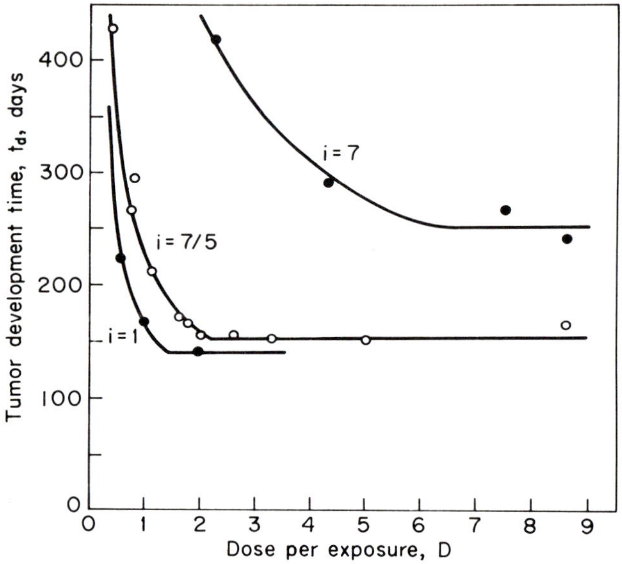

FIG. 10. Relationship between dose D and the development time t_d for three different intervals i. The drawn curves follow the theory of progressive acceleration of tumor growth (Blum, 1950).

prior to the appearance of visible tumors resulted in eventual cancer production if sufficient energy had been applied over a long enough period of time (Blum, 1950; Rusch *et al.*, 1941). Blum (1959) found that 50 exposures over 74 days was the smallest number of doses and the shortest time which would lead to cancer formation (total of 9×10^8 ergs/cm^2). Higher doses over a shorter period of time were ineffective. Further he reported that interruption of exposure periods for 30 days retarded tumor development rate but did not stop it. Resumption of irradiation again accelerated the process.

From this information and data derived from extensive experimentation, Blum (1959) surmised that UV-induced cancerization is a continuous cumulative process which begins with the initial exposure, and which cannot be divided into stages. Though some small degree of recovery appeared to occur it had little effect except perhaps when very low dose rates were used. He suggested that the inability to produce tumors with exposures carried out in less than 2½ months was due to insufficient time for growth acceleration to produce visible tumors within the lifetime of the experimental animals. In support of this concept, Rogers (1955) did produce adenomas in mouse lung tissue with a single UV exposure. Recently it has been shown that a single exposure will result in skin cancer formation in hairless mice if the process is stimulated with croton oil over the next several months (Section 5.2.3).

The mechanism of this acceleration of tumor growth is not clear. Blum postulated that the prime effect was intracellular in nature. However, recent studies indicate that removal of inhibitors of cell proliferation may play an extremely important role in cancer formation (Bullough and Laurence, 1968; Iversen, 1968).

4.2 Qualitative Changes Occurring with Tumor Development

4.2.1 Dermis

The role of the dermis in epidermal tumor formation has been emphasized by a number of investigators (Mackie and McGovern, 1958; Orr, 1963; Van Scott, 1964; Getzrow, 1966). Chronic exposure of human skin to the sun's energy over a number of years results in a striking change in the dermal connective tissue described as actinic elastosis (Lever, 1967). A proliferation of elastic tissue has also been produced experimentally in mice by repeated exposures of UV by Sams *et al.* (1964), who found that wavelengths between 290 and 320 nm were the most efficient in producing this effect. Similarly, Magnus and Johnson (1965) stimulated the formation of elastotic tissue following early destruction of elastic fibers with 300 nm rays using a monochromatic source. Nakamura and Johnson (1968) reported that dermal elastic tissue proliferation occurred in albino rats after chronic irradiation with UV only after discontinuance of the exposure. Johnson *et al.* (1968) postulated that this change was the result of photochemically induced alterations in fibroblast function rather than degradation of normal elastic fibers. In support of this concept, W. L. Epstein, *et al.* (1969b) recently noted that unscheduled DNA synthesis occurred in connective tissue cells of the upper dermis within minutes after exposure to UV shorter than 320 nm demonstrating a direct effect of UV on the dermal fibroblasts (Section 6.1.2).

Because of its frequent association with skin cancer formation, actinic elastosis has been considered to play an important role in the development of the tumors. However, Sams and his co-workers (1964) and Graham and Helwig (1965) have demonstrated that actinic elastosis is not essential for the development of epidermal malignancies. In addition, the experimental production of elastosis in animals has not been associated with cancer formation nor have UV-induced experimental cancers been dependent on the presence of this change (Winkelmann *et al.*, 1963; Epstein and Epstein, 1963).

Though the association of actinic elastosis and skin cancer formation appears to be fortuitous, other less obvious dermal changes may well influence tumor development. Connective tissue changes have been correlated with the development of basal cell epitheliomas in man and in experimental animals. These include normal collagen and elastic tissue dissolution, proliferation of argyrophilic fibers (most likely young collagen), increased amounts of acid mucopolysaccharides, increased numbers of mast cells and fibroblasts, deposition of periodic acid-Schiff positive material around the advancing epithelium and at times the formation of new elastic tissue (Sams *et al.*, 1963; Dobson, 1963; Zacheim, 1963; Mehregan *et al.*, 1964). Striking changes in the dermal vasculature have been also noted in experimentally induced as well as naturally occurring epidermal malignancies (Urbach, 1963; Graham and Helwig, 1965). Thus a number of dermal changes do occur which may be responsible, at least in part, for epidermal cancer formation.

4.2.2 Progressive Qualitative Changes Associated with Experimental Ultraviolet Light Induced Squamous Cell Carcinoma Formation

Using the hairless mouse model and repeated UV exposures from a hot quartz contact source, it has been possible to follow UV-induced dermal and epidermal responses progressing from benign hyperplasia through premalignant or actinic keratosis-like changes to frank invasive malignancy (Epstein *et al.*, 1966). Serial autopsies were performed over a 12 month period and the skin sections were stained with hematoxylin and eosin, Mowry's colloidal iron (MCI) which stains acid mucopolysaccharide blue (Mowry, 1958), luxol fast blue which stains mature collagen blue (Hale *et al.*, 1960) and periodic acid-Schiff (PAS) stains. The results indicated that an accumulation of acid mucopolysaccharides (AMP) and a loss of insoluble collagen in the upper dermis appeared by the stage of benign hyperplasia and persisted throughout the succeeding stages of cancer formation. In addition, proliferation of mast cells and fibroblasts also occurred during the early changes and became more prominent with the more advanced carcinogenic alterations. Though similar dermal changes have been associated with human basal cell epitheliomas (Sams *et al.*, 1963) and with chemically induced cancer formation (Dobson and Griffin, 1962) they have also been produced by a variety of noncarcinogenic stimuli and thus, like actinic elastosis, they may have no specific relationship to tumor formation.

The most striking histologic changes occurred in the basement membrane and epidermis. Following three to six weeks of repeated UV exposures, there was a regular hyperplasia of the epidermis with hyperkeratosis, hypergranulosis and acanthosis. At this stage, the basement membrane had changed from a delicate, thin thread into a thick ropelike structure. With further exposures, epidermal hyperplasia became more intense and the cells primarily of the basal layer became abnormal in size, shape, polarity, and chromatin pattern and mitoses became increasingly prevalent (Fig. 11). As these epidermal cell changes occurred,

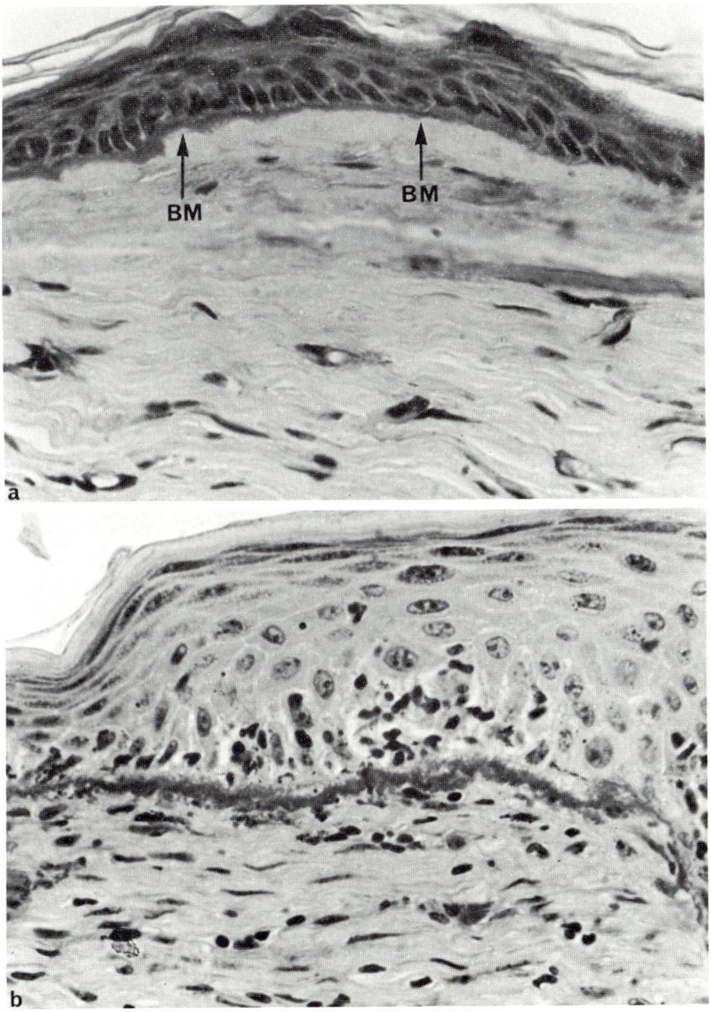

FIG. 11a. Histology of UV-induced benign hyperplasia showing hyperkeratosis, hypergranulosis, regular acanthosis, a thickened light microscopic basement membrane (BM), and hypertrophy of the dermis (Epstein *et al.*, 1966).

FIG. 11b. Histology of the premalignant-malignant stage of tumor development showing abnormality of size, shape, polarity and chromatin pattern in the epidermal basal cell layer. The light microscopic basement membrane has become clumped, frayed and discontinuous at this point (Epstein *et al.*, 1966).

the basement membrane became thicker, more irregular, clumped and frayed and then, with frank invasive malignancy, it disappeared altogether. Breaks in or loss of the light microscopic basement membrane have also been observed associated with invasion in naturally occurring skin cancers (Cawley *et al.*, 1966) as well as chemically induced experimental tumors (Dobson, 1963). Recent studies further suggest that microinvasion, through gaps in the electron microscopic basement membrane, represents the earliest stage of tumor invasion (Ashworth *et al.*, 1961; Frie, 1962). However, as Pinkus and Mehregan (1966) point out, temporary breaks do occur in simple inflammatory dermatoses such as psoriasis or seborrheic dermatitis. Therefore, the questions whether the basement membrane changes are due to abnormal dermal or epidermal signals or both and whether the disruption of this structure is important to cancer invasion remain to be answered.

Epidermal cellular kinetics were also studied at the various stages induced by chronic UV irradiation of hairless mouse skin *in vivo*, utilizing mitotic counts and tritiated thymidine (TdR-H^3) labeling of cells in DNA synthesis (Epstein and Fukuyama, 1969). As noted in Table I, only a few basal cells were in mitoses or synthesizing DNA prior to UV irradiation. With the development of benign hyperplasia, an increase in mitoses and cells in DNA synthesis occurred. Igarashi (1967) reported a similar response in guinea pigs following chronic UV exposures. A further increase in the mitotic count and cells synthesizing DNA was seen with the progressive development of the premalignant or benign-malignant stage. This proliferative activity was confined to the germinative cells of the epidermis until frank invasive malignancy had occurred, at which time cells throughout the tumor mass were involved in DNA synthesis and mitoses. Many of the mitoses in the cancer tissue were quite abnormal in appearance. In addition to proliferation, a reduction in the DNA synthesizing time and G-2 period and a marked acceleration in epidermal cell transit time was also noted with the progression

TABLE I
EPIDERMAL MITOSIS AND DNA SYNTHESIS IN NORMAL, HYPERPLASTIC AND PREMALIGNANT STATES

Stages	Mitosis/Basal cells[a]	Labeled cells[b]/Basal cells[a]
Normal	0.77%	2.3%
Hyperplasia	1.9%	5.4%
Premalignant/early malignant	3.2%	18.5%

[a] Only interfollicular areas were counted.
[b] Cells labeled with tritiated thymidine.

to malignancy. The number of cells produced exceeded those being eliminated, accounting for the hyperplasia and expansion of the tumor mass. The hyperplasia produced could not be ascribed primarily to a population shift to younger cells which normally take longer to mature (Iversen and Evensen, 1962; Skjaeggestad, 1964) or to the effects of delayed maturation (Setala *et al.*, 1962). The primary site of experimental squamous cell carcinoma in this study appeared to be in the germinative layer or basal cells of the epidermis. In contrast to a number of cancers (Evensen and Iversen, 1962; Johnson *et al.*, 1960; Baserga, 1965; Hoffman and Post, 1966) these UV-induced malignancies were characterized by acceleration of cell formation, maturation, and turnover.

4.3 Summary

Experimental UV-induced tumor formation has been examined from a quantitative and a qualitative point of view. Quantitatively the studies of Blum and Rusch and their co-workers indicated that under controlled circumstances skin cancers will develop if a sufficient amount of energy is delivered over a long enough time period. Shortening the interval of delivery and/or increasing the dose accelerated tumor formation. Discontinuing the exposures retarded the tumor development rate. From this data Blum postulated that UV-induced carcinogenesis began with the first exposure. Studies on mouse lung tissue and croton oil promotion studies on mouse skin support this concept.

Qualitative studies demonstrated progressive changes in the epidermis, the dermis and the epidermal/dermal basement membrane associated with the development of UV-induced squamous cell carcinomas. These experimental tumors appeared to develop from the epidermal basal cell and were characterized by a rapid cell turnover and decreased transit time. Among the most striking changes were those which occurred in the basement membrane which initially became thick, then clumped, frayed, and finally, with frank invasive malignancy, disappeared altogether. Whether these changes were related to an abnormal epidermal or dermal signal or both and whether changes in this structure related to control of cancer invasion remains to be determined.

5. Physical and Chemical Influences on Ultraviolet Light Carcinogenesis

5.1 Physical Factors Influencing Ultraviolet Light Carcinogenesis

Tumor promoting effects of physical stimuli, including freezing, scalding, and wounding have been reported in chemically induced cancer formation (Berenblum, 1930; DesLingneris, 1940; Pullinger, 1945). However, there is little information concerning the effects of these factors in UV-induced cancer formation. The influence of heat on this process is an exception. Though there have been some theoretical questions concerning the effect of temperature on acute UV effects (Clark, 1936; Blum, 1945; Van der Leun, 1966), the enhancement of erythema and cutaneous injury by heat has been well established (Hill and Eidenow, 1923; Freeman and Knox, 1964). Further, Freeman and Knox (1964) demonstrated that animals kept in a heated environment rapidly developed more UV-induced tumors than mice living in a temperate environment confirming the previous studies of Bain *et al.* (1943). These authors also showed that the increased temperatures had to be present at the time of the UV exposures to produce this acceleration. Whether this heat effect results from an influence on biochemical reactions following this photochemical response, an additive injury, or a number of other possibilities remains to be determined. However, it is likely that temperature does play a role in the epidemiology of human skin cancer formation.

5.2 Influence of Chemical Agents on Ultraviolet Light Carcinogenesis

5.2.1 Photosensitization

As noted previously, wavelengths shorter than 320 nm, especially between 280 and 320 nm constitute the carcinogenic action spectrum in experimental cancer formation (Section 2). However, with the introduction of a photosensitizer with a suitable action spectrum, acute phototoxic responses can be produced by rays longer than 320 nm (Blum, 1941; Buck *et al.*, 1960; Pathak, 1961). Repeated long wavelength exposures in the presence of such active agents as 8-methoxypsoralen, coal tar, eosin, hematoporphyrin and anthrasol result in a high cancer yield (Büngler, 1937; Blum, 1941; O'Neal and Griffin, 1957; Griffin *et al.*, 1958; Urbach, 1959). Thus there appears to be some correlation between early phototoxic effects and eventual cancer formation.

5.2.2 Chemical Carcinogens

The polycyclic hydrocarbon carcinogens are potent photosensitizers with action spectra primarily in the long UV range (Blum, 1941). Santamaria (1964) and S. S. Epstein *et al.* (1964) have demonstrated a significant relationship between the photodynamic reactivity and the carcinogenic potential of these compounds. Since the chemicals are photosensitizers, the addition of their inherent carcinogenicity to the cancer-inducing effects of photoxic insults should accelerate tumor formation. When this concept was tested, however, a variety of conflicting results were noted (Findlay, 1928; Kohn-Speyer, 1929; Schorr and Ssobolewa, 1930; Büngeler, 1937; Taussig *et al.*, 1938; Morton *et al.*, 1940; Rusch *et al.*, 1942, and Epstein *et al.*, 1961).

Recently Clark (1964) attempted to clarify the problem using 20-methylcholanthrene (MCA) topically followed by varying doses of long UV energy. She noted that exposure to lower energy levels ($4\text{–}8 \times 10^7$ ergs/cm^2) shortly after MCA application accelerated papilloma formation but did not influence cancer production. When larger amounts of long UV energy were given ($10\text{–}18 \times 10^7$ ergs/cm^2) an early acceleration of papilloma formation occurred, but most of these growths regressed. The final results showed a retardation of both papilloma and cancer formation. The mechanism of this inhibition has not been clearly established. Morton *et al.* (1951) suggested that light energy enhances absorption and leaves less carcinogen in contact with the skin surface. Miller, (1951) and Moodie *et al.* (1954) proposed that photooxidation of the chemical may alter its carcinogenic potential. More recently, the studies of Santamaria (1960) indicate that photooxidation of the substrate may limit the carcinogenic efficiency of a compound. Whatever the mechanism, it is clear that the direct influence of UV on the hydrocarbon must be eliminated if their combined carcinogenic effects are to be examined. Using this separation technique, Epstein and Epstein (1962) demonstrated that subcarcinogenic amounts of UV will stimulate cancer formation initiated by a single application of 7,12-dimethylbenz(a) anthracene (DMBA) (Table II). Each animal developed one or two large tumors, none regressed and most of them were malignant. Squamous cell carcinomas and/or sarcomas were produced (Epstein and Epstein,

TABLE II
TUMOR INCIDENCE (DMBA AND UVL)

Group	Survivors at 55 weeks (after DMBA)[a]	Mice with tumors	% Tumor bearing mice	
[b]Group I DMBA plus UVL (X)	6	5	83%	63%
	13	7	54%	
[b]Group II DMBA	12	2	17%	18%
	10	2	20%	
Group III acetone plus UVL (X)	23	0	0	0
	11	0	0	

Note: (X) Total UVL by 55 weeks = 12.06×10^7 ergs/cm^2.
[a] Includes dead mice with significant tumors.
[b] Difference between Group I and Group II is significant (Chi^2 $p = 0.01$). (Epstein and Epstein, 1962.)

1963). Multiple papilloma formation and tumor regression did not occur. Therefore, the pattern of tumor development differed from croton oil promotion of chemically initiated cancer formation (Shubik, 1961). These findings support the concept that tumor promotion by a potent carcinogen results in the progressive formation of malignant tumors (Shubik, 1950), which appears to hold true whether the promoter is a chemical carcinogen or UV.

Further studies using hairless mice and carcinogenic amounts of UV demonstrated that pretreatment with a single application of DMBA accelerated the rate of tumor appearance and growth and increased the incidence of tumors per mouse (Epstein, 1965) (Figs. 12 and 13). Almost all of the tumors were squamous cell carcinomas and occasional lymph node metastases were noted.

The pattern of tumor formation resembled that described for summation of carcinogenic responses (Saffiotti and Shubik, 1956; Poel, 1963). These results suggest that chemical and UV carcinogenic effects are additive in nature. Whether they act at the same sites or through similar mechanisms remains to be determined.

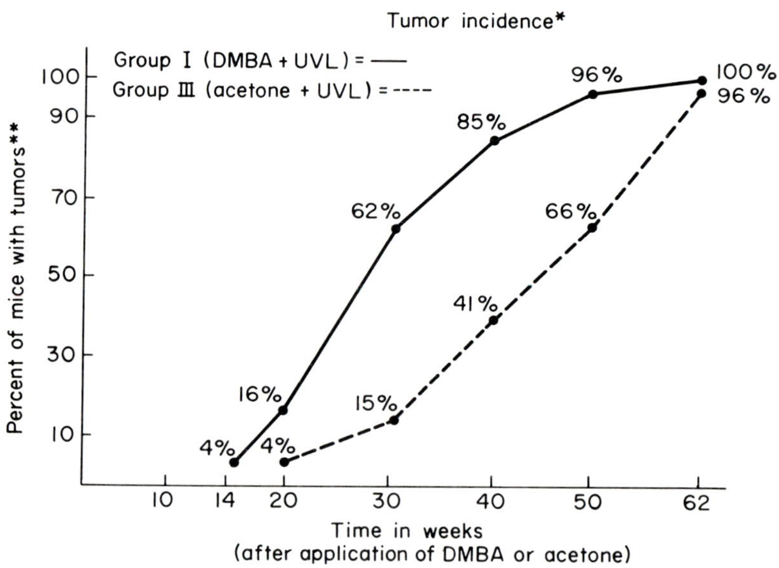

FIG. 12. Tumor incidence in random-bred, hairless mice exposed to UV plus 7,12-dimethylbenz(a)anthracene (DMBA) or acetone. The median test, comparing incidence of tumors in Groups I and III at time of 50 percent incidence in pooled Groups, gave a $x^2 = 11.768$ ($P < 0.001$), which confirms the significance of the DMBA effect (Epstein, 1965).

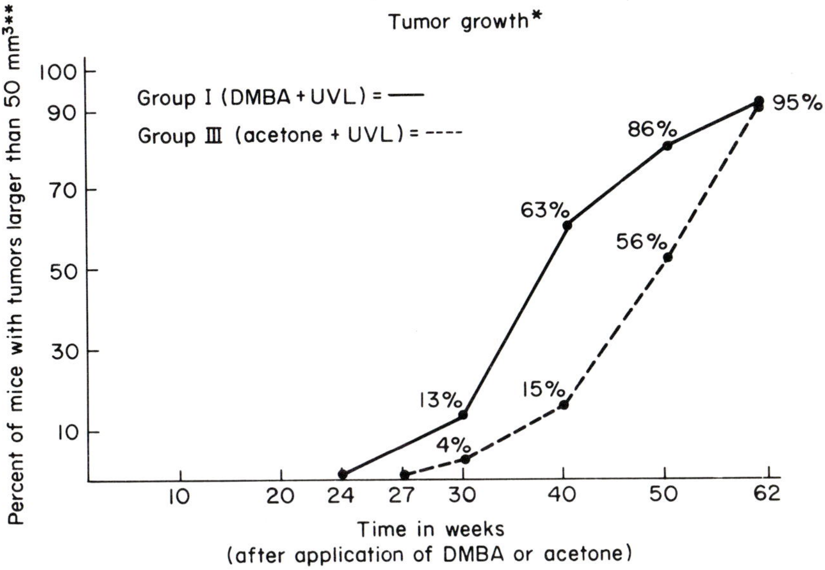

FIG. 13. Tumor growth rate in random-bred, hairless mice exposed to UV plus 7,12-dimethylbenz(a)anthracene or acetone. The median test comparing incidence of tumors in Groups I and III at time of 50 percent incidence in pooled groups gave a $x^2 = 7.27$ ($P < 0.01$), which confirms the significance of the DMBA effect (Epstein 1965).

5.2.3 THE EFFECTS OF CHEMICAL PROMOTION ON ULTRAVIOLET-INITIATED CARCINOGENESIS

Extensive studies using promotion techniques suggest that chemically induced cancer formation involves two or more stages beginning with the initial application of the carcinogen (Berenblum and Shubik, 1949; Boutwell and Rusch, 1951; Roe and Salaman, 1954; Berenblum and Haran, 1955). Boutwell's (1964) studies suggest that promotion actually encompasses two processes: specific conversion of the initiated cells to become dependent tumor cells and nonspecific propagation to a multiplying tumor cell population. In contrast, Blum's (1959) findings and calculations indicate that experimental UV carcinogenesis is in fact, a continuous process developing progressively and not in stages from the first exposure.

However, Blum (1950) could not produce tumors unless the exposures were carried out over a minimum of 2½ months regardless of the energy used. He suggested that shorter exposure periods did not accelerate tumor formation sufficiently for the growths to become visible within the lifetime of the experimental animals.

TABLE III
TUMOR INCIDENCE (UVL + CROTON OIL)

Group	18 Months T/S
Group I (UVL + Croton Oil)	8/44
Group II (UVL)	0/35
Group III (Croton Oil)	0/31
Group IV (no application)	1/21

T = number of mice with tumors >50 cu mm on posterior half of their backs; S = survivors including mice with tumors >50 cu mm autopsied or dying prior to the time noted. The difference between Group I and the Control Groups showed a P value of 0.001 using a 2 × 2 contingency table (Epstein and Roth, 1968).

In a recent study using a single UV exposure (1.3×10^7 ergs/cm^2 of UV energy between 280 and 320 nm) we noted that cancer formation did occur when followed by repeated applications of the noncarcinogenic promoting chemical, croton oil (Table III) (Epstein and Roth, 1968). Whether the croton oil applications in these studies served to accelerate tumor growth that had started with the single UV exposure or functioned as a promoter by transforming UV-initiated cells into a tumor forming population has not been determined. However, in contrast to the effects of initiation with a polycyclic hydrocarbon carcinogen, only one growth was produced per tumor-bearing mouse. Multiple papilloma formation and significant tumor regression typical of chemical initiation with croton oil promotion was not noted. Thus these findings indicate that not only does the process of cancer formation begin with the initial UV exposure, but also the characteristics of the tumor appear to be established by the light energy prior to the chemical stimulation.

5.3 Summary

Studies concerning the influence of physical stimuli on UV-induced carcinogenesis have been sparse with the exception of heat which accelerates tumor formation. In contrast there have been a number of studies evaluating the effects of various chemicals on this process. The carcinogenic effects of repeated exposures to noncarcinogenic longer UV in the presence of a photosensitizer with the proper action spectrum has been well established. However, when the hydrocarbon carcinogens were used as the photosensitizer, a variety of conflicting results were reported. Recently, separation of the photodynamic and carcinogenic properties of these chemical carcinogens revealed that the cancer producing effects of UV and the hydrocarbon carcinogens were additive in nature. Further it was observed that chemical promotion following a single UV exposure resulted in a significant number of tumors adding support to Blum's concept that UV-induced cancer formation begins with the first exposure.

6. Acute Reactions to Ultraviolet Light

The carcinogenic effects of chronic repeated exposures to UV energy shorter than 320 nm are well established (Section 2). Recent studies have demonstrated that a single UV insult can result in cancer formation when croton oil promotion techniques are utilized similar to findings with polycyclic hydrocarbon carcinogens (Section 5.2.3) indicating that the process of carcinogenesis begins with the initial insult. In general, no direct correlation between the early effects of either UV or the chemical carcinogens and eventual cancer formation can be made. However, certain similarities between their acute responses have been noted and will be discussed in this section. For a thorough review of the acute reactions of mammalian skin to UV, the reader is referred to the recent review in *Photophysiology*, Volume IV (Johnson *et al.*, 1968).

6.1 Comparison of Chemical Hydrocarbon and Ultraviolet Light Effects on Mitoses and Macromolecule Synthesis

Influences on the mitotic cycle have been among the most extensively studied aspects of the acute reactions to chemical carcinogens *in vivo*. Utilizing colcemide injections to arrest the mitotic cycle in metaphase and tritiated thymidine (TdR-H^3) tracer techniques, Iversen and Evensen (1962) and Elgjo (1966) noted an early inhibition of DNA synthesis and mitosis formation which lasted for about 12 hours. This was followed by an acceleration of these cellular functions associated with increased cell renewal and hyperplasia.

Recently, the early effect of UV on mitoses and macromolecule synthesis were evaluated utilizing similar techniques.

6.1.1 UVL on Mitoses

Using intraperitoneal injections of colcemide to arrest mitoses in metaphase, Epstein *et al.* (1968a) noted a marked depression in mitosis formation in the epidermal basal cells within one hour post irradiation of hairless mouse skin *in vivo* with a single exposure to 4.49×10^6 ergs/cm^2 of UV energy shorter than 320 nm. This depression persisted up to at least seven hours and was associated with a marked reduction in the mitotic rate and prolongation of the basal cell turnover time. By 24 hours, acceleration of the mitotic rate and basal cell turnover time had occurred, reaching a peak at 72 hours post irradiation. This acceleration of mitotic activity was associated with hyperplasia of the epidermis. Utilizing the formula developed by Leblond and Stevens (1948), no change in the metaphase duration was detected during post UV-induced changes.

6.1.2 Ultraviolet Light Effects on DNA Synthesis

Intraperitoneal injections of TdR-H^3 (specific activity 3 c/mM) were used to evaluate the influence of a single exposure of UV energy shorter than 320 nm (4.49×10^6 ergs/cm^2) on DNA synthesis in the hairless mouse skin *in vivo* (Epstein *et al.*, 1968a). A depression in DNA synthesis occurred in the epidermal basal cells within one or two hours and persisted for several hours post irradiation. Acceleration of this function occurred by 24 hours, reached a peak at 48 hours and persisted though to a lesser degree for seven days, again associated with hyperplasia of the epidermis. Similar acute changes have been reported in human skin *in vivo* post UV injury (W. L. Epstein *et al.*, 1969b).

The *in vivo* DNA synthesis response to UV simulates changes noted in micro-organisms and cultured mammalian cells (Kelner, 1953; Hanawalt and Setlow, 1960; Dendy and Cleaver, 1964; Klímek and Vlašínova, 1967). However, stimulation of unscheduled DNA synthesis as well as depression of normal, predivision, semiconservative DNA formation has also been noted in the early post-UV irradiation period (Rasmussen and Painter, 1964; Rasmussen and Painter, 1966; Cleaver, 1968). The unscheduled synthesis presents as a sparse labeling with TdR-H^3 of essentially all the irradiated cells in the tissue culture monolayer as noted autoradiographically. Using density gradient techniques, and tritiated bromouracil deoxyriboside (H^3BrUdR), this sparse labeling was found to represent the insertion of small numbers of bases throughout preformed DNA chains most likely due to repair replication or dark repair of UV injured DNA. In support of this concept Regan and Trosko (1967) and Horikawa *et al.* (1968) reported excision of thymine dimers from ultraviolet light damaged DNA in cell cultures.

Recently, unscheduled DNA synthesis has been demonstrated in hairless mouse and human skin *in vivo* post-UV-exposure utilizing intradermal injections of TdR-H^3 of a high specific activity (11c/mM) and in relatively high concentrations (J. H. Epstein *et al.*, 1968b; W. L. Epstein *et al.*, 1969b). This unscheduled synthesis occurred in the differentiated epidermal granular cell, Malpighian cell and dermal fibrocyte nuclei as well as in the germinative basal cell nuclei of the epidermis, and was demonstrated by a sparse nuclear labeling pattern with the radioactive tracer. Chemical extraction and DNAase incubation studies confirmed that the TdR-H^3 noted in the autoradiograph was incorporated into DNA. Whether this unscheduled synthesis represents an *in vivo* demonstration of the dark repair process remains to be determined.

No evidence of unscheduled DNA synthesis has been noted following the application of a polycyclic hydrocarbon carcinogen (Goldyne *et al.*, 1969).

6.1.3. Ultraviolet Light Effects on RNA and Protein Synthesis

Chemical extraction and autoradiographic studies utilizing radioactive cytidine and amino acids as tracers indicate that inhibition of RNA and of protein synthesis are also early post-UV irradiation effects in the mammalian epidermis *in vivo* (Baden and Pearlman, 1964; Fukuyama *et al.*, 1967; Epstein *et al.*, 1968a). As with the mitotic rate and DNA synthesis response, acceleration of these functions occurs by 24 to 72 hours after the UV exposure. Studies of the effects of chemical carcinogens on protein synthesis are sparse. However, some evidence of RNA synthesis inhibition by these substances has been reported (Sinclair and McCarter, 1964).

6.2 Summary

A comparison of the acute UV-induced responses with the early effects of chemical carcinogens revealed a number of striking similarities including the rapid marked inhibition of the mitotic rate, cell turnover, and DNA and RNA synthesis. The post-injury recovery and hyperplasia stage occurs after a number of traumatic procedures (Lobitz and Holyoke, 1954; Brophy and Lobitz, 1959; Daniels *et al.*, 1961; and W. L. Epstein *et al.*, 1969b). As Skjaeggestad (1964) points out, this recovery from chemical carcinogenic injury does not differ qualitatively from an irritant-induced response. Soffen and Blum (1961) reported that the post-UV epidermal hyperplasia and increased number of mitoses may last at least 40 days.

Since a single application of a chemical carcinogen or a single UV exposure can lead to eventual cancer formation, at least part of the acute response must be an initial step in carcinogenesis. Further, the additive effect of these two carcinogenic modalities might represent a possible similarity of their cancer producing mechanisms. However, there are a number of differences in the effects of these two types of injuries including the apparent mechanism of the early inhibition of DNA synthesis and mitoses. Chemical carcinogens prolong the synthesizing time and mitotic duration in contrast to UV which appears to stop these two cell functions. In addition, no proof of a correlation between any of the known early responses to either carcinogenic stimuli and eventual cancer formation is available at this time.

7. Human Skin Cancer Formation

As noted in the introduction, examinations of the sun's role in the production of human skin cancers does not lend itself to direct experimentation. However, extensive astute observations have strongly suggested the etiologic significance of light energy in the induction of these tumors. Skin cancers in Caucasians in general are most prevalent in geographical areas of the greatest insolation and among people who receive the most exposure, i.e., men who work outdoors (Magnusson, 1935; Segi, 1963; Haenszel, 1963; Urbach *et al.*, 1966). They are rare in Negroes and other deeply pigmented individuals who have the greatest protection against ultraviolet light injury (Dorn, 1944; Oettlè, 1963; Quisenberry, 1963). Further, the lightest complexioned individuals such as those of Scotch and Irish descent, appear to be most susceptible to skin cancer formation when they live in geographic areas of high UV exposure (Seldam, 1963; Urbach *et al.*, 1966). When skin cancers do occur in the darkly pigmented races, they are not distributed primarily in the sun-exposed areas as they are in light skinned people (Mulay, 1963). The tumors in these pigmented individuals are more commonly stimulated by other forms of trauma such as chronic leg ulcers (Oettlè, 1963), irritation due to the lack of wearing shoes (Pringgoutomo and Pringgoutomo, 1963), the use of a Kangeri (an earthenware pot which is filled with burning charcoal and strapped to the abdomen for warmth), the wearing of a Dhoti (loin cloth), and so on (Mulay, 1963). In contrast, the distribution of skin cancer in the Bantu albino and in patients with xeroderma pigmentosum follows sun-exposure patterns (Oettlè, 1963). Blum (1959) and Urbach *et al.* (1966) have summarized the arguments supporting the role of sunlight in human skin cancer formation as follows: (*a*) skin cancers do occur predominantly on the sun-exposed parts of the body; (*b*) they are more common in regions of the earth that receive the most sunlight; (*c*) pigmented races are much less susceptible to skin cancer formation than Caucasians who work outdoors.

Though these arguments do not constitute absolute proof and they are in themselves not wholly established, there is a considerable body of circumstantial evidence supporting the role of sunlight in at least three types of skin cancers: basal cell epitheliomas, squamous cell carcinomas and melanomas.

7.1 Basal Cell Epithelioma

The basal cell epithelioma is the most common human skin cancer that occurs in Caucasians. It is derived from epidermal and/or adnexal basal cells and frequently presents as a slow growing, shiny nodule composed of masses of cells with darkly staining nuclei which stimulate epidermal basal cells. These lesions invade locally but rarely metastasize.[1] Extensive epidemiologic studies (Magnusson, 1935; Urbach *et al.*, 1966) indicate that these tumors are found primarily on the head and neck—that is, in sun exposed areas. They are exceedingly rare in deeply pigmented races (Oettlè, 1963). In a recent study, Gellin *et al.* (1965) reported that there is a statistically significant tendency for patients with basal cell epitheliomas to have light eyes, light hair and light complexions, to sunburn easily and to spend more hours outdoors than a control group of people without such tumors. These findings add further support to the importance of sunlight in basal cell epithelioma formation. However, the distribution of this type of cancer on the head and neck does not correspond with sites of greatest UV exposure, i.e., they are concentrated in the inner and outer canthus, the postauricular areas and other relatively protected sites. Thus, factors in addition to sun exposure must play a role in the production of this type of skin cancer. The persistence of cell rests along embryonic lines of closure of the facial skin, or the presence of more reactive adnexal structures, have been suggested but there is no proof that such factors are of etiological significance.

Basal cell epitheliomas have been produced experimentally with chemical carcinogens in rats (Zacheim *et al.*, 1959; Dobson, 1963), and, as noted in Section 5.2.2, UV will stimulate cancer formation initiated by a chemical carcinogen. It is conceivable that such chemical mediators which are present in our environment (S. Epstein, 1967) could concentrate in areas of highest basal cell epithelioma formation. However, there is no evidence that this type of tumor can be produced in experimental animals by UV energy alone or in combination with a chemical carcinogen. Therefore, the evidence for the role of sunlight in the production of this tumor is confined to the aforementioned epidemiological findings.

7.2 Squamous Cell Carcinoma

The squamous cell carcinoma is the second most common skin cancer in Caucasians and is more common than basal cell epitheliomas in the

[1] The reader is referred to Rook *et al.* (1968) for the descriptions of clinical and microscopic variations of basal cell epitheliomas, squamous cell carcinomas and melanomas.

pigmented races (Pringgoutomo and Pringgoutomo, 1963; Oettlè, 1963; Shanmugaratnam and La'Brooy, 1963; Tansurat, 1963). It generally presents as a relatively slow growing keratotic tumor composed of masses of cells that simulate epidermal Malpighian cells and tend to form keratin. However, the individual cells frequently show bizarre forms and shapes as well as malignant dyskeratosis.[1] These lesions do metastasize at times.

The evidence for the role of sunlight in producing squamous cell carcinomas in Caucasians, albinos, and patients with xeroderma pigmentosum is even more convincing than it is for basal cell epithelioma formation. These growths are distributed primarily over the head and neck and to a lesser extent the exposed areas of the upper extremities (Magnusson, 1935; Huriez *et al.*, 1962; Urbach *et al.*, 1966; Eastcott, 1963; Seldam, 1963). As noted for basal cell epitheliomas, squamous cell carcinomas are found most abundantly in men who work outdoors in areas of the highest insolation. In addition, though both squamous cell carcinomas and basal cell epitheliomas are more prevalent in geographical areas of high sun exposure, there is a relatively greater increase in squamous cell carcinoma incidence with decreasing latitude and increasing sun exposure (Urbach *et al.*, 1966).

These lesions are uncommon in the more deeply pigmented races and when they do occur they are found primarily on lower extremities, abdomen and the like, depending on sites of chronic irritation and injury but not solar damage (Oettlè, 1963; Mulay, 1963). Further, in contrast to the basal cell epithelioma, the squamous cell carcinoma occurs almost exclusively in anatomical sites which receive maximal UV irradiation such as the rim of the concha in Caucasians (Huriez *et al.*, 1962; Urbach *et al.*, 1966). Thus evidence that sun plays an important role in squamous cell carcinoma formation in Caucasians is quite convincing. This concept gains additional support from experimental studies in which squamous cell carcinomas can be produced in appropriate animals to the exclusion of other types of malignancies by UV irradiation (Section 3.1.1).

7.3 Melanomas

Melanomas are uncommon skin malignancies which metastasize relatively frequently as compared to basal cell epitheliomas and squamous cell carcinomas. Thus these tumors have a much greater mortality rate than the more benign common skin malignancies. Clinically, they gen-

[1] The reader is referred to Rook *et al.* (1968) for the descriptions of clinical and microscopic variations of basal cell epitheliomas, squamous cell carcinomas and melanomas.

erally present as relatively rapidly growing, darkly pigmented nodules which often ulcerate and show evidence of inflammation. The presence of satellite lesions around the central tumor mass strongly suggests the diagnosis of malignant melanoma.[1] The melanotic freckle of Hutchinson or circumscribed precancerous melanosis of Dubreuilh represents a striking exception. These lesions are usually extremely slow growing tumors most commonly involving the face and neck and show the picture of a melanoma histologically when the nodular stage occurs but they appear to metastasize rarely.

The origin of the neoplastic melanocytic cell is in a state of controversy. Some investigators feel that most of these tumors develop from pre-existing nevi whereas others consider that the growths originate more commonly from the epidermal melanocytic system (Becker, 1930; Beerman *et al.*, 1955; Belisario, 1959; Quevedo, 1963; Mishima, 1967; Clark, 1967). In Helwig's (1963) extensive series, 80% of the patients claimed that the melanomas had developed from pre-existing nevi or pigmented growths. However, he was only able to find histologic evidence of nevus cells in 20% of the tissues.

The role of sunlight in malignant melanoma formation is much less clear than its association with basal cell epithelioma and squamous cell carcinoma production. Melanomas are not found primarily on the sun-exposed parts of the body, and the protective effect of melanin pigment is not so obvious as it is in the other forms of skin cancer (Helwig, 1963; Oettlè, 1963; Gellin *et al.*, 1969). However, a number of surveys do suggest that sunlight does influence the formation of some melanomas. Caucasians with melanomas statistically tend to have light hair, blue eyes, fair complexions and spend more time outdoors when compared to a control group of patients who do not have this type of cancer growth (Lancaster and Nelson, 1957; Cade, 1961; Helwig, 1963; Gellin *et al.*, 1969). Studies on geographic distribution of this tumor in the United States (Haenzel, 1963), Australia (Lancaster, 1956) and New Zealand (Lancaster, 1956; Lancaster and Nelson, 1957; Eastcott, 1963; Davis *et al.*, 1966) demonstrate a significant increased incidence in areas of highest insolation. A survey of genetically similar populations demonstrated a much greater melanoma prevalence associated with high sun exposure in Australia and New Zealand as compared to the low insolation found in England (Eastcott, 1963; Davis *et al.*, 1966).

The occurrence of sun-induced melanomas in patients with xeroderma pigmentosum (Moore and Iverson, 1954 and Lynch *et al.*, 1967), and

[1] The reader is referred to Rook *et al.* (1968) for the descriptions of clinical and microscopic variations of basal cell epitheliomas, squamous cell carcinomas and melanomas.

the production of melanomas from benign pigmented lesions by chronic UV exposures in experimental animals (Epstein *et al.*, 1967) (Section 3.1.2) further confirms the potential of sunlight energy to produce these tumors under the proper circumstances. In addition, the distribution of melanomas developing from circumscribed precancerous melanosis strongly suggests that the sun is at least in part responsible for such lesions (Ward, 1967). However, the anatomical distribution of melanomas in general indicates that other factors are more important than sunlight in the etiology of the majority of these tumors.

7.4 Xeroderma Pigmentosum

Xeroderma pigmentosum is a rare genodermatosis generally inherited as a mendelian autosomal recessive trait, though an apparent incomplete sex-linked recessive pattern has been reported (Macklin, 1944; Anderson and Begg, 1950; El-Hefnawi and Rasheed, 1963). The condition is characterized by hyperpigmentation, depigmentation, atrophy, telangiectasia, verrucous excrescences, actinic keratoses and ulcerations involving sun-exposed parts of the skin. The clinical picture resembles chronic x-ray damage. The patients frequently develop all variety of skin malignancies including basal cell epitheliomas, squamous cell carcinomas, sarcomas, and even malignant melanomas (El-Hefnawi and Rasheed, 1963; Attie and Khafif, 1964; Lynch, 1967; Lynch *et al.*, 1967; and Reed, *et al.*, 1969). Prognostically, patients with xeroderma pigmentosum usually do not survive past the first two decades of life due to metastases from the cutaneous malignancies.

Though photodermatoses simulating polymorphous light eruptions have been reported, the basic problem in these individuals is their marked sensitivity to the carcinogenic effects of the sun's rays. The action spectrum in xeroderma pigmentosum appears to fall in the sunburn range (290 to 320 nm) (Hausser and Vahle, 1922; Rothman, 1923; Lynch, 1934; Zoon, 1938; Nürnberger, 1968). Erythema, telangiectasia and persistent pigmentation have been produced (Rothman, 1923), and histological and electron microscopic evidence of chronic sun damage following a single exposure to these UV rays has been reported (Nürnberger, 1968).

The mechanism of this carcinogenic potential is not clear. Loss or lack of the protective effects of melanin pigmentation as noted in albinos or patients with light skin and red hair does not play a role since the condition occurs in deeply pigmented races as well as Caucasians. The acute response of these dark complexioned patients with xeroderma pigmentosum to the sunburn rays does not differ from noninvolved individuals with similar pigmentation (Findlay, 1968). Further, actinic degeneration is not present in the corium of the Bantu patients suggesting that the carcinogenic UV effects occur directly in the epidermal cell and are not related to stromal damage (Oettlè, 1963).

Recently a genetically inherited enzyme defect has been reported which may be responsible, at least in part, for the cancer forming potential in these patients with xeroderma pigmentosum. As noted in Section 6.1.2, UV irradiation of mammalian cells in culture and epidermal and dermal cells *in vivo*, will stimulate unscheduled DNA synthesis in cells which are not preparing to divide. Density gradient studies indicated that this unscheduled synthesis represents repair replication which involves the insertion of small numbers of bases throughout the DNA, pre-

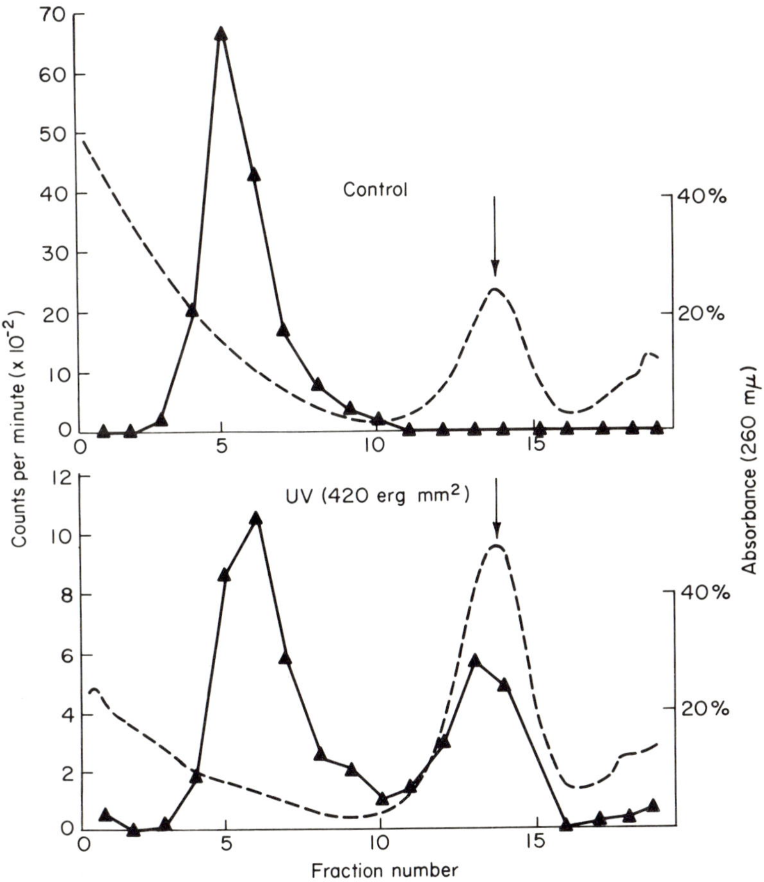

FIG. 14. CsCl equilibrium density gradient of DNA from normal human skin fibroblasts labeled for 1 hr with BrUdR, irradiated with UV light, and labeled for 4 hrs in H°BrUdr absorbance 260 mu --------, H^3 activity. Arrows mark position of normal density DNA (Cleaver, 1969a).

sumably to replace damaged regions which are excised. Though there is little evidence of excision steps in mammalian cells as yet, there is evidence to support the repair replication phenomenon. Recently Cleaver, utilizing the density gradient method (1968, 1969b) demonstrated that UV-irradiated fibroblasts from the dermis of patients with xeroderma pigmentosum in cell culture failed to demonstrate unscheduled DNA synthesis and showed no evidence of repair replication. Labeling the normal fibroblast DNA with H^3-BrUdR post irradiation results in radioactivity in the lighter fractions separated by the density gradient technique. These findings indicate that small amounts of the heavier

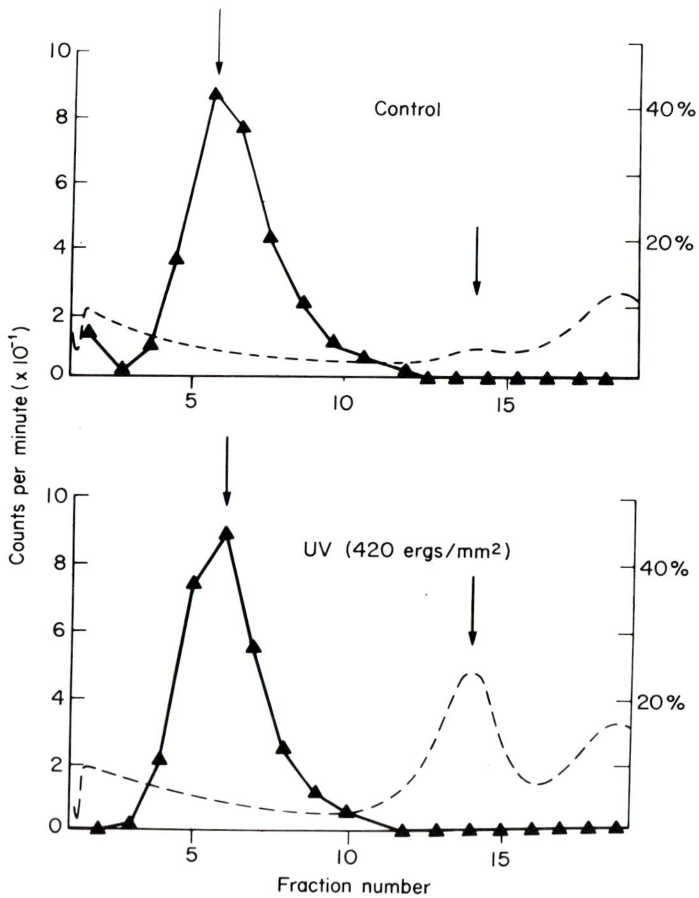

FIG. 15. CsCl equilibrium density gradient of DNA from homozygous xeroderma pigmentosum fibroblasts labeled for 1 hr with BrUdR, irradiated with UV light, and labeled for 4 hrs with H^3BrUdR, absorbance 260 mu H^3 activity. Arrows mark positions of normal density DNA and DNA with BrU in one strand (Cleaver, 1969a).

material are incorporated into the DNA apparently due to the removal of thymine dimers and replacement with normal bases in proper sequence (dark repair) (Fig. 14). In xeroderma pigmentosum this repair process is missing (Fig. 15).

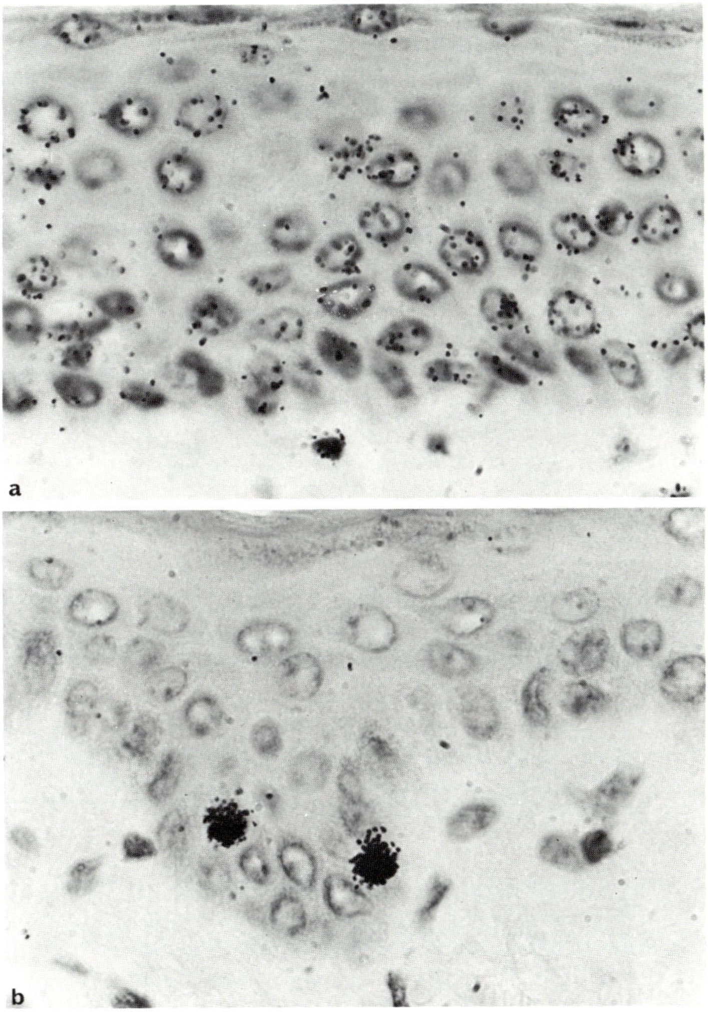

FIG. 16a. Autoradiograph of irradiated normal human skin *in vivo* showing sparse labeling with TdR-H^3 in the granular, Malpighian and basal cell nuclei (15 minutes post-UV irradiation).

FIG. 16b. Autoradiograph of irradiated xeroderma pigmentosum skin *in vivo* showing the absence of the sparse labeling with TdR-H^3 (15 minutes post-UV irradiation). The dense labeling in two basal cells represents premitotic, semiconservative DNA synthesis.

Examination of unscheduled DNA synthesis in epidermal cells and dermal connective tissue cells in xeroderma pigmentosum patients *in vivo* using TdR-H^3 as the radioactive tracer (Section 6.1.2) revealed results similar to the cell culture findings (J. H. Epstein *et al.*, 1969). As noted in Fig. 16, UV irradiation of normal human skin *in vivo* results in unscheduled DNA synthesis throughout the differentiated epidermal granular and Malpighian cell and the dermal connective tissue cell nuclei as well as the germinative basal cell nuclei. In xeroderma pigmentosum this process does not occur or does so at a very low rate. Thus, this *in vivo* response most likely reflects the same abnormality described in the cell culture system. In contrast to the pure UV effect, Cleaver (1969b) reported that x-radiation and UV sensitization by bromouracil (BrU) was followed by apparently normal DNA repair in the xeroderma pigmentosum fibroblasts in culture. He points out that x-radiation and UV irradiation of DNA containing BrU produce strand breaks. Repair of this injury requires enzymes for excision and patching but not for the initial chain breaks. In UV injury alone, the repair process apparently needs an enzyme to produce the initial chain break which will result in the removal of the damaged bases (presumably pyrimidine dimers especially of the thymine bases).

From this data, Cleaver postulates that cells of patients with xeroderma pigmentosum are deficient in an enzyme which cuts DNA (an endonuclease) to initiate repair of the UV damage. He further suggests that possibly this defect results in a high somatic mutation rate which eventuates in cancer formation in these individuals. However, correlation of this apparent enzyme deficiency and tumor formation will have to await further evidence since correlation of specific acute responses with eventual carcinogenesis has yet to be established.

7.5 Summary

The relationship of UV or sunlight to human skin cancer formation depends primarily on epidemiologic and anatomical distribution studies. Despite some minor discrepancies, the role of sunlight in the production of basal cell epitheliomas and squamous cell carcinomas is relatively well established. The association between melanoma formation and sun exposure is less clear. The production of these tumors by UV in experimental animals and clinical observations of light-induced melanoma formation in patients with xeroderma pigmentosum indicate that the sun's rays can stimulate this type of carcinogenesis. This concept gains further support from several geographic distribution surveys. However, anatomical location studies indicate that a number of other factors are important and that the sun's rays can play a role, if at all, in the development of only a minority of these lesions.

The inherited condition, xeroderma pigmentosum, presents a caricature of the problem of sunlight-induced carcinogenesis in humans. These patients develop all varieties of skin cancers including melanomas due to sun exposure. These growths appear within a matter of months or a few years in contrast to the usual tumor latent period of several decades in normal people. Recent cell culture and *in vivo* studies indicate that the cells of patients with xeroderma pigmentosum are deficient in an enzyme which is necessary for repair of UV damaged DNA. Whether this defect is responsible for the marked carcinogenic potential of these patients remains to be determined.

Acknowledgments

This work was supported in part by the U. S. Public Health Service Research Grant No. CA06386 from the National Cancer Institute, Department of Health, Education and Welfare, U. S. Public Health Service.

I am indebted to the Williams and Wilkins Co., Baltimore, Maryland, for permission to include in the text Figures 1, 2, 3, and 4, and Table II.

References

Anderson, T. E., and Begg, M. (1950). *Brit. J. Dermatol.* **62,** 402.

Ashworth, C. T., Stembridge, V. A., and Luibel, F. J. (1961). *Acta Cytol.* **5,** 369.

Attie, J. L., and Khafif, R. A. (1964). "Melanotic Tumors: Biology, Pathology and Clinical Features." Thomas, Springfield, Ill.

Baden, H. P., and Pearlman, C. (1964). *J. Invest. Dermatol.* **43,** 71.

Bain, J. A., Rusch, H. P., and Kline, B. E. (1943). *Cancer Res.* **3,** 610.

Baserga, R., Henegar, G. C., Kisieleski, W. E., and Lisco, H. (1962). *Lab. Invest.* **11,** 360.

Baserga, R. (1965). *Cancer Res.* **25,** 581.

Becker, S. W. (1930). *A. M. A. Arch. Dermatol. Syphil.* **21,** 818.

Beerman, H., Lane, R. A. G., and Shaffer, B. (1955). *Am. J. Med. Sci.* **229,** 583–444.

Belisario, J. C. (1959). "Cancer of the Skin." Butterworth, London.

Berenblum, I. (1930). *J. Exp. Pathol.* **11,** 208.

Berenblum, I., and Haran, N. (1955). *Brit. J. Cancer* **9,** 268.

Berenblum, I., and Shubik, P. (1949). *Brit. J. Cancer* **3,** 384.

Blum, H. F. (1941). "Photodynamic Action and Diseases Due to Light." Rhinehold, New York.

Blum, H. F. (1943a). *J. Natl. Cancer Inst.* **3,** 533.

Blum, H. F. (1943b). *J. Natl. Cancer Inst.* **3,** 538.

Blum, H. F. (1945). *Physiol. Rev.* **25,** 483.

Blum, H. F. (1950). *J. Natl. Cancer Inst.* **11,** 463.

Blum, H. F. (1959). "Carcinogenesis by Ultraviolet Light," Princeton Univ. Press, New Jersey.

Blum, H. F., and Lippincott, S. W. (1942). *J. Natl. Cancer Inst.* **3,** 211.

Boutwell, R. K. (1964). *Progr. Exptl. Tumor Res.* **4,** 207.

Boutwell, R. K., and Rusch, H. P. (1951). *Cancer Res.* **11,** 238.

Brophy, D., and Lobitz, W. C. (1959). *J. Invest. Derm.* **32,** 495.
Buck, H. W., Magnus, I. A., and Porter, A. D. (1960). *Brit. J. Dermatol.* **72,** 249.
Bullough, W. S., and Laurence, E. B. (1968). *Nature* **220,** 134.
Büngeler, W. (1937). *Ztschr F. Krebforsch.* **46,** 130.
Cade, S. (1961). *Ann. Roy. Coll. Surg. Eng.* **28,** 331.
Cawley, E. P., Hsu, Y. T., and Weary, P. E. (1966). *Arch. Dermatol.* **94,** 712.
Clark, J. H. (1964). *Cancer Res.* **24,** 207.
Clark, J. M. (1936). *Am. J. Hyg.* **24,** 334.
Clark, W. H., Jr. (1967). *In* "Advances in Biology of the Skin" (W. Montagna, ed.), Vol. VIII, pp. 621–647. Macmillan (Pergamon), New York.
Cleaver, J. E. (1968). *Nature* **218,** 652.
Cleaver, J. E. (1969a). Atomic Energy Commission Syposium on Biological Implications of the Nuclear Age (in press).
Cleaver, J. E. (1969b). *Proc. Natl. Acad. Sci.* **63,** 428.
Daniels, F., Jr., Brophy, D., and Lobitz, W. C. (1961). *J. Invest. Dermatol.* **37,** 351.
Davis, N. C., Herron, J. J., and McLeod, G. R. (1966). *Lancet* **2,** 407.
Della Porta, G., Rappaport, H., Saffiotti, U., and Shubik, P. (1956). *Arch. Pathol.* **61,** 305.
Dendy, P. P., and Cleaver, J. E. (1964). *Intern. J. Radiation Biol.* **8,** 301.
DesLingneris, M. J. A. (1940). *Am. J. Cancer* **40,** 1.
Dobson, R. L. (1963). *J. Natl. Cancer Inst.* **31,** 841.
Dobson, R. L., and Griffin, M. (1962). *J. Invest. Dermatol.* **39,** 597.
Dorn, H. F. (1944). *Public Health Rep.* **59,** 33, 65, 97.
Dubreuilh, W. (1907). *Ann. Dermatol. Syphil.* **8** (4), 387.
Eastcott, D. E. (1963). *Natl. Cancer Inst. Monograph* **10,** 141.
Elgjo, K. (1966). *In* "Norwegian Monographs on Medical Science." (Universitetsforlaget) Olso, Norway.
El-Hefnawi, H., and Rasheed, A. (1963). *J. Egypt. Med. Assoc.* **47,** 985.
Epstein, J. H. (1965). *J. Natl. Cancer Inst.* **34,** 741.
Epstein, J. H., and Epstein, W. L. (1962). *J. Invest. Dermatol.* **39,** 455.
Epstein, J. H., and Epstein, W. L. (1963). *J. Invest. Dermatol.* **41,** 463.
Epstein, J. H., and Fukuyama, K. (1969). (Unpublished data).
Epstein, J. H., and Roth, H. L. (1968). *J. Invest. Dermatol.* **50,** 387.
Epstein, J. H., Sullivan, F. J., and Epstein, W. L. (1961). *J. Invest. Dermatol.* **36,** 73.
Epstein, J. H., Fukuyama, K., and Dobson, R. L. (1966). Presented at the Fourth International Congress of Biometeriology, Rutgers, New Jersey.
Epstein, J. H., Epstein, W. L., and Nakai, T. (1967). *J. Natl. Cancer Inst.* **38,** 19.
Epstein, J. H., Fukuyama, K., and Fye, K. (1968a). "Photochemistry and Photobiology," presented at the 60th Annual Meeting of the Amer. Soc. Clin. Res., Atlantic City, New Jersey, May 6. (In press).
Epstein, J. H., Fukuyama, K., and Epstein, W. L. (1968b). *J. Invest. Dermatol.* **51,** 445.
Epstein, J. H., Fukuyama, K., and Epstein, W. L. (1969). Presented at the 61st Annual Meeting of the Amer. Soc. Clin. Res., Atlantic City, New Jersey, May 5.
Epstein, S. (1967). *J. Air Pollution Control Assoc.* **17,** 728.
Epstein, S. S., Small, M., Falk, H. L., and Mantel, N. (1964). *Cancer Res.* **24,** 885.
Epstein, W. L., Fukuyama, K., and Cortese, T. A. (1969a). *Arch. Dermatol.* **99,** 94.
Epstein, W. L., Fukuyama, K., and Epstein, J. H. (1969b). *Arch. Dermatol.* (In press).
Epstein, W. L., Fukuyama, K., and Epstein, J. H. (1969). *Arch. Dermatol.* (In press).
Evensen, A., and Iversen, O. H. (1962). *Nature* **96,** 383.

Findlay, G. H. (1968). Presented at the Fifth Intern. Congr. of Photobiology, Hanover, New Hampshire, August 26–31.
Findlay, G. M. (1928). *Lancet* **2,** 1070.
Forbes, P. D. (1965). *J. Invest. Dermatol.* **44,** 388.
Freeman, R. G., and Knox, J. M. (1964). *Arch. Dermatol.* **89,** 858.
Frei, J. V. (1962). *J. Cell Biol.* **15,** 535.
Fukuyama, K. Epstein, W. L., Epstein, J. H. (1967). *Nature* **216,** 1031.
Gellin, G. A., Kopf, A. W., and Garfinkel, L. (1965). *Arch. Dermatol.* **91,** 38.
Gellin, G. A., Kopf, A. W., and Garfinkel, L. (1969). *Arch. Dermatol.* **99,** 43.
Getzrow, P. L. (1966). *Arch Dermatol.* **94,** 44.
Ghadially, F. N., and Barker, J. F. (1960). *J. Pathol. Bacteriol.* **79,** 263.
Ghadially, F. N., Illman, O., and Barker, J. F. (1960). *Brit. J. Cancer* **14,** 647.
Goldyne, M., Epstein, J. H., and Fukuyama, K. (1969). (Unpublished data).
Grady, H. G., Blum, H. F., and Kirby-Smith, J. S. (1941). *J. Natl. Cancer Inst.* **2,** 269.
Grady, H. G., Blum, H. F., and Kirby-Smith, J. S. (1943). *J. Natl. Cancer Inst.* **3,** 371.
Graham, J. H., and Helwig, E. B. (1965). *In* "Advances in Biology of Skin" (W. Montagna and R. L. Dobson, eds.), Vol. II, pp. 277–327. Macmillan (Pergamon), New York.
Griffin, A. C., Hakim, R. E., and Knox, J. (1958). *J. Invest. Dermatol.* **31,** 289.
Haenszel, W. (1963). *Natl. Cancer Inst. Monograph* **10,** 225.
Hale, D. M., Cromartie, W. J., and Dobson, R. L. (1960). *J. Invest. Dermatol.* **35,** 293.
Hanawalt, P. S., and Setlow, R. B. (1960). *Biochem. Biophys. Acta* **41,** 283.
Hausser, K. W., and Vahle, W. (1922). *Strahlentherapie* **13,** 41.
Helwig, E. B. (1963). *Natl. Cancer Inst. Monograph* **10,** 287.
Heuper, W. C. (1941). *Cancer Res.* **1,** 402.
Hill, L., and Eidenow, A. (1923). *Proc. Roy. Soc.* (Biol.) **95,** 163.
Hoffman, J., and Post, J. (1966). *Cancer Res.* **26,** 1313.
Horikawa, M., Nikaido, O., and Sugahara, T. (1968). *Nature* **218,** 490.
Huriez, C., Lebeurre, R., and Leperre, B. (1962). *Bull. Soc. Franc. Dermatol. Syphil.* **69,** 886.
Hyde, J. N. (1906). *Am. J. Med. Sci.* **131,** 1.
Igarashi, Y. (1967). *Jap. J. Dermatol.* (Series A) **77,** 207.
Iversen, O. H. (1968). *Nature* **219,** 75.
Iversen, O. H., and Evensen, A. (1962). "Experimental Skin Carcinogenesis in Mice." Norwegian Univ. Press, Olso.
Johnson, H. A., Rubini, J. R., Cronkite, E. P., and Bond, V. P. (1960). *Lab. Invest.* **9,** 460.
Johnson, B. E., Daniels, F. Jr., and Magnus, I. A. (1968). *In* "Photophysiology" (A. C. Giese, ed.), Vol. IV, pp. 139–202. Academic Press, New York.
Kelner, A. (1953). *J. Bacteriol.* **65,** 252.
Kelner, A., and Taft, E. B. (1956). *Cancer Res.* **16,** 860.
Klaus, S. N., and Winkelmann, R. K. (1965). *J. Invest. Dermatol.* **45,** 160.
Klímek, M., Vlašínova, M. (1967). *Neoplasma* **14,** 385.
Kohn-Speyer, A. C. (1929). *Lancet* **12,** 1305.
Kuper, J. B., Brackett, F. S., and Eicher, M. (1941). *Rev. Sci. Instr.* **12,** 87.
Lancaster, H. O. (1956). *Med. J. Australia* **1,** 1082.
Lancaster, H. O., and Nelson, J. (1957). *Med. J. Australia* **1,** 452.
Leblond, C. P., and Stevens, C. E. (1948). *Anat. Rec.* **100,** 357.
Lever, W. F. (1967). "Histopathology of the Skin" (4th ed.). Lippincott, Philadelphia.

Lobitz, W. C., and Holyoke, J. B. (1954). *J. Invest. Dermatol.* **22,** 189.
Lynch, F. W. (1934). *Arch. Dermatol.* **29,** 858.
Lynch, H. T., (1967). *Cancer Res. Monograph Series* **12,** 1, New York.
Lynch, H. T., Anderson, D. E., Smith, J. L., Jr., Howell, J. B., and Krush, A. J. (1967). *Arch. Dermatol.* **96,** 625.
Mackie, B. S., and McGovern, V. J. (1958). *Arch. Dermatol.* **78,** 218.
Macklin, M. T. (1944). *Arch. Dermatol. Syphil.* **49,** 157.
Magnus, I. A., and Johnson, B. E. (1965). Cited by Johnson, B. E., Daniels, F., Jr., and Magnus, I. A. *In* "Photophysiology" (A. C. Giese, ed.), Vol. IV, pp. 139–202. Academic Press, New York.
Magnusson, A. H. W. (1935). *Acta Radiol. Suppl.* **19.**
Mehregan, A. H., Staricco, R. G., and Pinkus, H. (1964). *Arch. Dermatol.* **89,** 33.
Miller, E. C. (1951). *Cancer Res.* **11,** 100.
Mishima, Y. (1967). *Cancer* **20,** 632.
Moodie, M. M., Reid, C., and Wallick, C. A. (1954). *Cancer Res.* **14,** 367.
Moore, C., and Iverson, P. C. (1954). *Cancer* **7,** 377.
Morton, J. J., Luce-Clausen, E. M., and Mahoney, E. B. (1940). *Am. J. Roentgenol. Radium Therapy* **43,** 896.
Morton, J. J., Mider, G. B., Luce-Clausen, E. M., and Mahoney, E. B. (1951). *Cancer Res.* **11,** 559.
Mowry, R. W. (1958). *Lab. Invest.* **7,** 566.
Mulay, D. M. (1963). *Natl. Cancer Inst. Monograph* **10,** 215.
Nakai, T., and Rappaport, H. (1963). *Amer. J. Pathol.* **43,** 175.
Nakamura, K., and Johnson, W. C. (1968). *J. Invest. Dermatol.* **51,** 253.
Nürnberger, F. (1968). Presented at the Fifth Intern. Congr. of Photobiology, Hanover, New Hampshire, Aug. 26–31.
Oettlè, A. G. (1963). *Natl. Cancer Inst. Monograph* **10,** 197.
O'Neal, M. A., and Griffin, A. C. (1957). *Cancer Res.* **17,** 911.
Orr, J. W. (1963). *Natl. Cancer Inst. Monograph* **10,** 531.
Pathak, M. A. (1961). *J. Invest. Dermatol.* **37,** 397.
Pinkus, J., and Mehregan, A. H. (1966). *J. Invest. Dermatol.* **46,** 109.
Poel, W. E. (1963). *Natl. Cancer Inst. Monograph* **10,** 611.
Pringgoutomo, S., and Pringgoutomo, S. (1963). *Natl. Cancer Inst. Monograph* **10,** 191.
Pullinger, B. D. (1945). *J. Pathol. Bacteriol.* **57,** 467.
Quevedo, W. C., Jr. (1963). *Natl. Cancer Inst. Monograph* **10,** 561.
Quevedo, W. C., Jr., Cairns, J. M., Smith, J. A., Bock, F. G., and Burns, R. J. (1961). *Nature* **189,** 936.
Quisenberry, W. B. (1963). *Natl. Cancer Inst. Monograph* **10,** 181.
Rasmussen, R. E., and Painter, R. B. (1964). *Nature* **203,** 1360.
Rasmussen, R. E., and Painter, R. B. (1966). *J. Cell Biol.* **29,** 11.
Reed, W. B., Landing, B., Sugarman, G., Cleaver, J. E., and Melnyk, J. (1969). *J. Am. Med. Asso.* **207,** 2073.
Regan, J. D., and Trosko, J. E. (1967). *Radiation Res.* **31,** 548.
Roe, F. J. C., and Salaman, M. H. (1954). *Brit. J. Cancer* **8,** 666.
Roffo, A. H. (1934). *Bull. Assoc. Franc. Etude Cancer* **23,** 590.
Rogers, S. (1955). *J. Natl. Cancer Inst.* **15,** 1001.
Rook, A. J., Wilkinson, D. S., and Ebling, F. J. (1968). "Textbook of Dermatology." Blackwell, Oxford.
Rothman, S. (1923). *Arch. Dermatol. Syphil.* **144,** 440.

Rusch, H. P., and Baumann, C. A. (1939). *Ann. J. Cancer* **35,** 55.

Rusch, H. P., Kline, B. E., and Baumann, C. A. (1941). *Arch. Pathol.* **31,** 135.

Rusch, H. P., Kline, B. E., and Baumann, C. A. (1942). *Cancer Res.* **2,** 183.

Saffiotti, U., and Shubik, P. (1956). *J. Natl. Cancer Inst.* **16,** 961.

Sams, W. M., Smith, J. G., and Finlayson, G. R. (1963). *J. Invest. Dermatol.* **41,** 457.

Sams, W. M., Smith, J. G., and Burk, P. G. (1964). *J. Invest. Dermatol.* **43,** 467.

Santamaria, L. (1960). *Recent Contrib. Cancer Res. Italy* **1,** 167.

Santamaria, L. (1964). Presented at the Fourth International Photobiology Congress Oxford, England, July 26–30.

Schorr, G., and Ssobolewa, N. (1930). *Ztschr f. Krebsforsch.* **31,** 308.

Segi, M. (1963). *Natl. Cancer Inst. Monograph* **10,** 245.

Seldam, R. E. J. (1963). *Natl. Cancer Inst. Monograph* **10,** 153.

Setala, K., Nyyssonen, O., Lindroos, B., (1962). *Europ. J. Cancerol.* **1,** 92.

Shanmugaratnam, K., and La'Brooy, E. B. (1963). *Natl. Cancer Inst. Monograph* **10,** 127.

Shield, M. (1899). *Lancet* **1,** 22.

Shubik, P. (1950). *Cancer Res.* **10,** 713.

Shubik, P. (1961). *Acta Un. Int. Cancr.* **17,** 24.

Sinclair, N. R., and McCarter, J. A. (1964). *Nature* **203,** 521.

Skjeggestad, O. (1964). "Experimental Epidermal Hyperplasia in Mice, Relation to Carcinogenesis." Universitetsforlaget, Olso, Norway.

Soffen, G. A., and Blum, H. F. (1961). *J. Cellular Comp. Physiol.* **58,** 81.

Tansurat, P. (1963). *Natl. Cancer Inst. Monograph* **10,** 71.

Tausig, J., Cooper, Z. K., and Seelig, M. G. (1938). *Surg. Gynecol. Obstet.* **66,** 989.

Unna, P .G. (1894). "Die Histopathologie der Hautkrankheiten." Hirschwald, Berlin.

Urbach, F. (1959). *J. Invest. Dermatol.* **32,** 373.

Urbach, F. (1963). *Natl. Cancer Inst. Monograph* **10,** 539.

Urbach, F. Davies, R. E., and Forbes, P. D. (1966). *In* "Advances in Biology of the Skin" (W. Montagna and R. L. Dobson, eds.), Vol. VII, pp. 195–214. Macmillan (Pergamon), New York.

van der Leun, J. C. (1966). Ph.D. Thesis, Univ. of Utrecht.

van Scott, E. J. (1964). *In* "The Epidermis" (W. Montagna and W. C. Lobitz, Jr., eds.), pp. 573–586. Academic Press, New York.

Ward, W. H. (1967). *Australian J. Dermatol.* **9,** 70.

Winkelmann, R. K., Baldes, E. J., and Zollman, P. E. (1960). *J. Invest. Dermatol.* **34,** 131.

Winkelmann, R. K., Zollman, P. E., and Baldes, E. J. (1963). *J. Invest. Dermatol.* **40,** 217.

Zacheim, H. S. (1963). *J. Invest. Dermatol.* **40,** 283.

Zacheim, H. S., Simpson, W. L., and Langs, L. (1959). *J. Invest. Dermatol.,* **38,** 57.

Zoon, J. J. (1938). *Strahlentherapie* **61,** 640.

AUTHOR INDEX

Numbers in italics indicate the pages on which complete references are cited.

C

D

L

Pringgoutomo, S., 260, 262, *272*

T

U

V

W

Y

Z

SUBJECT INDEX